D0742610

EDITORS

BRIAN K. HALL

WENDY M. OLSON

Keywords and Concepts in Evolutionary Developmental Biology

Harvard University Press

Cambridge, Massachusetts

London, England

2003

Printed in the United States of America

Library of Congress Cataloging-in-Publication Data
Keywords and concepts in evolutionary developmental biology / Brian K. Hall and Wendy M. Olson, editors.
p. cm — (Harvard University Press reference library)
Includes bibliographical references (p.).
ISBN 0-674-00904-5
1. Developmental biology. 2. Evolution (Biology).
I. Hall, Brian Keith, 1941– II. Olson, Wendy. III. Series.
QH491 .K49 2003
571.8—dc21 2002192201

Contents

JOHN TYLER BONNER

Foreword

There has been a remarkable explosion of interest in evolutionary developmental biology in the last twenty years and one might well ask why this has occurred. The subject, after all, in a different guise was central to biology in the nineteenth century; it was a subject of importance to Charles Darwin, August Weismann, Ernst Haeckel, and many others, but to some degree it faded in the twentieth century. Embryology concerned itself primarily with the mechanism of development, a subject that goes back to Wilhelm Roux, Hans Driesch, and others; the study of evolution rather aggressively stepped to one side of development as though to keep itself uncontaminated by secondary details.

There were notable exceptions of individuals who brought development and evolution together during the last century, but their ideas never became the popular central issue of either developmental biology or evolution. To name a few who made significant contributions, in the beginning of the century Walter Garstang pushed the idea that it was not just adult animals that evolved; the larval stages of metamorphosing animals also could be independently modified by natural selection. He was followed by Gavin de Beer, who made a spirited effort to expose what he felt was the oversimplification of the problem by the great nineteenth-century propagandist Ernst Haeckel with his biogenetic law that ontogeny recapitulates phylogeny. The subject was far richer than that and de Beer did a fine job of exploiting that richness, although he did it by coining a large number of forgettable terms.

By the mid-twentieth century, we began to hear new thoughts on the subject from Ivan Schmaulhausen and Conrad Waddington. The latter's work is particularly interesting because he made a concerted effort to bring genetics and development together and thought of them in an evolutionary context. He felt quite strongly that the new synthesis of evolutionary biology—population genetics—seriously missed the mark. It was a hollow quest because it completely ignored how genes are involved in the formation of the adult: what genes do is direct development, and that is what evolves. From today's perspective he was certainly on the right track, though his ideas, while greatly admired by many, had only a modest impact.

In the 1950s and 1960s, as the popularity and importance of biochemistry increased, scientists began to examine the nature of the chemical messengers that account for the many events of development. In particular they wanted to know the nature of the substances that emanated from the dorsal lip of the blastopore that induced the amphibian embryo to form. This was the key to understanding embryonic induction, discovered in the 1920s by Hans Spemann and Hilda Mangold in their famous experiment showing the remarkable stimulating properties of this region, which they called the "organizer." The problem seemed manageable until Johannes Holtfreter discovered that implanted dead organizer tissue could also induce a secondary embryo, thereby showing that the region might be giving off chemical stimuli. This led a number of researchers to try to isolate and characterize those chemicals. The search was not successful and while "chemical embryology" in other forms was central for a while, it was an area of research of modest success.

The crucial change came with the advent of molecular genetics: suddenly there was a new way of examining the causes of development—the main object of experimental embryology for the previous one hundred years. Ultimately the chemical nature of the induction described by Spemann and Mangold and various other mysteries were solved, or are within reach of being solved. The key genes involved in pattern and many other processes in development could now be identified along with the proteins they specified; the chemical details of causal embryology could be explored at a far deeper level. One of the great outcomes of this new plan of study was the discovery that these genes existed in widely different groups of organisms, sometimes with the same function, sometimes with a different one. In this way, in a brilliant flash, development again hooked up with evolution. We could examine the evolution of development in a new and exciting way.

It is striking that the big jumps forward in the twentieth century all involved the coming together of two disciplines, and in each of these momentous collisions, genetics was one of the components. The first occurred in the 1900s when Mendel's genetics encountered chromosomal cytology and cytogenetics was born. In the 1930s genetics collided with evolution to produce population genetics, which had, and continues to have, a significant impact on our thinking. The greatest explosion of all was when genetics collided with biochemistry in the 1950s to produce molecular biology. Finally, in the 1980s genetics—now molecular genetics—quite accidentally produced wonders by joining with evolution to produce modern evolutionary developmental biology. This most recent wave not only has shown great promise for the future, but has its roots far back in the nineteenth century.

With the new interest in the marriage of evolution and development, nothing could be more timely and appropriate than this compendium of keywords and concepts. Over the last two centuries, words and ideas have continued to emerge. For Darwin, evolution and development were closely intertwined; now they have come together again in an important and integrated way.

BRIAN K. HALL
WENDY M. OLSON

Introduction: Evolutionary Developmental Mechanisms

This book deals with mechanisms of evolutionary change, especially those mechanisms involving embryonic development as the vehicle for evolutionary change. The field that embraces such studies is known as *evolutionary developmental biology* or *evo-devo*. Evo-devo forges a synthesis of those processes operating during ontogeny with those operating between generations (during phylogeny). Evo-devo includes, but is not limited to:

Analyses of how embryonic development arose and evolved
The role of embryonic development and developmental processes in the evolutionary modification of existing features and in the origin of new (often novel) features
The origin, modification, suppression, or loss of life history stages
How genotype and phenotype interact over generational and gestational time spans
How development and ecology interact and co-evolve

The range of topics in this book reflects both the breadth of evolutionary developmental biology and the increasingly important role this discipline is playing in modern evolutionary and developmental biology. By development we mean any and all stages of an organism's life cycle from egg to embryo, to larval stage (in those animals with indirect development and, consequently, with metamorphosis), to newly emerged (hatched, born) offspring, through to adulthood, cessation of growth, and senescence.

Development occupies the all-important middle ground between the genotype and the phenotype. Although developmental processes figure prominently both in the production and evolution of the phenotype, and in the origin of novelties and new species, the role of development in evolution is too often neglected. The issue is not that we do not understand developmental processes; we have an extraordinary knowledge of how development works. For example, we have an extensive understanding of how genes function, both in individuals and in populations. We can describe the features (structures, behaviors, life history stages) that comprise the phenotype, and we realize that the phenotype is not a one-to-one readout of the genotype. It is no longer a question of whether ontogeny recapitulates (Haeckel) or creates (Garstang) phylogeny; we now appreciate that relationships between ontogeny and phylogeny are in fact reciprocal. Ontogeny enables phylogenetic change; ontogeny also evolves, and evolutionary change can take place at all stages of development.

We are just beginning to understand the evolutionary and developmental processes that constitute what we call *evolutionary developmental mechanisms,* which are the processes of development that mediate descent, whether that descent involves modification or stasis. Our aim in producing this book is to present those mechanisms and approaches in a way that illustrates the current state of the discipline and illuminates the road ahead for evo-devo—indeed, we believe, the road ahead for biology as an integrated and integrative discipline. To that end, the book discusses concepts and approaches at multiple levels of biological organization.

Natural selection operates at any of the hierarchical levels of individuals, kin, or groups, and on genes, gametes, zygotes, cells, or populations. Although often regarded as the raw material of evolution, genes do not make structures; there is no direct, one-to-one relationship between genotype and phenotype. Genes exert their influence on evolutionary change through mutations, duplications, rearrangements, novel cascades and networks, conserved patterns of expression, and so forth. The types and mechanisms of mutations are many and varied, including nucleotide substitutions, point mutations, insertion of transposable elements, shuffling of exons, and inversions of chromosomes. Some mutations are neutral but many are potentially beneficial. All are subject to selection, either directly or indirectly. Similarly, phenotypes and the processes that produce them are subject to selection; cells, embryos, and modifications of genetic and developmental processes are as much the raw material of evolution as are genes and mutations.

Evolutionary developmental mechanisms exist as a hierarchy of processes and so can be studied at all structural levels from genes to populations, using the wide variety of techniques and approaches developed in other fields. This diversity of mechanism and approach is both a strength and a drawback. On the one hand, a multitude of mechanisms act either alone or synergistically to effect evolutionary change. On the other hand, understanding and synthesizing this diversity demands a truly integrative approach and increased communication among the traditionally separate fields of genetics, molecular biology, developmental biology, paleontology, phylogenetics, morphology, and ecology (to name a few). No unified theory of evo-devo exists. However, many of us assume that development will find a more inclusive place in the study of evolutionary biology, just as it will find a more central role in studies of life history strategies, phenotypic plasticity, and ecological communities, to name a few.

To move toward these goals, we have invited experts with a diversity of backgrounds to contribute to this volume. Arranged alphabetically, the entries exemplify the hierarchical approach that is so characteristic of evo-devo and that is required to understand, in their fullness, developmental, evolutionary, and evolutionary developmental processes.

Topics traditionally associated with the Modern Synthesis, such as evolution, genetics, environment, selection, speciation, phylogenesis, variation, evolvability, and the relationship between the phenotype and genotype, occupy a central position in evo-devo, and so are included in this volume. Also included are discussions at the level of the gene, including regulation, cascades, and gene-gene interactions. At the cellular level, central topics such as cell division, determination, differentiation, patterning, interactions, and the role of cells in embryonic inductions are discussed in the context of their roles as evolutionary developmental mechanisms. Entries on genome size, epigenetic processes, and genomic and extragenomic inheritance show how genetic and cellular levels are linked.

The links among cells, tissues, and organs are discussed in such contexts as modularity, segmentation, epithelial-mesenchymal interactions, patterning, complexity, growth, heterochrony, and heterotopy. Links between development and morphology at the organismal level are discussed in the context of morphogenesis, ontogenetic repatterning, canalization, genetic assimilation, and structural and functional accommodation. Changes in life history are treated with reference to innovations and novelties, life history evolution, larvae, constraints, plasticity, and polymorphism.

The authors were encouraged to present and develop their own perspectives on their topics. Each provides an overview of the current status, the central concepts and approaches of evo-devo, and where the field is moving. Some entries include historical or philosophical background. We are delighted that John Bonner accepted our invitation to write the foreword. His seminal studies, which span half a century, laid the foundations for the current field of evo-devo.

How to Use This Book

The entries are arranged alphabetically by subject. Gilbert and Burian's entry, "Development, Evolution, and Evolutionary Developmental Biology," may be a good place to begin. This entry provides an overview of evo-devo and places the field in a historical context.

We placed references for all entries in a single bibliography at the end of the volume. The comprehensive index also serves as a glossary. To find definitions of terms, concepts, or ideas, consult this index. We also used the index for cross-referencing between entries. Genetic nomenclature follows Appendix 1 in Wilkins (2002).

Although we believe that we have covered the major concepts and approaches in evo-devo, we welcome your comments.

Keywords and Concepts in Evolutionary Developmental Biology

GRAHAM E. BUDD

Animal Phyla

The Animalia comprise one of the great kingdoms of life, probably encompassing several million species. Despite this enormous diversity, however, animals are placed in one or other of the relatively few phyla with, at least traditionally, few morphological links between them. Thus defined, the phyla have posed several important and classical questions, including:

Why are the phyla so distinct?
Why are there so many (or so few) of them?
Are there, or could there be, extinct phyla?
What are the relationships between the phyla, and what are their origins?
Is the fossil record at all informative about their early evolution?
Why have phyla appeared to be such stable entities through time?

This entry explores some of these issues and shows which are based on conceptual misunderstandings and which are genuine problems that could, and should, be addressed by research into fossil and living taxa.

The Body Plan Problem

Animals are traditionally defined as multicellular and heterotrophic, and they are grouped into some thirty or so phyla that have highly contentious relationships with each other. Apart from the basal groups (Porifera, Cnidaria, and some minor taxa with controversial positions), the majority of the phyla are grouped together into a large clade called the Bilateria, which are animals with bilateral symmetry along an anterior-posterior axis. These animals usually have a head with concentrated nervous tissue and sense organs, musculature, and three embryonic germ layers *(triploblasty)*. Many also possess a mesodermally bounded body cavity called a *coelom.* Bilaterians are traditionally divided into two great superphyla: the Protostomia and the Deuterostomia, which are distinguished on the basis of developmental features such as the mode of coelom formation. These features, however, are not as consistent as the groupings demand, and as a result these groupings have often been questioned. In recent years, the use of molecular systematics has brought the groupings under serious assault. Almost all work in evolutionary developmental biology has been carried out on bilaterians, although attention has now also been drawn to the more basal cnidarians. The development of sponges, the putative most basal grouping, has been relatively poorly studied, although recent advances in systematics renders sponges of great interest for the study of the origins of eumetazoans (cnidarians plus bilaterians).

Metazoans have customarily been divided into several grades that can be parodied as the possession of cells (sponges), tissues (cnidarians), and organs (bilaterians). These simple views mislead to the extent that they attempt to draw clear lines between grades that are likely to be evolutionarily related, and ignore taxa (such as ctenophores) that do not fit easily into these characterizations. Indeed, they may even ignore such data as the pronounced bilaterality of some cnidarians or the high degrees of differentiation and regionalization in sponges. Even the defining features of the metazoans are unlikely to have applied in the past in the same way as they do now: multicellularity and heterotrophy, for example, are unlikely to have evolved simultaneously.

Such definitional problems point to a conceptual difficulty that has bedeviled animal research for centuries, if not millennia. The *body plan* or *Bauplan* concept (Woodger, 1945) suggests that underlying different types of animals is some sort of invariable design, with each phylum being a set of variations on a particular theme. Certainly examination of a large group of organisms such as the echinoderms shows a degree of morphological similarity. Furthermore, it is possible to extract from them a notion of this *Bauplan*—a set of shared features that is inherent to our notion of what an echinoderm is.

The question, then, is whether this undeniable set of features (the water vascular system, the stereom skeleton, and so on) is merely an *a posteriori* construct from the present diversity, or whether it is more fundamental, with the implication that variation within the group is limited to within set boundaries. The former of these two views is trenchantly defended by Williams (1992), who implies that the retention of certain characters in a phylum is merely a statistical quirk. Others (e.g., Gerhart and Kirschner, 1997) have regarded the body plan as a definite package of morphology together with a distinctive developmental program that has hardened to become resistant to change at a certain level.

Both of these ways of looking at a phylum seem at least partly reasonable, and this paradox suggests a subtle linguistic conflict in how the term *phylum* is used. Simply put, when we group animals such as echinoderms together, we are relying on the ancient practice of categorizing organisms on the basis of their shared similarities, and in this sense, a phylum is a purely morphological and descriptive shorthand. Echinoderms, for example, are (seen in retrospect) those organisms that have a certain set of features. This pure term, however, is not a phylogenetic one; in theory such a group might be *monophyletic* (a group consisting of an organism plus all of its descendents), *paraphyletic* (a group consisting of an organism, but not all of its descendents), or even *polyphyletic* (a group that does not contain its own last common ancestor). Historically, early groupings of organisms were indeed a mixture of these sorts; as well as echinoderms (monophyletic), there were also "vermes," a group of worm-shaped organisms (polyphyletic), and rotifers (paraphyletic, because they gave rise to acanthocephalans).

The rise of phylogenetic systematics, however, has driven a wedge between relatedness and morphological similarity. This approach seeks to eliminate mere similarity as the basis for systematics. Nested clades may have accumulated many autapomorphies of their own and thus may be morphologically very different from the (paraphyletic) assemblage from which they arose. Discovering that this has in fact happened is rendered very difficult by the classical morphological approach to phyla, because paraphyletic phyla are so readily accepted. As a result, the relationship between a paraphyletic grouping and the phylum to which it gave rise, together with the great modification of the body plan that would be involved, might never be recognized.

Following this line of thought the body plan concept is partly artifactual. The relationship between affinities and morphology is, however, a subtle one that does not seem to be fully captured by either of the simple views we have outlined. Phyla do seem to be, in general, more coherent and stable entities than the view espoused by Williams would suggest. Taking a strictly monophyletic view of the clade, all living echinoderms appear to be characterized by certain features, and there are no echinoderm-derived taxa (as far as is known) that lack them. Furthermore, the accumulation of these important features seems, even when tempered by some recent arguments to the contrary (Budd and Jensen, 2000), to have taken place relatively early in each group's history (e.g., Erwin, 1993). Is this early lability followed by stability really just attributable to chance? If not, there may be more to phyla than mere historical descent. The field of evolutionary developmental biology has tended to regard this suspected stability as essentially a developmental one, but there are many other possibilities. For example, it could be caused by ecological conservatism or indeed conservatism in general: as long as morphology functions in any particular environment, it will be retained (see Valentine, 1995). The fossil record should be of help in resolving this issue.

Origins of the Metazoa

The fossil record is virtually silent on the origin of the Metazoa itself. Molecular evidence mostly suggests that there was a radiation in the eukaryotes some 2 billion years ago. The stem lineage of the Metazoa probably emerged at this time, together with plants and fungi. These early stem-group metazoans are unlikely to have been animals in the sense of being multicellular, and as such they are unlikely to be readily recognized in the fossil record. Nevertheless, the fossil record of the period does record a notable increase in diversity of single-celled presumed eukaryotes. Furthermore, many basal metazoans are soft-bodied, and may not be expected to leave a significant fossil record. The absence of sponge spicules or any of the calcareous skeletons of the cnidarians before the terminal Proterozoic is puzzling, suggesting that the origins of at least some clades were comparatively late.

One further point of interest is that, although molecular and morphological data agree in placing sponges followed by cnidarians at the base of the Metazoa, forming successive sister groups to the Bilateria, some molecular evidence (e.g., Collins, 1998) suggests that extant sponges are paraphyletic, with the Calcarea as the sister group to the Eumetazoa (cnidarians plus bilaterians). If confirmed, this highly significant result would imply that all extant higher metazoan diversity arose out of an organism with the same organization as living sponges, and that the shared morphological and developmental features of extant sponges were present in stem-group eumetazoans. A critical study of the shared developmental features of living sponges is thus an urgent priority for understanding subsequent evolution of animal development.

Origins of the Bilateria

The perceived remarkable contrast of bilaterians with cnidarians and sponges in terms of organization has led to an almost comical proliferation of theories to explain their origin, involving many of the principal concepts of evolutionary developmental biology over the last 250 years (partially reviewed in Willmer, 1990). The Continental schools of speculative zoology, drawing on the ideas of Haeckel, consistently saw the ancestral bilaterian as a coelomate organism that evolved through a cnidarian-grade ancestor, with the body cavity of cnidarians as directly homologous to the coelom. A more sophisticated version of this recapitulatory model was presented by Jägersten (1972), with refinements added by Nielsen (e.g., Nielsen and Nœrrevang, 1985) in more recent years. Conversely, Anglo-American workers, influenced by Libbie Hyman's monumental but unfinished survey of the phyla, have often concluded that the more common acoelomate phyla, especially the relatively featureless flatworms, were primitive and derived from a planula-like ancestor—the creeping, nonfeeding larva of cnidarians. This is a model of paedomorphic heterochrony rather than the peramorphosis of the Haeckelian school.

More recently, proposals have been made that connect morphological radiation in the bilaterians with increase in developmental complexity. In particular, indirect development, with the action of body patterning genes confined to a small set of cells in the larva (Davidson et al., 1995), has been seen as of potential importance in understanding the stages in the origin of the bilaterians (an idea once again returning to recapitulationist views, and one that has been vigorously disputed).

Such models are testable in principle by reference to bilaterian phylogeny. Molecular systematics has, however, tended not to confirm at least the Hyman-derived view, because acoelomates and pseudocoelomates tend, in molecule-based trees, to be scattered throughout the bilaterian tree and not to lie in a basal position. The initially surprising discovery of the conservation of developmental genes throughout the entire clade of bilaterians has led some researchers to conclude that the morphology the genes regulate today (e.g., formation of limbs, heart, eyes, and segments) was present in the bilaterian ancestor. The resulting picture of the so-called Urbilaterian has sometimes been remarkably complex, including a head with complex tripartite brain and eyes, limbs of some sort, segments, a heart, presumably a coelom (implied by the presence of the heart), and other features. If so, then the many phyla that lack some or all of these features would have to have become secondarily simplified, sometimes on a grand scale. Even traditionally (although, it should be stressed, not universally) accepted groupings within the coelomate bilaterians have come under assault (e.g., Peterson and Eernisse, 2001).

Molecular phylogenies have had two other major effects. The traditional protostome/deuterostome split has been blurred by the transferal of lophophorate animals from the deuterostomes to the protostomes. This move is in agreement with some morphological features, although the core of deuterostomes, echinoderms, hemichordates, and chordates remains intact. Controversy has also arisen concerning the protostomes. The traditional Articulata grouping of annelids and arthropods has been broken apart, with annelids allied with molluscs, brachiopods, and other coelomate protostomes, and the arthropods placed, together with their relatives the onychophorans and tardigrades, as part of a large group of moulting animals termed the Ecdysozoa. This group includes several of the pseudocoelomates: the nematodes, priapulids, nematomorphs, loriciferans, and kinorynchs. These reassignments are far from fully accepted, especially among morphologists; but they clearly have important implications about character homologies in important bilaterian features such as the coelom and heart, and thus about ancestral bilaterian features. These results have been in some ways anticipated and amplified by evidence from the Cambrian fossil record (e.g., Conway Morris and Peel, 1995; Budd, in press).

Early fossils. The earliest fossil assemblages that probably contain animals date from late in the Proterozoic. These are all probably younger than about 570 million years and part of the worldwide and infamous "Ediacaran biota." Despite the controversy that surrounds these fossils, it is generally agreed that ediacarans are a highly heterogeneous assemblage united by preservational circumstance rather than by close phylogenetic affinity. In addition, some of them were mobile, and some must have had rather unusual lifestyles, for example, being permanently buried. Hauntingly and tantalizingly, the later assemblages contain a set of simple trace fossils that must have been made by bilaterally symmetrical organisms. Although some cnidarians can burrow, these early traces probably represent at least basal bilaterians, although the actual nature of the organisms that made these traces is unknown.

The end of Ediacaran time was previously thought to be marked by a mass extinction that removed the ediacarans and allowed bilaterians to flourish, but recent reassessments of the timing of the terminal Proterozoic and Early Cambrian have called this view into question. Indeed, Cambrian Ediacaran forms have now been reported, suggesting an uninterrupted biotic expansion through this time—a view in accord with the critical trace fossil evidence (and indeed, the skeletal fossil record). If so, then the end of the Proterozoic and the beginning of the Cambrian record may document the actual emergence of the phyla, a critical point if the developmental evolution of the period is to be reconstructed.

The most exciting area of Cambrian research has been on the exceptionally preserved faunas, especially the "big three" of the Burgess Shale, Chengjiang fauna, and Sirus Passet fauna (Conway Morris, 1998). Another more recent source of extraordinary paleontological data concerning the origin of phyla comes from what can loosely be termed (in contradistinction to "Burgess Shale preservation") "Orsten" preservation, named after Late Cambrian faunas from Sweden and elsewhere. These fossils are typically small (less than 2 mm) and phosphatized, preserving stunning details of micrometer-scale morphology. Potentially the most interesting of all these finds are those of animal embryos from the Early Cambrian, and perhaps (although less certainly) from the terminal Proterozoic. These fossils show, in general, rather early stages of development, and they seem mostly to represent direct development (but see Chen et al., 2000 for putative examples of indirect development), an interesting point in light of the many theories of animal evolution that demand indirect development to be primitive. Unfortunately, none of the putative bilaterian embryos can be unambiguously tied to a known adult; but if this linking can be achieved, the Cambrian fossil record may yet reveal direct evidence for the evolutionary developmental biology that would have been thought beyond the reach of the fossilization process until only a few years ago.

In general, both Burgess Shale and Orsten faunas reveal a wide array of taxa, many of which are controversial. Modern study initiated by Whittington (e.g., Whittington, 1985) and others reveals that many or most of the taxa described do not fit into extant groupings at certain levels. For example, of the many arthropods, very few can be assigned to the modern crustaceans, chelicerates, and so on. Many taxa do not fit even into phyla. Do these taxa count as extinct phyla, implying an enormous capacity for the generation of new body plans (creating high "disparity") in the Cambrian that shut down shortly afterward? Until the 1990s, the view that many of these taxa represented extinct body plans and phyla was generally accepted. Since the publication of Gould's *Wonderful Life* (1989a) which was this view's apotheosis, however, there has been, especially in paleontological circles, a decided reaction against the view.

Modern systematics, building on conceptual advances by Willi Hennig and Dick Jefferies, takes a very different view of early disparity. The basic problem with almost all of the usual views of phyla is that modern groupings, by their very nature, cannot encapsulate all of ancient diversity, simply because of the vast amounts of extinction. Indeed, modern groups appear distinct, not because they have always been so, but because the intermediates between them are now extinct. Fossils inevitably fall into these gaps at some level or another. The fact that fossils do not fit into extant classifications merely reflects systematics procedure and is of no interest in itself.

The best way to encapsulate this difference between fossils and living forms is through the "stem" and "crown" group concepts (Budd and Jensen, 2000). In this view, Cambrian oddities are members of stem groups of groups of phyla, of phyla, or of classes. Only when stem and crown groups are recognized can the true pattern of the Cambrian fossils be examined objectively. Cambrian fossils, it should be noted, are but one example of many such cases. Whenever a major radiation occurs, it follows that many new clades appear, and thus many new stem groups emerge. One would therefore expect radiations—such as the origin of the tetrapods and mammals, or the invasion of land by plants—to be represented by similar patterns. This systematics reorganization does not in itself show whether Cambrian taxa are more disparate than extant ones, but merely presents the correct pattern to analyze. However, to date, no studies have shown any higher disparity in the Cambrian compared to the recent era.

Systematization of fossils into the stem groups of various phyla may allow the order of assembly of phyla body plans to be followed. This in turn suggests that the origin of the phyla should be considered in the context of ecological radiation in a population and adaptive sense (e.g.,

Schluter, 2000). The ways in which the phyla were assembled, and their subsequent stability or otherwise, thus both seem to be important topics for evolutionary developmental biology that the fossil record can contribute to. If there were something special about the period of evolution that led to the Cambrian explosion, then increased elasticity in morphology in Cambrian taxa might be expected. This popular notion has rarely been tested, but some results (e.g., Hughes et al., 1999) do indicate that Cambrian taxa (some trilobites, for example) are morphologically more variable than their post-Cambrian successors. This sort of analysis is, however, fraught with difficulties. Does greater variation in populations imply looser ecological and selective control (more allowed morphologies), as these authors suggest, or lower degrees of canalization (more morphologies generated by development)?

Finally, and highly controversially at the time of writing, molecular clocks suggest that the deuterostome-protostome split occurred as early as 1 billion years ago or more, predating the bilaterian fossil record by 500 million years (Wray et al., 1996). This timing obviously implies a lengthy cryptic period of bilaterian evolution. While this "phylogenetic fuse" has been happily accepted by some paleontologists (e.g., Wills and Fortey, 2000), it raises several important difficulties, the most obvious of which is the necessity to account for a missing fossil record of such a long period of time. This problem has been tackled by a variety of more or less *ad hoc* appeals, mostly to the soft-bodied or otherwise unpreservable nature of early bilaterians (e.g., by being planktonic or meiofaunal); but many of these hypotheses are at odds with phylogenetic reconstructions of early bilaterians (Budd and Jensen, 2000).

Gaining a clearer understanding of the evolution of animal development will require attention to phylogeny and shared features between the phyla, rather than treating them as separate entities. The phyla themselves have begun to look less monolithic, partly because of molecular and morphological assaults on their distinctness, leading to new groupings of phyla such as the Cycloneuralia and Ecdysozoa, and partly because of the slow recognition of Cambrian fossils lying in the stem groups of phyla. An understanding of the step-by-step processes by which modern animals were assembled will in turn allow a much tighter theoretical constraint on the types of change in developmental processes that took place concurrently, with the ultimate aim of tying in evolutionary developmental biology and the origin of phyletic-level features to population processes.

MELANIE L. J. STIASSNY

Atavism

The term *atavism* commonly refers to the reappearance of a character state typical of a remote ancestor in an individual that really shouldn't have it. That is to say, the state has been lost, or more commonly transformed, long ago within the history of its lineage. Such features have long intrigued evolutionary biologists, not least because they demonstrate the potential for underlying morphogenic mechanisms to be conserved over long periods of time, even though they may not have been phenotypically expressed for millions of years. Central to the concept of atavism is the notion of an underlying continuity spanning considerable phylogenetic distances, and consequently an atavistic feature can be recognized as such only within the context of a phylogenetic hypothesis.

Hall (1984, 1995) provides an excellent review of the subject in which he emphasizes features common to the formation of atavistic morphologies. In general, two main types of atavism are recognized, with a third waiting in the wings. The first is *spontaneous atavism,* in which rare atavistic anomalies occur in individuals of natural populations. Among vertebrates, familiar examples are the sporadic occurrence of hind limb rudiments in modern cetaceans, certain atavistic digits in horses, and atavistic muscles in birds and mammals. The second type are atavisms that can be generated experimentally. Perhaps the most spectacular of these *experimental atavisms* is the apparent re-establishment of ancestral patterns in the hind limbs of experimentally perturbed embryonic chicks (e.g., Hampé, 1960; Müller, 1989), or the formation of tooth-like structures by embryonic chick epithelium under the inductive influence of embryonic mouse dental mesenchyme (Kollar and Fisher, 1980). Similarly illustrative are other perturbed inductive interactions in the formation and transformation of various epidermal structures (Dhouailly, 1973).

The observation that ancestral traits may occasionally be re-expressed as spontaneous atavisms, or invoked by experimental manipulation, suggests that a considerable degree of genomic or epigenetic potential lies latent in evolving systems. Phenotypic expression may be lost, but underlying potentiality remains. The potential of such phenomena to lend insight into the mechanisms of evolutionary transformation is generally recognized but usually sidelined; atavism has been viewed more as the "Barnum and Bailey" of evolution than as mainstream. A notable exception to this view is presented by Reilly and Lauder (1988), who recognize an important auxiliary role for atavism in the assessment of homology. Their study of a population of *Notophthalmus viridescens* in which atavistic epibranchials are expressed in most individuals helped resolve a long-standing controversy regarding the homology of branchial arch segments in salamanders. With anatomical and topological data gleaned from careful examination of these atavistic arch elements, they were able to conclude that the basic segmental arrangement of the visceral arches has been conserved across the transition from aquatic to terrestrial life in vertebrate evolution.

The view that atavisms are interesting but basically irrelevant to the evolutionary process is well expressed by Levinton (1986, p. 265). He argues that, while it may be possible to induce novel structures by experimental manipulation, such manipulation only induces what evolution has already created. The importance of atavism is thus overstated because of "the fallacy of looking at highly evolved systems, backwards towards ancestral states that have been suppressed in the evolution of new taxa, and then mistakenly inferring a creative force." In addition, Rachootin and Thompson (1981, p. 188) claim that "these ancestral potentials have little chance of expression in themselves, because their ancestral phenotypic expression is irrelevant to the selective pressure favoring their retention," and in the impeccable language of the 1980s: "The legacy exists, but the assets are not convertible." They conclude with the observation that "A genuinely labile epigenetic system ought to be of the greatest interest to an evolutionist, and it is a far, far better thing than an occasional atavistic slip in a mature epigenetic system." While this view may represent reasonable caution against the overelaboration of theories based on experimental or spontaneous atavisms, there is a third type of atavism that may be of considerably more significance in this context.

This is the far more common and widespread occurrence of phylogenetic character reversal, or *taxic atavism,* so familiar to practicing systematists.

Recognizing phylogenetic character reversals as taxic atavisms, and therefore as manifestations of the same basic phenomenon as the other types of atavism, counters many of the objections already outlined. Here ancestral traits have become established as normal phenotypic components of descendant groups and are expressed in all members of a given clade. Taxic atavisms are solidly embedded in evolving entities (clades, species, populations) and expressed as a normal phenotypic component of each and every individual. As such, they can no longer be dismissed as individually expressed or experimentally induced anomalies.

How pervasive is this phenomenon? Of the considerable "noise" accompanying the phylogenetic signal (Rieppel, 1989; Wake, 1991; Futuyma, 1998; Meyer, 1999), there is little doubt that a significant portion results from character reversal. Examples of taxic atavism abound in the systematic literature (e.g., Raikow et al., 1979; Wake and Larson, 1987; Wyss, 1988; McCune, 1990; Stiassny, 1992; Desalle and Grimaldi, 1993); and as phylogenetic analyses become increasingly sophisticated, considerably more will be revealed. Such widespread occurrence suggests that evolving populations can and do generate variation by drawing on past morphologies, and that the underlying mechanism for this most probably involves the reactivation of conserved morphogenetic systems. The reintegration of these ancestral character states does not present significant developmental problems or cause major genomic disruption, as might be expected from a characterization as "atavistic slips" or maladapted "throwbacks." Moreover, within any evolving lineage there is a potentially vast reservoir of plesiomorphic morphologies that may be re-expressed at various levels; and while it is true that most taxic atavisms appear to "go nowhere," all present an opportunity for adaptive remodeling. The prospective potential of taxic atavism to provide an important auxiliary source of variation for subsequent remodeling within clades is explored by Stiassny (1992).

One example serves to illustrate the point. Schultz (1986), in a review of the origin of the spinning apparatus of spiders, proposed that the first arachnids had opisthosomal appendages and that all but spiders (Araneae) have lost them subsequent to the divergence of the scorpions. However, a simpler alternative to Schultz's hypothesis of plesiomorphic retention in spiders and multiple independent loss in uropygids, amblypygids, palpigrads, and other apulmonates would be that the fourth and fifth opisthosomal segmental appendages of spiders arose instead by reactivation of ancestral opisthosomal appendage expression (Figure 1). The functional significance of the unique spinning apparatus of spiders is in little doubt. What is less obvious is that this critical innovation may originate as a rather simple taxic atavism, in this case the release of a retained potential to express opisthostomal appendages followed by their subsequent remodeling into spinnerets.

In concluding his consideration of atavism, Levinton (1986, p. 267) writes that while "hopeful developmental monsters" may be of significance for evolutionary study, they are too rare and too fragile to be of much use in the laboratory. He adds that "major developmental mutants are invariably sickly and show pervasive deformities. From both theoretical and empirical viewpoints, hopeful monsters have led only to hopeless mooting." The recognition of phylogenetic character reversal as another, in this case very common, manifestation of the same atavistic phenomenon should alter our perception considerably. The manner in which these taxic atavisms can and do reappear, often over protracted periods of time and over large taxonomic distances, should be of the

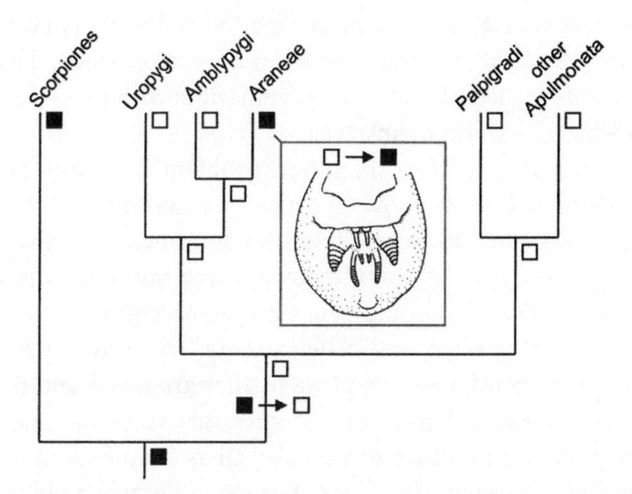

FIGURE 1. Intrarelationships of arachnid orders after Schultz (1986). Superimposed onto the scheme are the hypothesized character state changes. Black squares indicate the presence of opisthosomal appendages in adults, and white squares indicate their absence. Here the presence of opisthosomal appendages in Araneae is recognized as an example of taxic atavism. Inset illustrates opisthosomal appendages in ventral view of *Liphistius,* a primitive spider. Modified from Stiassny (1992).

greatest interest to evolutionary biology. It is in the bearers of complex taxic atavisms that are to be found a ready arsenal of viable "hopeful monsters" awaiting fruitful study.

GILBERT GOTTLIEB

Behavioral Development and Evolution

The problem of the origination of the behavioral actions, anatomical structures, or physiological functions that lead the way to evolutionary change—to the origin of new species—is a vexing one. The standard, neo-Darwinian solution has been genetic: a favorable change in genes and corresponding phenotype brought on by mutation, recombination, or drift. However, the distance between gene and phenotype is so great that there must be not only genetic compatibility but also developmental compatibility in order for the new phenotype to surface. That is why most mutations are lethal; they are incompatible with the developmental system in which they find themselves.

Another approach to the origination problem is to look at development itself—to determine how genetically compatible developmental changes can originate new physiological, anatomical, or behavioral features through the expression of already existing but otherwise quiescent genes. Many potentially coding genes are quiescent; and, in collaboration with messenger RNA and other factors, when activated they can participate in the production of proteins that are novel and thus result, with further epigenetic variations, in new structures or functions. We know this from tissue culture studies as well as from experimental reorganization of developing organisms. For example, when chick oral epithelial cells are brought into contact with mouse oral mesenchyme cells, the chick's oral epithelial cells are capable of participating in the formation of chimeric teeth (Kollar and Fisher, 1980). The contact reactivates a genetic potential that has remained dormant for 80 million years, at which time reptilian ancestors of birds possessed teeth. As a further example, G. Müller (1986) operated on chick embryo hind limbs to induce an enlarged, elongated fibula, one similar to that found in avian ancestors. The skeletal change resulted in secondary alterations of the later-forming zeugopod muscles that resembled the ancestral arrangement. Müller concluded: "The experimental reestablishment of an ancestral epigenetic condition within a developing organ system resulted in ancestral features of subsequently forming characters. This suggests that modification of epigenetic control of gene expression may be an important mechanism in evolutionary change" (Müller, 1986, p. 105).

Any effort to take this developmental or epigenetic view of evolution to the behavioral level is handicapped by the paucity of empirical examples. Behavior does not fossilize, and the idea that changes in behavioral development can bring about evolution is not popular, so few experiments are directed to this question. We do know that substantially changing species-typical developmental circumstances, in particular the physical or social environment, will bring out already-existing possibilities for behavioral change within a single generation. This entry builds on this fact to make a case for the role of behavioral development in evolution, an alternative originating pathway (not mutually exclusive) to the familiar gene-based originating pathway (mutation, recombination, drift) of the Modern Synthesis (Table 1).

TABLE 1. Developmental-behavioral evolutionary pathway

I. Change in Behavior	II. Change in Morphology	III. Change in Gene Frequencies
First stage in evolutionary pathway: change in ontogenetic development results in novel behavioral shift (behavioral neophenotype), which encourages new environmental relationships.	Second stage in evolutionary change. New environmental relationships bring out latent possibilities for morphological-physiological change. Somatic mutation or change in genetic regulation may also occur, but a change in structural genes need not occur at this stage.	Third stage of evolutionary change resulting from long-term geographic or behavioral isolation (separate breeding populations). It is important to observe that evolution has already occurred phenotypically before stage III is reached.

Modified from Gottlieb (2001), chap. 14.

Developmental Induction of New Behavioral Phenotypes (Behavioral Neophenogenesis)

In 1956, Seymour Levine performed a developmental manipulation that he thought would traumatize young rats: He removed them from the care of their mothers for a few minutes every day of the preweaning period, a procedure called *handling.* Rather than traumatizing the rat pups, this short-term stressor had a beneficent effect on their hypothalamic-pituitary-adrenal axis. As adults, the handled animals were more tolerant of serious stressors than unhandled control animals. Other experiments showed that the manipulation, in order to be effective, had to be carried out in the preweaning period (Denenberg, 1969). The handling procedure was similarly effective with laboratory strains of mice, but not with rabbits, in which the mother does not groom and care for her young as extensively as do rat and mouse mothers.

This early experience manipulation effected a dramatic change in the handled animals' ability to cope with stress in adulthood, while also enhancing their behavioral exploratory tendencies. These sorts of changes could have ramifications for natural selection because they would enhance the affected animals' capability of surviving and successfully reproducing in the face of a severe environmental alteration at any point after weaning. It might give such animals a slight reproductive advantage even in a relatively stable ecological circumstance, particularly if, e.g., the heightened exploratory tendency was deployed in the service of foraging. The main point is that an early experience manipulation of a non-species-typical kind brought out a new behavioral phenotype (a behavioral neophenotype), as had been predicted by the much earlier research of the Chinese psychologist Zing-Yang Kuo (summarized in Kuo, 1976). It is of considerable interest that the effects of the handling experience on the females carry over to the behavior of their unhandled progeny, mediated by changes in the maternal behavior of the handled females (Denenberg and Rosenberg, 1967).

Another early experience manipulation with implications for natural selection and evolution is the so-called environmental enrichment paradigm, in which laboratory animals, rather than being caged singly in relatively barren surroundings, are caged in social groups with other members of their species and exposed to a succession of novel toys and playthings that they explore and manipulate (summarized in Renner and Rosenzweig, 1987). This early experience had the effect of substantially elevating their learning ability in adulthood, particularly when challenged by the most demanding of tasks, the multiproblem Hebb-Williams maze. The enriched early experience also enhanced brain size.

Thus, the species-atypical experience of handling and enriched early experience lead to increased resistance to stress, enhanced exploratory behavior and problem-solving ability in adulthood, and an increase in brain size. These behavioral changes would facilitate adaptation should an organism's usual environment change drastically and would also support the seeking out of new habitats in the absence of environmental change, setting the stage for evolutionary change.

A third empirical example of the potential for "deviant" early experience to foster the development of behavioral novelties with evolutionary significance is the phenomenon of exposure learning (Sluckin, 1965). This form of learning takes place through mere exposure of young organisms to physical and social objects; the exposure leads to subsequent attachment, preference, or acceptance of these objects. This familiarity mechanism is pervasive. It is found in many invertebrate and vertebrate species (Szentisi and Jermy, 1989), and it is often referred to as imprinting—although, strictly defined, imprinting involves a sensitive period early in life and induces an enduring preference, not merely acceptance (nonavoidance) of an object or an experience. While particularly striking in young animals, exposure learning operates throughout life in some species such as our own (Hebb, 1946; Rheingold, 1985; Zajonc, 1971).

According to Mayr (1963, pp. 604, 605): "A shift into a new niche or adaptive zone is, almost without exception, initiated by a change in behavior. The other adaptations to the new niche, particularly the structural ones, are acquired secondarily. This is not the place to discuss how the behavior changes themselves originate." The preceding paragraphs sketched a theory of how such behavioral changes could originate. An empirical example of incipient speciation mediated by a transgenerational, developmental change in behavior is provided by the apple maggot fly, *Rhagoletis pomonella.* The original native (United States) host for the female apple maggot fly's egg laying was the hawthorn, a spring-flowering tree or shrub. Domestic apple trees were introduced into the United States in the seventeenth century. Haws and apple trees occur in the same locale. The first known infestation of apple trees by apple maggot flies was in the 1860s. There are now two variants of *R. pomonella,* one that mates and lays its eggs on apples and the other that mates and lays its eggs on haws (Table 2). The life cycles of the two variants are

TABLE 2. Developmental Behavioral Basis of Evolution: Incipient Speciation in Two Variants of Apple Maggot Fly *(Rhagoletis pomenella)*

Time	Apple Host	Hawthorn Host
Year 1	Eggs Laid	Eggs Laid
	↓ Fruit matures earlier than haw	↓ Fruit matures later than apple
Year 2	Hatch Late Summer	Hatch Early Fall
	↓ 5–12 days	↓ 5–12 days
	Offspring court and mate on or near host, and female lays eggs on same host	Offspring court and mate on or near host, and female lays eggs on same host
	↓	↓
Year 3	Cycle Repeats	Cycle Repeats

Abstracted from Prokopy and Bush (1993); Bush and Smith (1998).

now desynchronized because apples mature earlier than haws. Incipient speciation has been maintained by a transgenerational behavior induced by early exposure learning: an olfactory acceptance of apples for courting, mating, and ovipositing based on the host in which the fly developed (reviews in Prokopy and Bush, 1993; Bush and Smith, 1998).

We can only speculate on the cause of the original shift from hawthorns to apples as the host species for egg laying. Perhaps the hawthorn hosts became overburdened with infestations or, for other reasons, died out in a part of their range, bringing about a shift to apples in a small segment of the ancestral hawthorn population that did not have such well-developed olfactory sensitivity or an olfactory aversion to apples. This supposition is supported by behavioral tests, in which the apple variant accepts both apples and haws as hosts, whereas only a small percentage of the haw variant accepts apples and most show a strong preference for haws (Prokopy et al., 1988; Luna and Prokopy, 1995). As indicated by single-host acceptance tests, the apple-reared flies show a greater percentage of egg-laying behavior on the apple host than do the hawthorn-reared flies. Thus, the familiarity-inducing rearing experience (exposure learning) makes the apple-reared flies more accepting of the apple host, although they still have a preference for the hawthorn host. Given the ecological circumstances, the increased likelihood of acceptance of the apple host, even in the face of a preference for hawthorn, would perpetuate the transgenerational courting, mating, and laying of eggs in apple orchards.

Apple maggot flies hatch out at the base of the tree in which their mother laid eggs the previous summer (Prokopy and Bush, 1993, p. 6). As they mature sexually, the flies may wander tens or hundreds of yards, but they remain in the vicinity of the apple orchard, if not in the orchard itself. The scent of the apples attracts them. Due to this early rearing experience rendering the apple scent acceptable, the cycle renews itself because of the high probability that the early maturing apple maggot fly will encounter the odor of apples rather than hawthorns (see Table 2). In support of incipient speciation, the two variants are now genetically somewhat distinct and do not interbreed freely in nature, although they are morphologically the same and remain interfertile.

In contrast to the transgenerational behavioral scenario being put forward here, conventional evolutionary-biological thinking holds that "most likely some mutations in genes coding for larval/pupal development and adult emergence" brought about the original divergence and maintains the difference in the two populations (Ronald Prokopy, personal communication, August 2000). Although we cannot know with certainty, present evidence suggests a genetic mutation was not necessary. This is not a behavior versus genes argument. The transgenerational behavioral initiation requires genetic compatibility; otherwise, it would not work. The question is whether the original interaction (switch to the apple host) required a genetic mutation. The developmental timing change in the life histories of the two forms (see Table 2) has resulted in correlated genetic changes in the two populations that are indicated by allele frequency differences (Feder et al., 1997). That finding is consonant with the evolutionary model presented here (i.e., gene frequencies change sometime *after* the behavioral switch). From the present point of view, another significant feature of the findings is that, when immature hawthorn flies (pupae) are subjected to the prewintering environment of the apple flies (pupae), those that survive have a genetic makeup that is similar to that of the apple flies. Environmental selection is thus acting on already existing developmental-genetic variation. Most importantly, this result shows that there is still sufficient individual developmental-genetic variation in the hawthorn population, even at this late date, to support a transgenerational behavioral initiation of the switch from hawthorns to apples without the necessity of a genetic mutation.

To summarize, a developmental-behavioral change involving the apple maggot fly's choice of oviposition site puts it in a situation in which it must be able to withstand prewintering low temperatures for periods of time that differ between the apple and hawthorn forms (see Table 2). This situation sets up the natural selection scenario that brings about changes in gene frequencies that are correlated with the prewintering temperature regimen, as demonstrated in the Feder et al. (1997) experiments. The change in egg-laying behavior leads the way to genetic change in the population. The genetic change is thus a consequence of the change in behavior.

The Baldwin Effect

One of the earliest hypotheses bringing adaptability in behavior into the evolutionary process is the Baldwin Effect (Baldwin, 1896). The Baldwin Effect (BE) holds that (1) a severe change in the environment exerts a drastic selective effect, leaving only a fragment of the original population in place, i.e., those having sufficient behavioral or morphological adaptability to survive; (2) as the progeny of the survivors selectively breed among themselves for some number of generations in the new environment, (3) eventually the original behavior or morphological change brought out by the new environment becomes "congenital." It is fascinating that, although there was no empirical evidence for such a process, two other scientists put forward essentially the same scenario in the same year (Conwy Lloyd Morgan, 1896, and Henry Fairfield Osborn, 1896). The proposed process came to be called "organic selection" to differentiate it from natural selection and neo-Lamarckism.

Many years later, Waddington's (1953a) well-known "crossveinless" experiments concerning the eventual hereditary effect of heat shock on pupal development in fruit flies provided evidence for the "genetic assimilation of an acquired character" (his term, not Baldwin's). After fourteen generations of selective breeding, some individuals became crossveinless without the heat shock. The BE and genetic assimilation, and their relation to I. I. Schmalhausen's (1949, p. 175) notion of the importance of ontogenetic adaptability in the initial stage of the evolution of adaptations, are described more fully in Gottlieb (2002, pp. 126–136). The essence of the BE (organic selection) is that in due course, under the conditions described, an initially "acquired" character can become hereditary. This old-fashioned terminology obscures the hard-won fact that all phenotypes are a consequence of gene-environment interaction. Obviously, in Waddington's experiment the genes were there from the start in the organisms that responded with the crossveinless phenotype. Breeding among themselves could have heightened their temperature sensitivity so that, after fourteen generations, some crossveinless individuals did not require the original large heat shock but required only exposure to the inevitable small cyclic increases in temperature fluctuation that occur during normal incubation.

Thinking of the original response as nongenetic is one of the reasons why the BE has floated in a curious conceptual limbo to the present day. The idea that behavioral or morphological adaptability is not genetically based or correlated makes it difficult to incorporate the BE (and so-called genetic assimilation) into discussions of evolutionary mechanisms. For example, Ernst Mayr (1958) defined the BE as "the hypothesis that a nongenetic plasticity of the phenotype facilitates reconstruction of the genotype" (p. 354). Elsewhere in the same volume the BE is defined by another author as "the hypothesis that nongenetic adaptive modification of the phenotype may be replaced by genetically controlled modification" (Emerson, 1958, p. 318). The idea that behavior or any other organismic manifestation can be nongenetic in any sense of the word is incorrect. While population geneticists may sometimes say that the lack of heritable variation is "nongenetic" in the sense that the lack of phenotypic differences implies the absence of genetic differences, they clearly believe genes are correlated with traits. In other words, there must be a genetic basis for acquiring any character.

The present approach has in common with the BE and Schmalhausen's ideas the importance of behavioral adaptability in the early stages of the evolutionary process.

As has been shown, the first stage in an evolutionary pathway can be a change during ontogenetic development that results in a novel behavioral shift (a new behavioral phenotype) that recurs across generations, encouraging new organism-environmental relationships. In our hypothesized second stage, the new environmental relationships can bring out latent possibilities for anatomical or physiological change. Somatic mutations or changes in genetic regulation may also occur, but a change in structural genes need not occur at this stage. A change in gene frequencies may occur in the third stage of the evolutionary pathway as a result of long-term geographic or behavioral isolation creating separate breeding populations. It is important to observe that, in this theory, evolution has already occurred phenotypically at the behavioral, anatomical, and physiological levels before the third stage is reached.

Thus, new variations and adaptations in behavior can arise before they are selected for and are, therefore, not a consequence of natural selection. New variations and adaptations in behavior are a consequence of changes in individual development mediated by a transgenerationally persistent (consistent) change in the developmental milieu. As is widely appreciated, natural selection is not the cause of the new adaptations but acts only as a filter through which the new adaptations must pass. Changes in behavior create the new variants on which natural selection works. As noted by Mivart (1871, p. 240), "Natural Selection favours and develops useful variations, though it is impotent to originate them." As elaborated by Endler (1986, p. 51), "Natural selection cannot explain the origin of new variants and adaptations, only their spread." And, as Ernst Mayr remarked in 1963, novel behavioral shifts antedate the anatomical changes, which arise secondarily, in the evolution of new species. (And, of course, it can also go the other way around. There are multiple evolutionary pathways.)

The contribution of the present theory is to offer a behavioral scenario to bring about novel changes in development and an explanation of how such changes can be maintained across generations (through the transgenerational persistence of the changed developmental milieu) without any initial change in genotype.

The role of individual development in the evolutionary scenario is sometimes recognized, but there is resistance to accepting evolutionary pluralism, that is, alternative evolutionary pathways that do not involve genetic origination in the sense of mutation, recombination, or drift, the mainstays of the Modern Synthesis (Løvtrup, 1987; Futuyma, 1988).

Even Darwin himself was an evolutionary pluralist: "I am convinced that Natural Selection has been the main, but not exclusive, means of modification" (Darwin, 1859, p. 6).

Acceptance of the role of behavior in evolution (in the sense of seeing behavior as the leading edge of evolution) has a following beyond the more than thirty authors who contributed to the volume edited by Leonovicová and Novák (1987), *Behavior as One of the Main Factors of Evolution.* Along with Ernst Mayr, a number of biologists and a few psychologists have recognized that novel shifts in behavior are involved in speciation (e.g., Bonner, 1983; Hardy, 1965; Larson et al., 1984; Parsons, 1981; Piaget, 1978; Plotkin, 1988; Reid, 1985; Sewertzoff, 1929; Wyles et al., 1983). But what has not been recognized is the extragenetic developmental conditions that allow, prepare, or dispose animals to make novel shifts in behavior.

The "mechanisms" described earlier (handling, enrichment, exposure learning) are the means that have already been experimentally delineated; obviously, there are likely other similar behavioral mechanisms that have not yet been described. The developmental causes of novel behavior have been too little explored. Perhaps it is not too much to hope that this will change now that there is a theoretical rationale (see Oyama et al., 2001).

ADAM WILKINS

Canalization and Genetic Assimilation

At first glance, *canalization,* which is a property of developmental systems, and *genetic assimilation,* which is a type of adaptive evolutionary process, appear to be separate and distinct phenomena. Canalization refers specifically to the intrinsic robustness that developmental processes often display in response to perturbations, whether external (environmental) or internal (genetic). Genetic assimilation, on the other hand, denotes the evolutionary phenomenon in which an environmentally induced developmental potentiality within a population is selected and eventually becomes a fixed characteristic of that population, appearing in the absence of the original inductive stimulus.

Yet the two terms are linked historically and conceptually. The historical linkage is that they were both first characterized by the same two individuals, C. H. Waddington (1905–1975) and I. I. Schmalhausen (1884–1963). The principal underlying connection is more important, however. States of canalization must be broken or overcome during an episode of genetic assimilation in which a new developmental propensity becomes characteristic of a population. Once a new developmental pathway has achieved that new status, however, it becomes stabilized (canalized). In effect, genetic assimilation replaces one canalized state with a new one. Furthermore, by buffering against the expression of genetic variants, canalization allows such variants to accumulate in the population. That store of genetic variation, in principle, can lay the grounds for episodes of genetic assimilation.

The two phenomena share an additional feature: the precise evolutionary significance of both is a matter of controversy. Thus, while canalization, in dampening down the expression of genetic variation, should act as a brake on developmental evolution, it is not clear how much it actually retards such evolution in general. The debate over this point is connected to the difficulty of quantifying degrees of canalization, making it hard to compare different canalized states, especially between different organisms.

Genetic assimilation has received less attention than canalization in recent years, but its importance for developmental evolution remains equally problematic. First, whether genetic assimilation is a distinct phenomenon in its own right, different from the well-known selection of polygenically based characters that appear above a certain threshold of cumulative effect, has been questioned (Mayr, 1963). Second, Williams (1966) argues, in contrast to the claims of Waddington (1961) and Schmalhausen (1949), that genetic assimilation is unlikely to have been a significant source of genetic innovation in adaptive (and morphological-developmental) states. Resolving this debate will be difficult: ascertaining whether *any* particular evolutionary innovation originated through genetic assimilation or another evolutionary route is virtually impossible (Hall, 1992a).

This entry examines basic characteristics of canalization and its possible genetic and molecular foundations, discusses the evidence for genetic assimilation and discusses the debates about its character, and assesses the significance of both phenomena for the evolution of developmental systems.

A definitional ambiguity to note, at the outset, is that canalization (or "auto-regulation" in Schmalhausen's terminology) can refer either to a *state* of developmental resistance to perturbation or to the *process* by which such states are acquired. Although there are numerous variant definitions of the term (see Debat and David, 2001), most refer to canalization as a state. Furthermore, all emphasize that this state involves a damping down of possible phenotypic variation, whether the perturbations are environmental or genetic.

No state of canalization, however, has ever been directly demonstrated or measured. Instead, the existence of such states is inferred from comparisons between a reference strain, usually a wild type, and one or more genetically different strains. Thus, within a wild-type strain, the operation of canalization against an environmental insult (such as a temperature shock or exposure to sublethal doses of a toxic compound) is usually deduced from the fact that one or more other strains show a greater degree or frequency of developmental abnormality (of one or more kinds) in response to that treatment. Similarly, the existence of some built-in genetic "buffering" of wild-type developmental systems against genetic perturbation is inferred from the fact that mutants often show a greater range of phenotypic variability, as either variable penetrance or expressivity or both, than the wild type. (Whether the extent of fluctuating asymmetry, namely the difference between left and right sides of an animal, is an inverse measure of canalization strength has long been contentious. The developing consensus is that, in general, it is not [Debat and David, 2001]).

An important point about states of canalization is that they are specific to traits, not to organisms as a whole. Thus, for instance, those life history traits most central to evolutionary fitness appear to be more strongly canalized than other traits (Stearns and Kawecki, 1994). A more subtle point is that, in any assessment of the mechanisms of canalization, the particular way in which one defines an individual trait can influence how one interprets the basis of its canalization.

Take the example of scutellar macrochaete bristles in *Drosophila melanogaster.* Wild-type fruit flies have four of these large bristles on the dorsal thorax. Various mutant states and genetic backgrounds can alter the numbers significantly, from zero to eight. A great deal of early work went into elucidating the mechanism of canalization of scutellar bristle number (see Rendel, 1959, 1967). Yet, it has become increasingly apparent that these bristles do not develop as a function of a homogeneous field. Rather, it seems that the development of particular bristles at particular sites is differentially determined—and canalized—at these sites (Scharloo, 1991; Whittle, 1998). In effect, treating the scutellar bristles as a single trait, and their state of canalization as reflected in their numbers, "sacrifices important spatial information" (Whittle, 1998). Waddington (1961) made a similar point concerning the canalization of cross-vein development in fruit fly wings. Defining a trait precisely is important as a first step in attempting to understand the basis of its canalization.

While states of canalization are assessed from developmental and morphological outcomes, the property itself is genetically based and, for any specific trait, is characteristic of a particular strain. This consideration is germane to whether there are two distinct general categories of canalization. Much of the early work on canalization dealt with the ability of wild-type strains to resist environmental shocks *(environmental canalization),* while more recent discussion concentrates on robustness to genetic perturbation *(genetic canalization).* Yet it is by no means certain that different genetic elements are involved in these two operationally distinguished forms of canalization (Scharloo, 1991). Indeed, experimental evidence supports the idea that canalizing mechanisms operative against environmental and genetic perturbations of particular traits involve the same genetic architectures (Stearns et al., 1995).

Ultimately, if discussions of canalization are to be fruitful, they must move from phenomenological characterization to the nature of its molecular and genetic foundations. In principle, there would seem to be only three general properties that can convey resilience of developmental systems to perturbation: (1) self-correcting or self-balancing biochemical processes, (2) elements of redundancy, and (3) error reduction systems.

In the first category are feedback loops and efficient enzyme fluxes (Whittle, 1998). Although such biochemical behaviors are ultimately genetically based, they take place well away from the level of the gene. The second category consists of the well-documented phenomenon of partial genetic redundancy (Wilkins, 1997), based, in particular, on coexpressed paralogous genes whose activities can partially substitute for another. Functional redundancy, however, may also include genes that are not sequence-related but that perform the same biochemical function

(Koonin et al., 1996). Such similar activities may act as backup for each other and contribute to the canalization of those traits in whose development those gene products participate.

The third category of built-in genetic robustness, error reduction systems, might seem unlikely *a priori,* but evidence supports the existence of these systems. Rutherford and Lindquist (1998) reported that flies heterozygous for the gene *Hsp83,* which encodes a chaperone protein, are far more likely than wild-type to express otherwise hidden (recessive) genetic variation for many morphological (developmental) traits. In subsequent generations, however, that expression ceases to be dependent on *Hsp83* heterozygosity, as homozygosity for the "uncovered" mutations becomes frequent. Evidently, an initial weakening of the strength of the various signal transduction pathways in which the *Hsp83*-encoded protein participates reduces the effective wild-type gene activity of many proteins. The result is the unmasking, and expression, of many normally recessive alleles.

Because canalization masks the expression of genetic variability, and only expressed genetic variation can be directly acted upon by natural selection, canalization seemingly has the potential to restrict patterns and rates of evolutionary change. The critical question provoked by this inference concerns the strength and significance of such constraints. Can canalization, in effect, seriously retard rates of developmental evolution? Conversely, do rapid rates of organismal or trait evolution indicate reduced strength of canalized states? The idea that weak canalization might have contributed, in particular, to the rapid diversification of animal forms in the "Cambrian explosion" has been discussed by Gould (1989a) and Bard (1991). This idea can be extended and generalized further: perhaps all early stem groups exhibit less canalization, hence, more variability, than later, more derived groups. If this idea is correct, the fossils of stem group forms or of transitional species should, in general, show more phenotypic variability than later, crown groups.

In a recent test of this idea, using trilobite fossils, Hughes et al. (1999) concluded that there was no irreversible trend toward greater developmental constancy during trilobite evolution. Older evidence also provides little support for the idea that rates of evolution are inversely correlated with "strengths" of canalization, measured in terms of degrees of phenotypic variability. Thus, from his extensive surveys of mammalian fossils, Simpson (1953) concluded that during strong directional selection, such as the kind that might underlie many adaptive radiations, traits often appeared less variable than during periods of relative stability. While fossil evidence of this kind is always subject to more than one interpretation, the available evidence suggests that canalization does not comprise a straight jacket on evolutionary developmental biology and may not even be a particularly strong retarding force.

Genetic assimilation denotes the process in which an initially environmentally induced developmental/morphological trait that is favored by natural selection comes to be generally expressed by the population as a whole in the absence of the initial environmental stimulus. The term was coined by Waddington (1953a); but the process was also described by Schmalhausen (1949), who called it both "stabilizing selection" and "autonimization."

The basic *a priori* argument for the existence of genetic assimilation was that if an adaptive change were triggered by an environmental stimulus, selection for that trait should lead to canalization for that trait—and hence its appearance in the absence of the inducing stimulus. Both Waddington and Schmalhausen described in general terms how an infrequently occurring "plastic" response in a population could be converted to a fixed permanent feature of the population's developmental repertoire by strong adaptive selection.

In addition, Waddington experimentally demonstrated that the induction of certain developmental abnormalities by environmental shocks in *D. melanogaster,* if followed by selection, could ultimately lead to the expression of those traits, in the absence of the inducing stimulus, in the selected lines. The requisite conditions were selection—either by the experimenter or the environment—and pre-existing genetic variation in the populations favoring the response. His examples involved induction of the crossveinless phenotype by heat shock and of bithorax phenocopies by ether (Waddington, 1953a, 1956a), and development of enlarged anal papillae in response to osmotic shock (Waddington, 1959). In all populations in which there was an initial positive response to the regime, artificial selection increased the percentage of individuals showing the response "spontaneously." For a review of some apparent cases of genetic assimilation under natural conditions, see Hall, 1998, pp. 313–315.

Williams (1966) took issue with the claim that genetic assimilation can be an important source of adaptive evolution and, therefore, of developmental innovation. He pointed out that the developmental alterations produced by environmental shocks were, in themselves, unrelated to traits having adaptive value and criticized Waddington for not distinguishing between susceptibilities to environmental shocks and responses (which have an inherent adaptive component). He also noted that any developmentally plastic response, evoked in response to a change in an environmental parameter, was itself the product of prior adaptive conversion. Hence, the conversion of a sometime plastic response to an invariant developmental outcome actually involves a narrowing of adaptive response rather than a source of innovation *per se*. The development of enlarged anal papillae in response to high osmotic pressure in the medium would be an example of such a narrowing of response.

Eshel and Matessi (1998) supported the general validity of Williams's objections but argued that the breakdown of canalization under extreme environmental conditions can itself be adaptive. Specifically, they proposed that the environmental shocks might occasionally unmask genetic variations that favored adaptation in unusual circumstances in the past. In their quantitative genetic model, such external perturbations might allow expression of rare "preadaptations," permitting new evolutionary directions. In effect, they suggest, Williams's objections to genetic assimilation as a general means of adaptive change basically are valid but may not pertain to rare bursts of evolutionary change in morphology and development.

An additional possibility may be worth considering. In addition to all-or-none traits, there are meristic traits. If there is a latent capacity for induction of such traits, such as certain sensory modalities, and if acquisition of expression of one or a few elements has adaptive value, continued selection could, over time, increase the number of such elements. Such changes would be both adaptive and innovative, at the developmental level, leading to a novel morphological pattern. For such "plastic" traits, genetic assimilation might have genuine potential for evolutionary innovation.

In conclusion, the precise significance of both canalization and genetic assimilation for the evolution of developmental processes remains controversial. At present, it seems probable that the principal biological significance of canalization is to ensure developmental fidelity, a prerequisite of species' identity and fitness. Although it dampens expression of genetic variation, canalization does not seem to produce a tight restraint on developmental evolution. Conversely, while genetic assimilation is unlikely to have been a major general source of morphological-developmental innovation in evolution, several considerations suggest that it may come into play in certain situations. As the molecular mechanisms of canalization and the genetic foundations of particular traits are increasingly illuminated, we should be able to better understand the circumstances in which states of canalization break down and those in which genetic assimilation plays an evolutionary role.

ANDRES COLLAZO

Cell Determination and Differentiation

Defining terms such as "determination" is difficult because of possible confusion with other concepts such as "specification" and "commitment." Slack (1991) provides one of the most cogent discussions of these terms, and his definitions are used in this entry. Developmental *commitment* represents the intrinsic character of a cell or region in an embryo. Slack (1991) distinguishes between three types of commitment: specification, determination, and potency.

A cell or tissue is *specified* to become a particular structure if it develops autonomously into that structure when placed in a neutral environment outside of the embryo. Needless to say, this experiment cannot be done for many species because cells or tissues do not survive or differentiate in isolation. These experiments are most common with invertebrate and amphibian embryos in which cell and tissue explants can survive in physiological buffers.

Determination refers to a further state of commitment in which the specified cell or tissue develops autonomously even when grafted into a different environment in the embryo. Determined cells or tissues are considered to be specified (even though often this has not or cannot be tested directly), while the reverse is not necessarily true.

The *potency* of a cell or tissue is the complete range of derivatives these cells or tissues can form if placed in the appropriate environment.

This is another type of commitment that can be difficult to test experimentally. How is it possible to know whether the cell or tissue has been challenged in all possible environments? The creation of chimeric mouse embryos is relatively straightforward (Gardner, 1968); and because the labeled cells introduced this early in development (blastocyst stage) can go on to form any cell type in the adult, the potency of any cell type can be tested.

Any of these concepts of commitment can be applied to a tissue or region of an embryo as well as to a single cell. A tissue and a given cell within it are not necessarily committed at the same point in development. For example, a tissue that is determined to form an organ consisting of several cell types can consist of individual cells that are not yet determined to form a particular cell type. This difference can even be seen at the nuclear level, such as in the early syncytial blastoderm stage of insects (Slack, 1991).

A related term, *differentiation,* refers to the development of specialized cell types (Gilbert, 2000a). This is the final result of commitment; some differentiated cell types, however, can dedifferentiate and form other cell types during processes such as regeneration (Brockes, 1994; Jones and Corwin, 1996). The developmental *fate* of a cell or region represents what that cell or region differentiates into during development. A diagram representing the fate of multiple cells or regions during an embryo's normal development is called a *fate map.* Such fate maps have been generated for many species and are critical for interpreting gene expression patterns and experimental manipulations (Woo et al., 1995). However, fate maps do not give any direct information on the state of commitment of a cell or region (Slack, 1991).

Davidson (1991) uses similar concepts but slightly different terminology in his categorization of three different types of animal development: (1) the canonical invertebrate embryo, (2) vertebrates, and (3) insects. In defining these three types of development, he discusses two modes of commitment, autonomous and conditional specification. Slack would consider these two modes as dealing more with the acquisition of commitment. Davidson defines specification more broadly to include Slack's concept of determination as well as specification.

Autonomous specification results in the specification of the cell or tissue of interest, independent of its surroundings. This process typically occurs by the segregation of cytoplasmic determinants (such as proteins or mRNAs) during cell cleavages. In *conditional specification,* interactions between cells are essential for specification and determination. These cell-to-cell interactions can result from signaling molecules from adjacent or distant cells.

Autonomous specification results in a developmental pattern that classical embryology calls *mosaic* or determinative development, while conditional specification results in a pattern called *regulative* development (Gilbert, 2000a). Just as it is possible to construct a fate map for different regions of a developing embryo, it is possible to generate a specification map by culturing different pieces in isolation. Identical fate and specification maps indicate that an embryo has undergone mosaic development (Slack, 1991); different and independent cells and/or tissues are specified separately.

Regulative development is more complex because it encompasses potentially more mechanisms than mosaic development. Slack (1991) describes four different types of regulative behavior, which could represent very different underlying mechanisms. Though earlier developmental studies characterized different species' development as either mosaic or regulative, elements of both have been found in the same individual's development (Gilbert, 2000a).

The nematode *Caenorhabditis elegans* is a classic example. Early work describing the complete cell lineage and laser ablation experiments of single cells revealed a pattern of development that seemed to be highly mosaic (Sulston et al., 1983). Now, however, with a better understanding of development based on new experiments and a molecular context, cell-to-cell interactions are recognized as critical to cell commitment even at the earliest cleavage stages such as the 2-, 4-, 8-, and 16-cell stages (Schnabel and Priess, 1997; Bowerman, 1999). Many cell differentiation events later in development, such as the induction of the vulva and surrounding tissues by gonad cells (Sulston and White, 1980; Kornfield, 1997), also depend on conditional specification.

The *C. elegans* example is not meant to argue against the importance of autonomous specification, which certainly occurs in most, if not all, species. For example, the germ line in *C. elegans* is autonomously specified (Bowerman, 1999). Another classic example of mosaic development is the ascidian, a marine invertebrate that is the basal chordate taxon (Satoh, 1994). The primary muscle lineage of ascidians is autonomously specified, and recent studies in the ascidian *Halocynthia roretzi* have identified a molecule involved, the mRNA for the gene *macho-1* (Nishida and Sawada, 2001). However, conditional specification is also important for the differentiation of other tissue types such as the notochord (Nakatani and Nishida, 1999; Takahashi et al., 1999a; Minokawa et al., 2001).

While generalizing across the diversity of eukaryotes is difficult, modern developmental biology does offer some interesting observations. The utility of defining an embryo's development as mosaic or regulative is questionable, not only because many organisms display elements of both, but also because many elements of conditional specification have been found in embryos that were once thought to be primarily mosaic. In general, autonomous specification is most often associated with maternal determinants and relatively early developmental events (Pourquie, 2001), while conditional specification appears to be the more common mode of commitment for most cell and tissue types. Conditional specification is thought to act later in development than autonomous specification because it requires a cellularized embryo, but it can occur as early as the two-cell stage.

Given the importance of conditional specification during development, the differentiation of most cell types appears to depend on signals from surrounding cells. *Induction* is the process by which a cell or tissue influences the fate of another cell or tissue. *Competence* is the ability to respond to an inductive signal and is a subset of that cell's or tissue's potency (Slack, 1991). Induction can occur by direct cell contact or at a distance. Induction at a distance is typically thought to involve a *morphogen,* which is a long-range signaling molecule that acts in a concentration-dependent manner. It is also possible for a cell to signal another cell at a distance by a cell relay mechanism (Vincent and Perrimon, 2001) or long cytoplasmic projections (Teleman et al., 2001).

Diffusible morphogens, and the resulting morphogenetic gradients, are an appealing mechanism for induction because of the ample evidence for differential responses of cells or tissues to different concentrations of morphogens *in vitro* (Gurdon et al., 1994; Smith, 1995; Jones et al., 1996). However, with the notable exception of *Drosophila* development (Zecca et al., 1996; Strigini and Cohen, 1997; Teleman et al., 2001), there has been surprisingly little evidence for morphogenetic gradients acting *in vivo,* especially in vertebrate embryos. The best evidence of a morphogen in vertebrate development is the nodal protein Squint (member of the transforming growth factor-beta family of signaling molecules) in the zebrafish (Chen and Schier, 2001). The problem with most studies (except for those done in *Drosophila*) has been that the amount of exogenous morphogen used is probably much higher than the endogenous amount, although endogenous concentrations have been difficult to measure (Teleman et al., 2001).

Signaling molecules are important factors in the specification and determination of cells and tissues. Other factors are required for imparting identity. The same signaling molecules can be used in the development of different structures. Garcia-Bellido (1975), working in *Drosophila* with its clearly compartmentalized development, called genes that impart identity to a structure *selector genes.* Since then, many selector genes have been identified, the most famous being the homeobox containing genes of the Hox complex.

Patterning the different parts of the embryo involves an interaction between signaling pathways and selector proteins. Two models outline how signaling pathways and selector proteins interact. In the first, selector genes and signaling pathways have different target genes that interact to provide positional information and identity. The second model proposes that signaling pathways and selector proteins have the same target genes, with the signaling effector proteins interacting with the transcription factor protein encoded by the selector gene to regulate the target gene's transcription. Current evidence, including a study of the role of the selector gene *Scalloped* in *Drosophila* wing development, supports the second model as the most likely mechanism by which signaling pathways and selector genes pattern embryonic structures (Affolter and Mann, 2001; Guss et al., 2001).

Potency, the third type of commitment, can be used in defining determination, for example, as a restriction of potency (Slack, 1991). A cell or tissue that can form more than one thing is called *multipotent* or *pluripotent,* while one that can form everything is called *totipotent* (Slack, 1991). *Stem cells* are multipotent or even totipotent cells that can self-replicate and produce differentiated progeny. Stem cells are important not only because of their role during development (Slack, 2000; Watt and Hogan, 2000), but also because they can be found in some adult tissues (Gussoni et al., 1999; Pittenger et al., 1999). Embryonic cells, however, probably have the greatest range of potency (Vogel, 2000).

Developmental biologists can test the potency of single cells *in vivo* and in multiple species by labeling them with a lineage tracer that is passed on only to that cell's progeny (Bronner-Fraser and Fraser, 1988; Collazo et al., 1993; Raible and Eisen, 1994). This methodology has revealed, for example, that neural crest cells are multipotent even after they have begun to migrate (Fraser and Bronner-Fraser, 1991). Single-cell injections can demonstrate multipotency but cannot demonstrate that a cell is determined even if it gives rise to only one cell type. To test determination, the single cell needs to be transplanted to challenge its potential, and this is a more difficult experiment. These types of experiments are being undertaken in more species, providing comparative material for studying the evolutionary changes in potency and lineages of single cells during embryonic development (Boyer et al., 1996; Martindale and Henry, 1999; Henry et al., 2000).

All the terms discussed in this entry provide a framework for understanding the underlying mechanisms that organisms use during development. Though many of these terms were defined before modern molecular techniques revolutionized our understanding of development, they maintain their relevance by providing an all-important context. Modern developmental biology uses many cellular and molecular approaches that need to be better incorporated into studies of development and evolution (Zimmer, 1994; Collazo and Fraser, 1996). Learning how cells become committed to a particular fate and differentiate into the numerous cell types that make up eukaryotes will help provide a mechanistic basis for the morphological and physiological changes seen during organismal evolution.

JAMES W. VALENTINE

Cell Types, Numbers, and Body Plan Complexity

Following the founding of the cell theory in 1838, and especially after Schwann's famous treatise (1839), the notion that cells are building blocks of all living things became absolutely fundamental to our understanding of how life is organized. Defining the cell, however, proved to be difficult. Some early biologists even disputed that unicellular organisms were organized as cells; and certainly the great diversity of cell organizations, including prokaryotic cells, multinucleate eukaryotic unicells, and syncytial tissues in multicellular forms makes generalizing about types of cells extremely risky. Most early attempts at defining a cell were simply lists of properties, but most of the properties were found to be lacking in one or another cell type. A minimalist definition was offered by Alberts et al. (1989, p. 3): Cells are "small membrane-bound compartments filled with a concentrated aqueous solution of chemicals" that form living organisms. Though many cells are organized in distinctive ways, essentially all living things are cellular; and the multicellular kingdoms have arisen from unicellular ancestors through the evolution of developmental processes that produce and integrate multiple cells of differentiated types into complex bodies.

Cell Types and Cell Morphotypes

Definitions of the cell types in multicellular organisms depend on the methods by which cell properties are evaluated. Light microscopes can only resolve objects larger than about 0.5 μm, permitting visualization of most cell organelles and many cell inclusions such as granules and crystals. Electron microscopes can routinely resolve objects about 100 times smaller, permitting visualization of large molecules. Each of these methods produces pictures of cell morphologies—of cell sizes and shapes, of surficial structures, and of internal cytoarchitectural features that include the kinds, numbers, and arrangements of organelles and the presence of inclusions, pigments, and other notable features. Cells characterized by such features are *cell morphotypes*.

Electron microscopy has added vast and important detail to our knowledge of cell morphology and is clarifying the nature of some tissue types. For example, tissues in flatworms that were thought to be syncytial and that have played a role in phylogenetic theories have been shown to be cellular (Smith and Tyler, 1985; Rieger et al., 1991). Also, the cellular linings of the coelomic cavities of representatives of some invertebrate phyla have been shown to be quite diverse in cell types and perhaps not homologous (e.g. Bartholomaeus, 1994), a finding with important phylogenetic implications. Although cells belonging to the same cell type population are similar, a significant amount of morphological variation is commonly encompassed within a given type. For a simple example, the shape of otherwise-similar cells can vary gradually from relatively short and broad to tall and thin along an epithelial lining, and otherwise-similar muscle cells can have different dimensions in different regions. In practice, cell morphotypes are delineated at the more abrupt cytomorphological boundaries identified by histologists, rather than within such continua.

Cells are also studied by many nonmorphological techniques. Cell types can be sorted and their contents assayed by methods of chromotography or gel electrophoresis. For studying cell type differentiation in development, the most important techniques that do not rely on morphology per se are now those of developmental genetics. The activities of numbers of key developmental genes have been followed as cells differentiate during pattern formation. Cells that seem morphologically identical and are found in the same tissues, or in seemingly identical tissues in different regions, can have different developmental histories. These differences are usually simply subsumed under a given cell morphotype; but they indicate that, in a morphological evaluation of cell type numbers, the underlying developmental complexity must be significantly underestimated.

The classification of cell types is not formalized. Cells are usually described in the context of the tissues or organs in which they reside, which results in a partially functional arrangement—although many cell types (e.g., mucous cells) can be found in different tissue types and organs. For humans, Alberts et al. (1989) present a functional tabulation that recognizes 210 cell types arranged in 20 cell classes that are in some cases subdivided into smaller groups. The class of sensory transducers, for example, is subdivided into transducers for temperature or for touch. Alberts et al. note that some of their cell type categories are somewhat generalized and can be split even further, if desired.

Similar cell morphotypes that appear in different tissue types raise the question as to whether those cells are homologous. Many cell morphotypes found in different phyla are quite similar and function in similar ways, and they are often treated tacitly or openly as homologues, though the meaning of homology in cell types is not straightforward. Some differentiated cell types can develop from a variety of progenitor cell lines. However, developmental, morphological, and physiological aspects of biological organization do not stand in one-to-one relationship with each other, and homology in one of these aspects does not necessarily imply corresponding homology in another aspect (Striedter and Northcutt, 1991; Hall, 1992b; Wagner, 1994).

This is particularly clear for gene-to-phenotype relations (Wray, 1999). For example, Striedter and Northcutt point out that the archenterons of lampreys, sharks, teleost fish, salamanders, frogs, and lungfishes—while all homologous as embryonic guts—employ different, nonhomologous genes in their developmental pathways. In principle, dissociations of homology may also occur between the levels of the somatic hierarchy, such as cells and tissues or organs. Because cell types are basic morphological building blocks, they are available for use in the development of nonhomologous structures. Lack of homology among structures does not therefore necessarily imply a lack of homology in cell types. The evolutionary history of cell types within and across taxa, the history of their genetic bases, and the history of the structures in which they are embedded can all be different and can vary from case to case.

Cell Morphotype Numbers

Delimiting cell types is obviously difficult, and the same holds true for all evolving entities from genes to taxa. Nevertheless, the number of distinctive cell morphotypes clearly varies greatly among metazoans. For somatic (nongametic) cell types, parasitic metazoans may have as few as 3 (Myxozoa) and free-living forms as few as 4 (Placozoa), while humans have about 210 by a conservative count (variations in neurons are not counted). The number of cell morphotypes that can be found in all members of a phylum is usually much greater than that found in any one individual; thus, among all living sponges, Simpson (1984) discusses over thirty distinctive sponge cell morphotypes in the phylum as a whole, while most individual sponges have only five or six. No doubt many other sponge cell types have evolved, but have become extinct.

Table 1 lists the numbers of cell morphotypes, based on very conservative counts, reported in individuals belonging to each generally recognized metazoan phylum. Highly specialized and presumably derived cells, such as secretory cells found in restricted branches of a phylum, were not counted. In general, the counts represent minimalist individuals with the body plan of the phylum. For arthropods and chordates, a range of cell counts is given, the lower count representing the minimalist individual, such as might have existed as an early diverging member of the phylum (based on branchiopods and cephalochordates, respectively), the higher indicating the number of cell types reported among a more highly derived crown group member (based on flies and humans,

TABLE 1. Cell morphotype numbers estimated to be required for the basic body plans (stem members) of nominal metazoan phyla. Numbers for derived crown taxa are also given for arthropods and chordates

Phylum	Cell Morphotypes	Phylum	Cell Morphotypes
Acanthocephala	12	Mollusca	38
Annelida	37	Myxozoa	3
Arthropoda	42–90	Nematoda	14
Brachiopoda	34	Nematomorpha	8
Bryozoa	25	Nemertea	35
Chaetognatha	21	Onychophora	30
Chordata	38–215	Orthonecta	3
Cnideria	10	Phoronida	23
Ctenophera	17	Placozoa	4
Cycliophora	15	Platyhelminthes	20
Echinodermata	41	Pogonophora	20
Echiura	21	Porifera	4
Entoprocta	13	Priapulida	20
Gastrotricha	23	Rhombozoa	4
Gnathostomulida	16	Rotifera	15
Hemichordata	25	Sipuncula	25
Kinorhyncha	17	Tardigrada	18
Loricifera	18	Urochordata	38

respectively). Most of the invertebrate counts have been taken from the descriptions in Harrison and coeditors (1991–99).

Plotting the maximum number of cell types estimated for any metazoan living at a given time produces a curve that increases monotonically throughout the Phanerozoic Era (Valentine et al., 1994). Starting as they did with, say, two cell types at the founding of Metazoa, cell type numbers had nowhere to go but up. It is possible to interpret the temporal increase in the upper bound of cell morphotype numbers as simply reflecting an increase in the variance of cell type numbers associated with lineage diversification (see Fisher, 1986). Certainly, the trend of increase was far from universal, for a reduction in cell type numbers almost certainly occurred in a number of taxa. For example, cell type numbers appear to have been reduced in Acanthocephala (evolving from within Rotifera, Garey et al., 1996; Mark Welch 2000), in Pogonophora (evolving from within Annelida, McHugh, 1997; Kojima, 1998; Boore and Brown, 2000), and in such obligate parasitic phyla as Rhombozoa, Orthonecta, and Myxozoa (so reduced that their ancestral groups are uncertain). Other phyla with suspiciously low numbers of cell morphotypes include the parasitic Nematomorpha and the free-living but minute Tardigrada.

Cell Morphotype Number and Morphological Complexity

The morphological complexity of an organism can be defined as the size of its minimum description (Hinegardner and Engleberg, 1983). The complexity of a body plan thus increases with the number of different kinds of parts and with their degree of disorder (McShea, 1991; Valentine et al., 1994). An organism with body parts that are morphologically similar is less complex than one in which they are highly differentiated, other things being equal.

Because cells are basic building blocks of bodies, the complexity of a body is at base sensitive to the number of cell types and to their arrangement in tissues, organs, and so forth. Because cell morphotype numbers can be more or less reliably compared across all metazoan phyla, they can be used as an index of comparative body plan complexity, an approach suggested by numbers of researchers (e.g. Sneath, 1964; Bonner, 1965; Raff and Kaufman, 1983; Valentine et al., 1994). Because the counting of cell morphotypes alone does not address their ordering at higher morphological levels or evaluate the morphological heterogeneities to be found there, their numbers by no means represent a scalar metric of complexity. However, cell morphotype numbers should serve a useful purpose as rough-and-ready indications of relative complexities among body plans, and therefore of the evolution of levels of developmental sophistication appropriate to those complexities.

Using this criterion, changes in metazoan complexity can be traced through the fossil record, and some generalities emerge that seem to be free from the problems of delimiting cell morphotypes. The earliest fossils of body parts, from near 570 Ma, are of sponges and possible cnidarians, which presumably had few cell types, perhaps fewer than ten for early cnidarians. The earliest trace fossils, probably from a bit later, are of minute forms creeping on or just below the sea floor. They can be interpreted as bilaterians with acoelomate and pseudocoelomate body plans that might have had fifteen or so cell types.

Near the beginning of the Cambrian, at about 543 Ma, trace fossils indicate that body sizes increased significantly, and that more active and presumably more complex organisms had appeared. These trends continued until the Cambrian explosion, around 530–520 Ma, when advanced invertebrate bilaterian phyla, including mollusks, arthropods, echinoderms, and cephalochordates, make their first appearances as fossils. Early-branching members of these more complex forms have cell morphotype numbers in the mid-thirties to high forties. Agnathan fish might also be represented during the Cambrian explosion; living agnathans have about sixty cell morphotypes (Hardisty and Potter, 1971). Members of more derived branches of some phyla, especially those that later invaded the land, can have cell morphotype numbers several times greater than those inferred for their stem groups (see Table 1). This history, which must be somewhat speculative, suggests that the period just before and during the Cambrian explosion was characterized by a particularly high rate of developmental evolution as numbers of novel, complex body plans were produced.

Cell Morphotype Number and Developmental Complexity

The genomic architecture responsible for metazoan development is under intensive study and its basic framework is rapidly emerging. Most of the ligands, signaling pathways, and transcriptional regulators that perform the molecular mechanics of cell differentiation and body plan development throughout Bilateria seem to have been present in cnidarians, and indeed many of them are found in sponges. However, one might hypothesize that metazoans that are simpler in terms of cell type number have less complex developmental systems and thus simpler genomes than more complex forms. By analogy with morphological complexity, genomic developmental complexity can be defined as the size of the minimum description of its processes.

In practical terms, this definition suggests that the number of genes in a genome represents a rough index of its developmental complexity. But this has not proven to be the case. We can assess relations between gene number and complexity from four organisms whose genomes have been sequenced and gene number estimated.

A single-celled fungus *(Saccharomyces)* has about 6340 genes (Goffeau et al., 1996).

A nematode *(Caenorhabditis)*, with about 16 cell morphotypes, has about 18,400 genes (*C. elegans* sequencing consortium, 1998).

An arthropod *(Drosophila)*, with about 90 cell morphotypes, has only about 13,600 genes (Adams et al., 2000).

And a mammal *(Homo)*, with about 210 cell types, has about 39,000 genes (Ventner et al., 2001).

Clearly, gene number is not a good predictor of the scale of morphological complexity even as defined by cell type number, and thus it is also not a good predictor of developmental complexity (Valentine, 2000; see also Brookfield, 2000).

Distinct cell types are produced by the differential expression of genes, each type involving different fractions of the genome (Davidson, 2001). The expression of a metazoan gene is regulated by factors that bind to a promoter complex that is upstream from the portion of the gene that is transcribed—in *cis* position. The promoter complex includes numerous binding sites, the enhancers, which are organized in modules; the gene may be expressed or repressed depending on combinations of factors that occupy the enhancer sites. Such combinatorial systems have the ability to produce vast numbers of outcomes from relatively few variables. Different combinations of signals to the *cis*-regulatory portion of a gene may thus cause its transcription at many different times and places during development, influencing differentiation within numbers of cell types. Furthermore, transcribed gene sequences can produce more than one gene product owing to alternate splicings, and these alternate products often function at different times and in different cell types.

The inference from these data is that there is more than one way for a lineage to become more complex: It can add more genes or alternate splicings and thus increase the number of regulatory genes, enzymes, and structural proteins with which to construct a more complex body; or it can repeatedly use a given set of genes in different combinations and developmental contexts, adding, instead of genes, enhancer sites or modules (Valentine, 2000). While the number of genes represents genomic diversity, the number of gene expression events represents genomic richness, and probably reflects developmental complexity more closely than any other measure.

Even with gene expression events taken as a metric of developmental complexity, problems arise in relating this metric to morphological complexity. For example, some organisms pass through numbers of body plans during ontogeny, and presumably employ far more gene expressions than if they developed more directly into their adult body plans. The ratio of gene expression events to adult body plan complexity in forms with multiphasic life cycles is presumably higher than for, say, their direct-developing relatives. It may be possible to devise a measure of body plan complexity that would include the developmental stages, integrating all of the morphologies produced during ontogeny. Whether that measure would bring developmental and morphological complexities into close correlation is an interesting and unanswered question.

FRIETSON GALIS AND BARRY SINERVO

Conserved Early Embryonic Stages

Embryonic development of metazoans begins in a stereotyped fashion. The zygote rapidly divides into many smaller cells. The blastula, which results from this cleavage, then undergoes a series of dramatic morphogenetic movements, a process called gastrulation, which results in a multilayered embryo, the gastrula.

It is usually assumed that similarity among early embryonic stages is the result of evolutionary conservation (Seidel, 1960; Sander, 1983; Buss, 1987; Raff, 1996; Hall, 1996a). Even though Von Baer did not believe in evolution by descent, he was the first to propose that early changes in ontogeny were more likely to have cascading consequences than later changes. Although this evolutionary constraint is undoubtedly real (Buss, 1987), its importance has recently been challenged by the revelation of considerable variation in early embryonic stages. In addition, evolutionary conservation of early embryonic stages may be less than previously thought because of similarity due to convergent or parallel evolution *(homoplasy)*. Similarity is almost unavoidable in the early stages of metazoans because of the complete reset that occurs at the initial single-celled stage. Only a limited number of permutations are possible in embryos with few undifferentiated cells. The diversity of multicellular clonal propagules, which lack such a reset, is indeed remarkably high.

Similarity arises because of either similar selection pressures or stringent constraints. In the case of constraints, conservation of morphological patterns arises because no genetic variation is produced, or invariably negative pleiotropic effects are associated with changes. Similarity in a phylogenetic context, thus, can have four causes: (1) *plesiomorphy,* conserved ancestral traits; (2) *synapomorphy,* shared derived traits; (3) homoplasy due to common adaptational pressures, or 4) homoplasy due to developmental constraints (convergent derived adaptive traits promoting trait corrrelation). These four causes reflect, in increasing order, null (1) to more complex adaptational hypotheses (adaptation is implicated in 2–4). We provide examples of each.

Diversity and Similarity in Cleavage Patterns

Mechanistic causes of cleavage patterns. Patterns of cleavage are determined by a small number of mechanistic factors, variation in which allows a high diversity of cleavage patterns. One mechanism is the amount and distribution of vitelloprotein within the zygotic cytoplasm (Gilbert, 2000a). Yolk impedes cleavage. In zygotes with little yolk, cleavage usually separates the embryo into distinct cells (*holoblastic cleavage,* as in sponges, sea urchins, lancelets, most amphibians, and mammals).

In zygotes with a large yolk component, cleavage occurs only in the part of the cytoplasm with little or no yolk *(meroblastic cleavage),* leading in cephalopods, fishes, reptiles, and birds to an embryonal disc on top of a large yolk mass (*epiblastic cleavage,* Figure 1) and in crustaceans and insects to a superficial layer of embryonal cells with yolk in the center *(periblastic cleavage).* For example, most amphibians exhibit holoblastic cleavage. However, eleutherodactyline frogs with derived yolky eggs also have secondarily derived epiblastic cleavage (Elinson, 1987). Given the adaptive value of high yolk volume for progeny survival and the extreme convergence of epiblastic or periblastic cleavage among diverse taxa, cleavage mechanisms are constrained to evolve in response to selection on embryonic nutrition (cause 4 in previous list).

The second mechanism is the angle and timing of mitotic spindle formation and, thus, orientation of cleaving cells: spiral (protostomes), radial (deuterostomes), and rotational (mammals) cleavage. Coordinated movement arising from cleavage and the sequestration of cytoplasmic

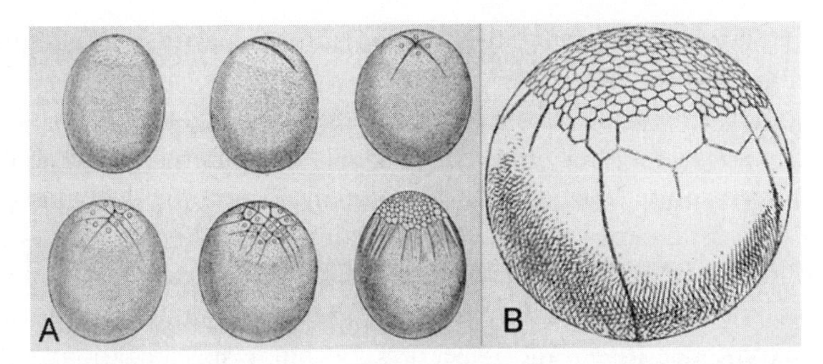

FIGURE 1. Convergence. Meroblastic, epiblastic cleavage in (A) the cephalopod *Loligo pealii* (after Claus and Grobben, 1917) and (B) the longnose gar *Lepisosteus osseus* (after Balfour, 1881).

factors that determine cell fate are undoubtedly important in many groups but are most clearly illustrated during ascidian larval development, in which mesoderm is derived from one blastomere of the four-cell embryo.

The third mechanism is differential cell adhesion. Differential adhesion can cause important shape differences, for instance in compaction in mammals and polyembryonic wasps (in which blastomeres are not loosely packed but very closely joined together) or in coeloblastulae (in which the blastula is a hollow ball of cells instead of a solid ball, as in stereoblastulae).

Diversity in cleavage patterns can be understood in terms of adaptations for embryonic life such as nutrient uptake, locomotion, and maternal determination.

Nutrients can be obtained from yolk, maternal tissues, or the environment. Specializations for embryonic nutrition uptake can arise as a direct consequence of constraints (e.g., yolk constraints) or as adaptations to specialized conditions. For example, evolution of mammalian viviparity has led to compaction (Gilbert, 2000a), a conspicuous specialization in which loosely arranged blastomeres suddenly huddle together and form a compact mass. Compaction and associated intracellular changes function during embryonic implantation in the uterus. Compaction triggers separation of the inner cell mass (from which the embryo will develop) from the trophoblast, which provides fetal contributions to the placenta. Interestingly, polyembryonic insects have converged on compaction, together with early separation of embryonic and extraembryonic cell lineages (Grbić et al., 1998), presumably because endoparasitism imposes similar spatial constraints as viviparity, an example of cause 3 in previous list.

Locomotor demands also influence cleavage (Buss, 1987). Free-living dispersal stages have external ciliated cells and internal dividing cells. This configuration appears to be caused by locomotor demands combined with a universal metazoan constraint, example of cause 1—cells cannot divide when ciliated (Buss, 1987). Metazoan cells have only one microtubule center, which can be used either for a mitotic spindle or for cilia, axons, dendrites, and other microtubular specializations. Thus, cilia are concentrated on the surface and dividing cells without cilia are inside propagules to prevent interference with locomotion.

Maternal determination. Buss (1987) argued that selection for maternal determination has shaped the evolution of cleavage. Determination of early development via maternal cytoplasmic factors provides a powerful means for the mother to prevent proliferation of one cell line at the expense of others, thus helping establish selection at the level of the individual. The amount of development under maternal control is necessarily limited; only a finite number of maternal cytoplasmic factors can be provided to the zygote. Unequal cleavage potentially increases effectiveness of maternal control. In unequal cleavage, some cells differentiate and begin zygotic transcription, marking the end of maternal control. These cells lose the capacity to become germ cells, whereas others continue undifferentiated and divide under maternal control. Differentiated cells usually lose the capacity to divide, which is why they can no longer become germ cells. The switch between maternal control and zygotic gene expression occurs at the mid-blastula transition in both anurans and *Drosophila.*

Polyembryony. Cleavage is radically different in polyembryonic parasitic wasps relative to other insects (Grbić et al., 1998). Cleavage is holoblastic, and the zygote produces not one but many small embryos. Cellularization occurs immediately, whereas in most other insects (including drosophilids) nuclei first divide into a syncytium and cellularization occurs later. Further specializations of compaction and early separation of embryonic and extraembryonic tissues are probably caused by the endoparasitic lifestyle, which is to a certain extent convergent with viviparity in mammals.

In conclusion, cleavage patterns are diverse, but important similarities occur because of evolutionary convergence due to similar embryonic adaptations and constraints. The most striking examples of convergent embryonic adaptations are compaction in mammals and polyembryonic insects and epiblastic cleavage in yolk-rich embryos.

Diversity and Similarity in Gastrulation Patterns

Morphogenetic processes of gastrulation that lead to germ layer differentiation are bewilderingly diverse. Processes that contribute to gastrulation are:

1. *Invagination,* in which a sheet of cells moves actively into the blastocoel toward the opposite side (Figure 2a)
2. *Unipolar ingression,* in which individual cells migrate inward at one end of a blastocoel (Figure 2a)
3. *Multipolar ingression,* in which individual cells migrate inward from all points of the blastocoel
4. *Epiboly,* the migration of apical cells over the other blastula cells, thus forming an external layer of cells (Figure 2b)
5. *Involution,* the moving in of a layer of cells so that it spreads over the inner side of the outer layer of cells (Figure 2b)
6. *Primary delamination,* in which the cells of a blastocoel divide such that one daughter cell remains at the surface and one moves to the interior (Figure 2c); and
7. *Secondary delamination,* in which a solid tissue of cells segregates into two layers.

Usually combinations of these processes form part of gastrulation, e.g., invagination and multipolar ingression (Figure 2a), and involution and epiboly (Figure 2b).

The combinatorial possibilities have led to a wide variety of morphogenetic processes among and within phyla and other higher taxa (below the level of phyla). Cnidarians, for example, have seven different types of gastrulation (Gilbert and Raunio, 1997). In insects there are drastic differences between short, intermediate, and long germ band insects, with polyembryonic wasps being even more derived. Within teleosts and mammals, gastrulation is highly variable (Collazo et al., 1994; Viehbahn, 1999); for example, the embryonic disc of rodents is cup-shaped instead of flat, and germ layers are inverted with endoderm outside and ectoderm inside.

Gastrulation is diverse but also characterized by similarity due to similar constraints and selection pressures. The important selection pressures

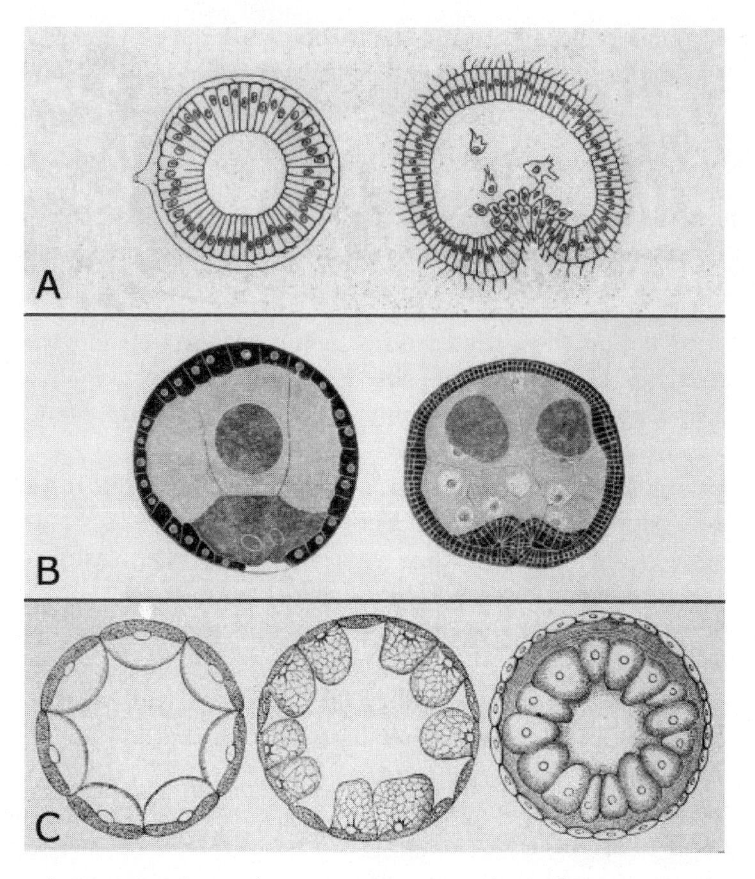

FIGURE 2. A. Diversity of gastrulation. Invagination and unipolar ingression during gastrulation in the echinoderm *Holothuria tubulosa* (after Balfour, 1881). B. Epiboly and involution in the echiuran *Bonellia viridis* (after Spengel in Balfour, 1881). C. Primary delamination in the cnidarian *Geryona* spec. (after Claus and Grobben, 1917).

are, as in cleavage, mainly related to locomotion and nutrient uptake. Viviparous and oviparous gastrulating embryos have numerous adaptations to living in eggs and other maternal tissues (Gilbert and Raunio, 1997; Grbić et al., 1998). The constraint on simultaneous ciliation and cell division of blastulae also holds for free-living gastrulae and leads again to ciliated outer cells (Buss, 1987). Yolk also constrains movements during gastrulation such as in the organizer region (Arendt and Nübler-Jung, 1999). The blastoporal canal of reptiles and primitive streak of birds and mammals are presumably specializations associated with this constraint. Further constraints and selective pressures follow from the type of blastula present at the start of gastrulation.

In blastulae with an outer layer of ciliated cells, for instance, epiboly is not possible; dividing cells would cover ciliated cells and preclude locomotion (Buss, 1987). Invagination is possible only in hollow blastulae (coeloblastulae). In solid blastulae, such as blastodiscs and stereoblastulae, invagination is not possible, and epiboly plays an important role. This is an example of cause 4, in which adaptative evolution of embryonic nutrition promotes correlated adjustments in other developmental processes.

Diversity of gastrulation and cleavage patterns is mirrored by diversity of genetic interactions at the molecular level. Comparisons of well-characterized taxa (insects versus vertebrates) show that gene networks underlying early development allow for substantial molecular reorganization (e.g., Holland, 2000; Damen et al., 2000; Stauber et al., 2002). However, evolutionary conservation is probably also important (De Robertis et al., 1994). Holland (2000) concludes, for example, that there are similarities in gene expression during protostome and deuterostome gastrulation, but it is not clear to what extent they arise from conservation versus convergence.

The larger number of cells during gastrulation, increased levels of differentiation, and large permutations of interacting processes generate almost infinite variation in gastrulation patterns. Yet, paradoxically, gastrulation is far more diverse than the end product; gastrulae invariably have two or three germ layers and never more (Hall, 1999a). Organ systems emerging from germ layers are similarly conserved: skin and nervous system from ectoderm, and digestive tube from endoderm. A key outcome is that gastrulation allows sheets of cells to come into contact, thus enabling embryonic inductions; neurulation is a classic example.

The infinite diversity of subsequent organic forms arguably begins with such inductions of tissue layers and organ systems. However, in order to develop organ systems, inductive events require topologically adjacent cell populations—a requirement that seems to form a stringent spatiotemporal constraint on the outcome of gastrulation.

Diversity and Similarity in Early Organogenesis

After germ layers are formed during gastrulation, cells interact with one another and rearrange themselves during organogenesis. This stage among metazoans is far more diverse than earlier stages. However, embryologists have noticed for a long time that within most high-order taxa, early organogenesis is far less variable than earlier stages (Seidel, 1960; Sander, 1983; Elinson, 1987; Hall, 1996a). Examples of high similarity of postgastrulation stages occur during the segmented germ band stage of insects, the nauplius stage of Crustacea, and the neurula/pharyngula stage of vertebrates (Sander, 1983; Grbić et al., 1998; Dahms et al., 2000). Recent data on genetic interactions confirm that within-taxa similarity is highest right after gastrulation (e.g., Damen, 2000; Holland, 2000). For example, expression patterns of segment polarity and Hox genes in insects are more similar than those of earlier-acting genes.

The phylogenetic distribution of these early stages of organogenesis suggests that similarity is due to conservation and not convergence. Sander (1983) introduced the term *phylotypic stage,* because these stages are typical for the phylum or higher taxa to which they belong. Although phylotypic stages are more conserved than earlier or later developmental stages, there is still important variation; what matters is that there is far less variation than at earlier and later stages. Apparently developmental stages evolve at different rates, and constraints to evolutionary change are particularly strong at stages shortly after gastrulation when axis patterning occurs.

Similarity and Conservation of Early Embryonic Stages

What causes conservation of these stages? There is abundant intraspecific genetic variation for phylotypic stages of vertebrates and insects; therefore, conservation must result from strong stabilizing selection against mutational change (causes 1 and 2, as defined earlier). Raff (1996) and Sander (1983) hypothesized that phylotypic stages are conserved due to global interactivity occurring during these stages: the web of intense interactions among organ primordia such as somites, neural tube, and notochord of vertebrates causes small mutational changes to promote widespread pleiotropic effects elsewhere in the embryo. At earlier stages, there are fewer inductive interactions. At later stages, there are many more inductive interactions, but they take place within semi-independent modules. Mutational change should thus be restricted to modules (Raff, 1996).

Slack et al. (1993) made the intriguing observation that the most conserved developmental stage within vertebrates is also the stage during which Hox genes are sequentially activated, mirroring their order on chromosomes (temporal and spatial colinearity). Comparison of the phylotypic stage among insects suggests that sequential gene expression is a highly conserved plesiomorphic trait (an example of cause 1). However, since then, evidence has accumulated that temporal and spatial activation of Hox gene expression at other stages coincides less well with chromosomal gene sequences. Apparently, keeping Hox gene regulation well organized matters less than the specific moment at which this should be the case.

Based on these observations, Galis and Metz (2001) proposed that selective mechanisms behind conservation of colinearity, at least in vertebrates, should primarily be sought in the high interactiveness during the phylotypic stage. They recently tested hypotheses of Sander (1983) and Raff (1996) by analyzing teratological studies on vertebrates that reported phenocopy effects of mutational change. Changes in number of cervical vertebrae, a highly conserved vertebrate trait patterned by Hox genes, also leads to a pleiotropic increase in susceptibility to cancer and stillbirths. Changes in Hox organization can thus have multiple and therefore major and detrimental effects. This argument places the general interactiveness of the phylotypic stage at the root of the conservation of the Hox organization, rather than placing the tight regulation of the Hox gene expression per se at the root of the phylotypic stage conservation. Examples of conserved traits determined during the phylotypic stage include numbers of mammalian cervical vertebrae as well as limb and digit number in tetrapods—and possibly many other features of the vertebrate body plan, such as the number of eyes and ears (Galis et al., 2001).

In conclusion, the rapid progress that has been made in understanding genetic and morphogenetic patterning of several invertebrate and vertebrate model species lets us determine whether similarity arises from conservation (constraint) or convergence. Unfortunately for many invertebrate taxa, we lack crucial information for meaningful interpretations. Emergent patterns suggest that similarity during cleavage and gastrulation not only results from conservation but also from convergence. In contrast, while early organogenesis is characterized by more divergence *between* phyla, *within* several phyla conservation of phylotypic stages is more dramatic than during cleavage and gastrulation. After an initial phase of rapid evolutionary change, there appears to have been prolonged stasis within phyla. An important cause of conservation appears to be stabilizing selection against negative pleiotropic effects. The causes for the rate of evolutionary change among taxa are not clear and remain a challenge for future evolutionary developmental research, as do the diverse forms of uninvestigated taxa of invertebrates.

KURT SCHWENK AND GÜNTER P. WAGNER

Constraint

Constraints are mechanisms or processes that limit the ability of the phenotype to evolve or bias it along certain paths. Beyond this general statement, however, little is agreed upon or clearly understood about constraints (see overviews by Gould and Lewontin, 1979; Alberch, 1982; Maynard Smith et al., 1985; Gould, 1989b; Wake, 1991; Amundson, 1994; Raff, 1996; Arthur, 2001; Burd, 2001; Schwenk, 2002). Ambiguity about constraint stems, paradoxically, from its very popularity. Spurred by Gould and Lewontin's (1979) influential critique of "adaptationism," evolutionists have embraced constraint enthusiastically, but not always critically. Thus, to a large extent, the facile assumption of constraint has merely replaced the facile assumption of adaptation. Coupled to an explosive proliferation of terminology, such uncritical usage has led to increasingly idiosyncratic meanings for constraint so that the concept is of diminishing value in scientific discourse (Antonovics and van Tienderen, 1991; McKitrick, 1993; Schwenk, 1995; Schlichting and Pigliucci, 1998).

If, as some claim, the constraint concept is moribund, it is only because the concept is so difficult to operationalize. Nonetheless, in capturing the growing sense that not all evolutionary outcomes are equally probable, constraint remains a central, even critical, element of evolutionary theory. It is therefore worth asking *why* the constraint concept is so elusive and how we might more effectively apply it. We believe that the central problem of constraint is a general failure to recognize its relativism. By relativism, we mean that constraint is only sensible within a local context (Stearns, 1986; Gould, 1989b; Schwenk, 1995; Fusco, 2001). Its meaning and appropriate invocation therefore depend on the specification of several factors that define the circumscribed realm within which it is applicable. Specifying factors include a designated clade (an in-group), a particular character or set of characters, a focal life stage, and most importantly, a null model. As these factors can change case by case, both the nature of constraint and the validity of its invocation can change as well. Attempts to define constraint in absolute or universal terms, as is typically done, are thus foredoomed.

Relationship to Selection

Contention and confusion surrounding constraint's relationship to selection illustrate the importance of relativism in characterizing constraint. Starting with Gould and Lewontin (1979), one view sees constraints as manifested in their effects on selection, implying that constraint is one thing and selection another. This dichotomous approach has been amplified to suggest that selection must be disallowed altogether as a cause or mechanism of constraint (e.g., Schwenk, 1995; Schlichting and Pigliucci, 1998). If not, it is argued, a well-understood process (selection) is replaced with a cryptic label (constraint), thus opening the door to "pan-constraint" and the dilution of the concept.

Another school of thought, however, does exactly this. By identifying constraint with a particular evolutionary outcome or pattern, the responsible mechanism is, by definition, "constraint." For example, a pattern of phenotypic stasis is often taken to imply that something has "constrained" phenotypic evolution in the characters of interest. If the cause of this limitation is stabilizing selection or conflicting environmental selection pressures, these are labeled "selective" or "ecological constraints," respectively. Constraint and selection are thus conflated.

Because this latter approach could quickly get out of hand, we should eliminate redundancy by making constraint and selection mutually exclusive. This goal is not so easily achieved, however. Maynard Smith et al. (1985), for example, adopted a dichotomous view in their classic treatment of developmental constraints. They wrote: "We would like to know whether [stasis] reflects developmental constraints limiting the possibility of change or, conversely, the maintenance of uniformity by stabilizing selection." How can the effects of constraint, they wondered, "be distinguished from the action of natural selection" (p. 266)? This partitioning of selection and constraint is significant in light of their definition of developmental constraints as "biases on the production of variant phenotypes or limitations on phenotypic variability caused by the structure, character, composition, or dynamics of the developmental system" (p. 265). This definition implies that constraints act during development to limit the generation of phenotypic variation, thereby limiting the pool of variation upon which selection can act. This approach is compelling because, by distinguishing the generation of variation from the operation of selection on that variation, Maynard Smith et al. seem to provide a logical basis for conceptually separating constraint and selection.

At least two mechanisms, however, probably cause "biases on the production of variant phenotypes": (1) Variations fail to arise in the first place because underlying genetic or epigenetic variation is not expressed phenotypically. Development, in this case, acts as a buffer ensuring consistency in a character's phenotype despite variation in the background information or developmental milieu—a process that is referred to as *canalization.* (2) Variations arise during development, but disruptions to developmental dynamics cause the death of the embryo and the variant characters never enter the population. This case represents nothing more than viability selection acting on early life stages. Thus, Maynard Smith et al.'s definition is not consistent with their dichotomous view; the failure to generate phenotypic variation during development (a putative developmental "constraint") cannot be temporally or mechanistically separated from the action of selection. Once again, constraint and selection are conflated.

Our inability to reconcile selection and constraint in this example stems from adherence to an absolute definition of constraint—in this case, one that would universally exclude selection. Followed to its logical conclusion, this dichotomous approach leads to a highly restrictive notion of constraint (Schlichting and Pigliucci, 1998) that does not satisfy the needs of many evolutionists for a constraint concept. With a relativistic approach, however, a potential solution emerges: selection might sometimes act as a constraint, but only in particular circumstances. These circumstances are determined by specification of the local context in which constraint is to be applied, a task to which we now turn.

Context of Constraint

Owing to constraint's relativism, a set of "fixed" points must be established around which constraint can be organized. These points, or specifying factors, serve to anchor the concept so it can be meaningfully defined. As a metaphor, consider the example of an unrooted phylogenetic tree. Until the tree is rooted by specification of an outgroup branch, several patterns of historical relationship are possible. The root anchors the tree and determines the order of the branching. Similarly, the following specifying factors anchor the context of constraint so it can be characterized in a given situation. Although some of these factors are implicit in the constraint literature, most are rarely, if ever, treated explicitly.

Time frame. Time frame is relevant to constraint in two ways. First, evolutionary constraints are manifested over long time periods (geological rather than ecological scales). Thus, a quantitative genetic approach, for example, may give only limited insight into the long-term potential for constraint (Arnold, 1992). On the other hand, given enough time, most constraints are eventually broken or circumvented. Constraints are therefore historically stable, but ultimately mutable (Schwenk, 1995; Wagner and Schwenk, 2000).

Historical pattern. Because constraints are manifested over geological time, a mechanistic hypothesis of constraint is best tested by reference to a historical pattern predicted as its consequence (McKitrick, 1993; Schwenk, 1995; Schwenk and Wagner, 2001). Patterns putatively associated with the past action of constraint include long-term phenotypic stability or stasis in characters, the observation that only a subset of theoretically possible phenotypes has evolved, the repeated independent evolution of a certain phenotype within a clade (parallelism), and the failure to achieve an "optimal" phenotype according to mechanical or engineering analyses.

Historical patterns are only weak tests of constraint hypotheses; alternative mechanisms can lead to similar patterns (e.g. Schwenk and Wagner, 2001). To be meaningful, the test requires comparison of "constrained" to related, "unconstrained" clades in which evolution has proceeded according to the expectations of a null model. There are, however, philosophical objections to any inference of past processes and selection pressures based on current conditions and patterns (Reeve and Sherman, 1992; Leroi et al., 1994). Given that hypotheses of constraint require exactly such inferences, their support is necessarily circumstantial—an operational reality that need not invalidate the concept.

Character specificity. A hypothesis of constraint applies to a designated character or set of characters, not to whole organisms (Schwenk, 1995). This is necessary both for precision and to avoid the paradox of simultaneous constraint and adaptation. Character constraints are thus decoupled from their ultimate effect on the state of organismal adaptation and lineage success. In other words, characters can be evolutionarily limited, but the effect of this limitation on organismal function and evolution can be neutral, positive, or negative (Wagner, 1988; Roth and Wake, 1989).

Clade specificity. Constraints are attributes of living systems and as such, they arise, evolve, and eventually disappear. In other words, constraints have phylogenetic continuity. At any given time, therefore, the organisms within a clade encompass a unique set of character constraints. A constraint hypothesis thus applies to characters within a particular clade at a particular time. This excludes from consideration universal attributes of life, such as a nucleic acid–based information system and conformation to physical laws. While such factors undoubtedly circumscribe the range of possible phenotypes—elephants cannot have spindly legs—they have less explanatory value for interpreting one evolutionary outcome over another within a particular clade.

This observation has several implications. First, for a given clade in which a constraint is hypothesized to exist, there must be a related clade without the constraint. Without the ability to compare character evolution in constrained and unconstrained clades, there is no possibility of falsifying the hypothesis. Second, given clade specificity, the historical stability of constraints, and their eventual dissolution in descendant taxa, constraints are distributed monophyletically or paraphyletically. Mechanistically similar constraints might arise polyphyletically, but these would be nonhomologous. Finally, so-called phylogenetic constraint, as typically used, merely implies historical contingency, not character constraint. It calls attention to the starting condition, which certainly influences evolutionary outcomes given that selection acts on existing phenotypic variation; but it does not speak specifically to the question of whether a particular character is constrained. Certainly constraints, like other organismal attributes, are part of a lineage's legacy, but historical contingency is a universal attribute of character evolution and does not, therefore, satisfy the requirement of clade specificity for hypotheses of constraint.

Focal life stage. The focal life stage is the designated life stage in which character evolution is putatively constrained. Typically, this is the adult stage, as implicit in most studies; but it can be any stage of ontogeny. Recognition that constraints are manifested at a particular life stage permits the decoupling of constraint from a particular mechanism because any mechanism that limits the ability of focal life stage characters to evolve according to the null model is potentially a constraint. For example, if the null model is evolution by natural selection, constraints are manifested by *their effects on selection acting during the focal life stage.* Designation of a focal life stage would help to resolve the problem of circularity encountered in the Maynard Smith et al. (1985) example.

Internal vs. external selection. The fitness consequences of character variation are traditionally seen as determined by the biotic and abiotic environment surrounding an organism. Environmentally imposed selection pressures act on characters to maintain the fit, or state of adaptation, between phenotype and environment. However, character variants are also tested within the internal arena of a functioning organism. Stabilizing selection pressures in this case stem from the dynamic interactions among characters, either in the instantaneous sense of function or in the sequential sense of development. Variants that disrupt the internal dynamics of functional or developmental systems have negative fitness consequences for the organism.

These components of selection are distinguished as external and internal selection, respectively (Whyte, 1965; Riedl, 1977; Cheverud, 1984, 1988; Arthur, 1997, 2001; Wagner and Schwenk, 2000; Fusco, 2001). The critical distinction for our purposes is that external selection changes with environment, whereas internal selection remains essentially constant because it travels with the organism. Thus characters are likely to be subjected to long-term, internal stabilizing selection imposed by intrinsic, organismal, functional integration, even as the external environment changes. If internal selection opposes external selection, the potential for constraint is clear. Such long-term stabilizing selection is, furthermore, likely to limit the pattern of variation available to external selection through its influence on the genetic variance/covariance matrix (Cheverud, 1984, 1988).

Null model. A null model represents an expectation concerning how character evolution would proceed in the absence of constraint (Gould, 1989b; Antonovics and van Tienderen, 1991; Schlichting and Pigliucci, 1998). Different models lead to different notions of constraint (Amundson, 1994; Schwenk, 20002). For example, implicit in most studies of nucleotide sequence evolution is a null model of neutral evolution, implying random mutation and fixation by drift. The expected pattern is a constant rate of sequence evolution, equal to the neutral mutation rate, and sequence divergence among taxa. A sequence that is highly conserved among taxa deviates from the null expectation and is said to be "constrained," meaning that it is under stabilizing selection (e.g., a change in nucleotides would entail harmful functional effects).

In contrast, a null model of evolution by natural selection is implicit in most organismal studies. The failure of the phenotype to have evolved in response to selection is then taken as evidence of constraint. Selection, in this case, is logically excluded as a mechanism of constraint (the dichotomous approach). Conflicting conclusions about the nature of constraint thus result from different null models. What is the best null model for constraints on phenotypic evolution?

It is reasonable to assume that, in the absence of constraints, phenotypic evolution would proceed according to the Darwinian notion of adaptive evolution by means of natural selection. An adaptive fit between phenotype and environment is driven by the external component of selection. Therefore, an appropriate null model for constraint is the adaptive evolution of characters by means of external selection.

A Definition of Constraint

Given the specifying factors, we can define constraint as a mechanism or process that limits or biases the evolutionary response of a character to external selection acting during the focal life stage. "Limits" and "biases" are usually identified post hoc by comparing patterns of character evolution in putatively constrained and unconstrained clades. In the latter, character evolution should conform to the null model; in the former, it should deviate from the model as predicted (e.g., Schwenk and Wagner, 2001).

This definition reconciles constraint and selection because its relativism accommodates the dichotomous view by suggesting that constraints are manifested in their effects on selection while allowing selection itself to serve as constraint. This is possible because internal selection acting at earlier life stages can limit the variation available to selection at the focal stage, and because internal selection can oppose external selection during the focal stage.

Types of Constraint

We have emphasized a context for constraint that leads to a workable definition. We are less concerned with compiling a list of possible constraints because these must be argued case by case. However, we consider briefly the potential nature of these constraints. The typological labels associated with most treatments of constraint are problematic due to definitional ambiguity and absolutism. Are genetic correlations caused by the epigenetic system genetic or developmental constraints? Does functional integration of developmental processes lead to developmental or functional constraint? We therefore stress mechanisms over labels. Our list may not be exhaustive.

Constraints are mechanisms that affect the evolutionary response of a trait to external selection acting at the focal life stage such that, over the long term, it deviates from null model expectations. Most constraints act at earlier life stages to restrict phenotypic variation of the trait within a population at the focal life stage, but some act during the focal stage to limit the efficacy of external selection acting on the trait. Possibilities include:

1. *Variational inaccessibility.* Despite mutations, certain character variants are never produced. These variants are therefore developmentally impossible to achieve and are never introduced into a population. This is implied by canalization and has been called both genetic and developmental constraint.

2. *Developmental lethality.* In this case, character variants can be produced, but their introduction into the developmental stream is lethal to the embryo. Lethality is caused by disruption to developmental dynamics and is therefore a case of internal selection. This has been called *developmental constraint,* but it is rarely attributed to selection.

3. *Deleterious pleiotropic effects.* In contrast to (2), individuals with variant character phenotypes survive until the focal life stage, but always suffer lower fitness because of deleterious changes in correlated characters (Galis, 1999; Galis and Metz, 2001). This has been called *genetic, epigenetic* and *developmental constraint.*

4. *Functional integration.* Functional integration implies internal stabilizing selection stemming from the dynamic interaction among characters during system function (Witte and Döring, 1999; Wagner and Schwenk, 2000). This can be manifested as constraint in two ways. First, a character variant is developmentally produced, but causes lethality at a later life stage when the relevant system becomes functional. For example, a modified dental character in a mammal might so reduce the efficacy of mastication that the afflicted individual starves to death, but only after weaning when it must eat solid food. This character variant fails to reach the adult focal life stage. Alternatively, variant individuals survive to the focal stage, but the variant character is always associated with lower fitness. Both situations have been called *functional constraint.*

Finally, our definition would seem to allow a final notion of constraint as the action of external selection during earlier life stages to limit variation available at the focal stage. In most species, for example, there is strong external selection on juveniles and relatively few survive to adulthood. A great deal of variation is therefore eliminated from a population during early life stages, possibly constraining the effects of external selection acting during the adult stage. However, this concept violates the important premise that constraints have phylogenetic continuity. External selection acting on juvenile stages results from a particular phenotype-environment interaction and is therefore not a property of the evolving clade. To the extent that juvenile adaptations evolve, including developmentally and functionally integrated character complexes, the efficacy of external selection acting at later stages will be reduced through the processes outlined in this entry. Metamorphosis is considered to be an evolutionary escape from such life stage-specific constraints (e.g., Moran, 1994).

Our goal has been to derive a biologically defensible and internally consistent concept of constraint that is applicable to a range of evolutionary interests. There can be no claim of a single, correct solution to this problem. The definition and types of constraint that we outlined follow directly from the factors we specified, but other definitions are possible if one or more of these factors are changed—for example, if a different null model is specified. The strength of this formalization is its ability to accommodate other needs or viewpoints.

By translating random genetic mutations into usable variation and ensuring the maintenance of a functionally coherent internal milieu, constraints act as regulators that tune or modulate the types of variation available to selection over time in a way that may facilitate adaptive evolution (e.g., Cheverud, 1984; Wagner, 1988; Wagner and Altenberg, 1996). Thus, while constraints imply evolutionary limitations on characters, their ultimate effect on organismal evolution may not only be positive but essential.

SCOTT F. GILBERT AND RICHARD M. BURIAN

Development, Evolution, and Evolutionary Developmental Biology

Evolutionary developmental biology (evo-devo) seeks to discern how the mechanisms of development effect evolutionary change and stasis, how changes in development generate evolutionary novelty, how development might constrain certain phenotypes from arising, and how developmental mechanisms themselves evolve. Evolutionary developmental biology links genetics and evolution via the agency of development. Because development connects genotype with phenotype, heritable changes in development can yield new phenotypes that can be retained through natural selection. Moreover, certain conservative features of evolution can be explained through the developmental mechanisms that enable the production of some morphological types but not others.

Evolutionary developmental biology is both a synthesis between evolutionary and developmental biology and an ongoing negotiation between these two disciplines. The developmental approach to evolution, drawing on traditional embryology, genetic studies of development, phylogenetics, and other disciplines, emphasizes conserved developmental pathways in evolution and the conditions and causes of evolutionary innovation. It seeks to complement and interact with population genetic approaches to evolution, which emphasize variations between individuals within species, particularly as they affect differential survival and reproduction. In addition, the developmental approach can illuminate the mechanisms of phenotypic plasticity, which enable the environment to generate specific phenotypes from an inherited genotype.

Macroevolutionary Concerns

The macroevolutionary component of evolutionary development biology attempts to discover the mechanisms by which evolutionary innovations are brought about. These macroevolutionary concerns of evolutionary developmental biology can be traced to Darwin and his contemporaries. After reading Johannes Müller's summary of von Baer's laws in 1842, Darwin (1859) saw that embryonic resemblances would be a very strong argument in favor of the genetic connectedness of different animal groups. "Community of embryonic structure reveals community of descent," he concluded in *On the Origin of Species.*

While von Baer never accepted homologies across phyla, evolutionary biology made it possible—on the principle of a monophyletic origin for the animal kingdom—to seek the links between the types (Bowler, 1996). Kowalevsky (1871) discovered that tunicate larvae have notochords and form their neural tubes and notochords in a manner very similar to the primitive chordate *Branchiostoma* (amphioxus). Moreover, Darwin (1874, p. 160) publicized this linkage between the invertebrates and vertebrate realms of the animal kingdom. Comparative developmental anatomy became evolutionary as questions of the origin and homologies of the germ layers in various animals became paramount (e.g., Lankester, 1877; Metchnikoff, 1879; Balfour, 1880, 1881).

A key issue in the macroevolutionary concerns of evo-devo is the origin and evolution of embryonic development. How, for instance, did the mesoderm first come into being, how did the various modes of gastrulation evolve, how did the first signal transduction cascades form, and how did the various body plans *(Baupläne)* originate? These investigations examine the origins of multicellularity during the pre-Cambrian and Cambrian radiations, and they often look at the physical and chemical properties by which cells interact.

Most of the other programs of evolutionary developmental biology concern the inherited modifications of existing developmental processes and the extent to which developmental mechanisms account for evolutionary innovations. How do changes in development lead to the production of novel structures? This is especially important when the novel structures are used to define clades, as the neural crest is used to define vertebrates. Such research attempts to answer the age-old questions of how birds formed feathers (Prum, 1999), turtles got their shells (Loredo et al., 2001), vertebrates their heads (Holland and Chen, 2001), arthropods their jointed legs (Carroll et al., 1995), and the elephant its trunk (still no answers, but see Gaeth et al., 1999, for an interesting developmental speculation). One of the main themes of evolutionary developmental biology is the formation of new structures through changes in the spatial and temporal expression of developmental regulatory genes (especially genes encoding transcription or paracrine factors), and the spatial and temporal occurrence of developmental phenomena (including cell-cell interactions, for example).

Jacob (1977) called this reutilization of pre-existing materials and processes "bricolage" (i.e., evolutionary tinkering; see also Duboule and Wilkins, 1998) and suggested that evolutionary biologists look to mutations of patterning genes in embryos to find the mechanisms of macroevolutionary changes. We now know, for example, that changes in Hox gene expression patterns correlate with the presence of maxillipeds in crustaceans, the formation of limbs from fins, the loss of snake forelimbs, and the numbers of thoracic vertebrae in vertebrates (Gaunt, 1994; Burke et al., 1995; Averof and Patel, 1997; Cohn and Tickle, 1999).

Similarly, changes in the timing and placement of paracrine factor expression have been correlated with the evolutionary changes of cusp locations in mammalian teeth (Jernvall et al., 2000). Changes of the timing and placement of developmental processes relative to those of ancestral organisms are called *heterochrony* and *heterotopy*, respectively. The differences in the formation of jaws in different strains of mice, for example, has been linked to heterochronic changes in epithelial-mesenchymal interactions (MacDonald and Hall, 2001), and the formation of the turtle carapace appears to be the heterotopic use of bone-forming pathways in the trunk dermis. Heterotopy and heterochrony have been used to describe changes in gene expression patterns (see Ferkowicz and Raff, 2001; Loredo, et al., 2001).

In addition to the redeployment of pre-existing genes, evolutionary developmental biology also studies those mutations of regulatory genes that give their protein products new functions. The clade of insects, for instance, is characterized by an Ultrabithorax protein that differs from the Ubx proteins of other arthropods in its ability to repress expression of the *Distal-less* gene. This new function appears to result from a mutation in the region of the gene encoding the C-terminus of the Ubx protein. This mutation may therefore be responsible for the lack of abdominal limbs in insects and for facilitating morphological diversity of the insect abdomen (see Galant and Carroll, 2002; Ronshaugen et al., 2002).

Macroevolutionary developmental biology also deals with the pressing issues of *homology* and *homoplasy* (Hall, 1994; Wake, 1994; Laubichler, 2000). Homology is usually defined as similarity due to descent with modification and can be applied at numerous levels—for example, to bones, organs, proteins, genes, and even signal transduction pathways. Similarity can also arise through the convergence of phenotypes due to their being selected for the same characters. Analogous structures (such as bird wings and insect wings) are selected independently for the same function, but they are not derived from the same structure. Homoplastic structures (such as sea turtle and dolphin flippers) share a common appearance due to being selected for similar properties, but they are independent modifications of organs in two different groups of animals.

One of the major findings of evolutionary developmental biology has been that analogous structures can be generated through homologous processes initiated by homologous genes. Homologous genes between taxa are called *orthologues,* within taxa paralogues. The *Pax6* orthologues, for instance, are involved in forming the eye in insects, vertebrates, molluscs, and planarians.

Moreover, not only is this one gene orthologous in these various taxa, but so are several other genes that interact with *Pax6* in the process of eye formation in these different clades (Xu et al., 1997; Czerny et al., 1999). This means not only that homologous genes *(Pax6)* can initiate the formation of analogous organs (fly eyes and mouse eyes), but also that similar pathways have been deployed repeatedly to produce eyes (Erwin, 1999; Hodin, 2000, Pineda et al., 2000). Instead of having to evolve a totally new way of developing an eye, which comparative anatomy suggests would have to be done forty-one separate times (Salvini-Plawen and Mayr, 1977), the eyes develop as modifications of the same basic theme. This theme involves interactions among several orthologous genes and gene products that have been conserved between protostomes and deuterostomes. Similarly, induction and specification of the ventral insect nervous system and the dorsal vertebrate nervous system are based on homologous pathways composed of homologous genes (Sasai and De Robertis, 1997; Oelgeschlager et al., 2000).

Conversely, distinct processes can generate homologous structures, such as the anterior and posterior neural tube in vertebrates (primary and secondary neurulation in this instance). Because cladistics is grounded on judgments of homology, it is essential to "get it right." As the above examples show, evolutionary developmental biology is crucial for homology assessment. Accordingly, evo-devo plays a major role both as a proponent and as a critic of various attempts to determine homology.

Related to homology is the concept of *modularity* (Gilbert et al., 1996; Raff, 1996; Wagner, 1996; Bolker, 2000). Evolutionary developmental biology asks: What are modules, how do they interact within the embryo, and how do they evolve? Numerous developmental modules are anatomical structures, some of which (e.g., rhombomeres, somites, parasegments, the enamel knot) play key roles in development but have no anatomical counterpart in the adult. Other modules, such as limb and heart fields, are regions in which particular interactions occur though the regions have no obvious anatomical boundaries. The modularity of development is commonly thought to be a prerequisite of evolution (see, e.g., Opitz and Gilbert, 1993; Raff, 1996). Modularity allows for dissociation, i.e., for one part of the embryo to change without causing major and probably deleterious changes in the other parts.

Dissociation is the developmental counterpart of the quasi-independence of traits in evolutionary biology (see Lewontin, 1978, 2001; Wimsatt, 1981; Brandon, 1999); unless traits can vary with considerable independence of one another, stepwise alteration of characters would be very difficult due to the need for coordinated changes. In development, modularity is required for heterochrony and allometry in the production of organs; it also allows for duplication, divergence, and reuse of modules—as when the duplication of somites enables some vertebrae to have ribs and others not. Modularity also enables some features within a module to diverge in concert, as in the evolution of Meckel's cartilage from fish gill cartilage to the mammalian middle ear ossicles. The co-option (recruitment) of developmental modules exemplifies how Jacob's bricolage facilitates evolution. Eyespots in the butterfly wing appear to have evolved through the recruitment of a battery of genes that had earlier been used for wing formation and, before that, for segmentation of the blastoderm (Keys et al., 1999).

Modules have also been considered critical to evolution because they mediate the transmission of genotypic information. A gene that is wild type in one individual may be mutant in another individual, because of the other alleles expressed in these modules (Eicher and Washburn, 1989; Wolf, 1995; Nijhout and Paulsen, 1997). By analyzing the regulatory gene networks that constitute these modules, evo-devo attempts to explain how changes in evolutionary phenotypes are initiated, maintained, altered, or constrained (see Raff 2000; von Dassow et al., 2000).

Another central problem of evo-devo is whether—and, if so, how—development constrains the morphologies that occur in evolutionary history (Maynard Smith et al., 1985). For instance, in the 300 million years of tetrapod evolution, no species has developed with a duplicated zeugopod (radius-ulna/tibia-fibula) or with eyes on its tail. Many morphologies that selection might favor seem never to appear and thus cannot be tested.

The notion of a *developmental constraint* used in evolutionary developmental biology is that of a "constraint on form." It should not be confused, as it often is, with a "constraint on adaptation," which is what many population geneticists mean by constraint (Amundson, 1994). Mathematicians and modelers are now looking at development to see whether certain developmental patterns are based on algorithms that restrict the permissible variations to parameters compatible with specific equations. For instance, the coat and shell pigmentation patterns of mammals, fish, and molluscs have been modeled on the Turing reaction-diffusion equations. These equations predict the existing variants and suggest that certain variants are not possible (Goodwin et al., 1993, Kauffman, 1993; Meinhardt, 1998; Painter et al., 1999; Meinhardt and Gierer, 2000). Developmental constraints can also become phylogenetic constraints for two reasons: It is difficult to modify a module, once in place, to suit it to another function, and certain opportunities are preferentially facilitated (e.g., eye formation) by the combinations of modules already in place (Raff, 1994).

Microevolutionary Concerns

Evolutionary developmental biology also focuses on processes that enable eggs, embryos, and larvae to survive in particular environments (Hall, 1998; Gilbert, 2001). Darwin's tangled bank is not for adults only. While Wilson (1898) pointed out that the homologies between spiralian embryos demonstrated their common origin, Lillie (1898) pointed to specific instances in which modifications of these common features permitted organisms to live in certain environments. Lillie showed how changes in the cleavage patterns of river mussels allowed them to generate a larval form adapted to life in swift-moving currents. Recent studies have shown that the eggs of amphibians and echinoderms laid in direct sunlight have evolved DNA repair enzymes and often have inducible pigments that block UV irradiation (Blaustein et al., 1994; Mead and Epel, 1995; Adams and Shick, 1996; Kiesecker et al., 2001). The fish and amphibian species that have evolved Wolffian regeneration (the regeneration of the lens from the iris) are those same species wherein trematode larvae can parasitize the lens cortex (Okada, 2000). Conversely, some developing organisms undergo various forms of symbiosis and mutualism. McFall-Ngai (2002) has argued that developmental symbioses, especially symbioses involving microbes, may constitute the rule rather than the exception.

Phenotypic plasticity is a critical adaptive trait in life history strategies and is important in the developmental responsiveness of organisms to their respective environments. As Woltereck (1909) and Waddington (1956b) pointed out, the genome is not only active; it is also reactive. Developmental (phenotypic) plasticity is the notion that the genome allows the organism to produce a range of phenotypes. A particular genotype does not necessarily yield a unique phenotype or even a single phenotype. The structural or behavioral phenotype induced by a particular environmental stimulus is referred to as a *morph*. When developmental plasticity manifests itself as a continuous spectrum of phenotypes expressed by a single genotype across a range of environmental conditions, this spectrum is called the *norm of reaction* or *reaction norm* (Woltereck, 1909; Schmalhausen, 1949; Schlichting and Pigliucci, 1998). Important areas of research concern whether the reaction norm is a property of the genome, the extent to which it can be selected, and whether different genotypes within a species differ in the direction and amount of plasticity that they can express within the available range of environments.

A special case of developmental plasticity, *polyphenism,* refers to the occurrence in a single population of discontinuous (either/or) phenotypes elicited by the environment from a single genotype (Mayr, 1963; Nijhout, 1991). One of the best-studied polyphenisms is that of the temperature-induced morphs of *Bicyclus anynana,* a Malawian butterfly. Here the evolutionary selection pressures for and against eyespots at certain seasons have been studied, as well as the gene expression changes induced by temperature (Brakefield et al., 1996; Beldade et al., 2002). In addition to temperature-dependent polyphenisms, there are polyphenisms induced by nutrition, population density, and the presence of predators and competitors. Such developmental responses to environmental inducers are the raw material from which an organism might generate pathways of response that could become independent of the environment and become the normative phenotype of the species (Waddington, 1942, 1953a; Schmalhausen, 1949; e.g., genetic assimilation, canalization).

Developmental plasticity also is seen in the life history strategies of many organisms. These include temperature-dependent sex determination, insect caste determination, and other developmental phenomena such as larval settlement cues, diapause, and symbioses.

Trade-offs are another important component of the interplay between development and evolution. In a predator-induced polyphenism, the induced phenotype can better survive the predator, but the phenotype may be less adaptive in other ways. In *Daphnia,* for instance, the production of helmets, which make *Daphnia* less palatable (an adaptive response to the presence of chemicals from a major predator), appears to lessen the resources allocatable for provisioning eggs (Riessen, 1992).

Evolutionary developmental biology is both a new discipline and an ongoing negotiation between disciplines. It seeks to amplify and extend the modern synthesis of evolutionary biology and genetics to include developmental genetics as well as population genetics. In so doing, evolutionary developmental biology is attempting to integrate developmental biology into an evolutionary framework and to enrich our explanations of speciation, phylogenetics, and the generation of morphology.

SCOTT F. GILBERT AND RICHARD M. BURIAN

Developmental Genetics

For much of the twentieth century, Mendelian genetics and evolutionary biology ignored ("black-boxed") the developmental mechanisms that produce variation (Harrison, 1937; Hamburger, 1980; Allen, 1986;

Sapp, 1987; Gilbert et al., 1996; Burian, 2000; Amundson, 2000). As a result, embryology and developmental biology were not included in the so-called Modern or neo-Darwinian Synthesis in evolutionary biology. Thus, in a major founding document of that synthesis, Theodosius Dobzhansky (1937) redefined evolution as change of allele frequency. Although several founders of the population genetic approach to evolution did not accept Dobzhansky's subsumption of evolution within population genetics, their objections did not turn on the need to include development within the synthesis, but on the need to account for population distributions, population sizes, and lineage splitting.

Population genetics became the predominant explanatory mode for evolutionary biology. By 1951, Dobzhansky (p. 16) could confidently declare, "Evolution is a change in the genetic composition of populations. The study of mechanisms of evolution falls within the province of populations genetics." Similarly, Ernst Mayr (1980, pp. 9–10), who rejected the definition of evolution in terms of change of gene frequency, wrote that "the clarification of the biochemical mechanism by which the genetic program is translated into the phenotype tells us absolutely nothing about the steps by which natural selection had built up the particular genetic program."

Until recently, developmental genetics had little to offer evolutionary biology except some ideas. However, since the 1980s, the mechanisms of differential gene expression have been elucidated, and the far-reaching effects of changes in gene expression patterns have been explored. Developmental biology is now providing evolutionary biology with tools to understand the contexts and identities of genes that are important for generating variation and to explore the contextual factors that regulate the genes' expression. Developmental genetics has also cleared up some misconceptions about evolution. For example, there was a consensus that "the search for homologous genes is quite futile except in very close relatives" (Mayr, 1966, p. 609, quoting Dobzhansky). Developmental geneticists disproved this view with the discovery that many of the genes encoding the developmental toolbox of paracrine and transcription factors were conserved between protostomes and deuterostomes.

Evolutionary developmental biology is now analyzing the proximate mechanisms responsible for some of the inherited variations studied by population geneticists. It has shown that some of the mechanisms that produce developmental variations have been co-opted into novel contexts, where they have played a significant role in evolutionary innovation (an idea called *bricolage;* see Jacob, 1977; Duboule and Wilkins, 1998). For example, differences in the expression of Hox genes can determine bristle pattern differences in *Drosophila* and axial skeleton differences in vertebrates and arthropods (Gaunt, 1994; Averof and Akam, 1995a; Burke et al., 1995; Averof and Patel, 1997).

Genes responsible for primitive characters can be used for new traits, such as when genes responsible for segmentation and neurogenesis are used for making eyespots of butterfly wings (Keys et al., 1999). Thus, it is reasonable to expect that the differences that transformed a flightless insectivore into a bat or a shell-less reptile into a turtle were produced by regulatory genes acting within the embryo. The identities of these genes and the manner of their expression, studied by developmental genetics, currently constitute a central theme in evolutionary developmental biology.

There are important historical differences between the account of genes used by the population genetics approach to evolution that was employed in the Modern Synthesis, and the account of genes in developmental genetics that is employed in evo-devo (Gilbert, 2000b). First, the genes of classical population genetics were uncovered by transmission patterns, typically captured in mathematical abstractions. The physical structure of a gene was not crucial for the identification of alleles A and a, hypothesized or detected by their effects, as long as they were stably inherited according to specific rules and mutated at particular rates. In contrast, the genes of developmental genetics are defined by conserved DNA sequences with specific regulatory regions, coding regions, and intron/exon boundaries.

Second, the genes of population genetics are recognized by the phenotypic differences caused by their various alleles. The very meaning of the term "allele" is tied to variants of a gene producing different phenotypes within a population. Developmental genes were recognized by the roles they played in controlling specific developmental processes, by the interactions of their products with DNA, and by their structural similarities in different taxa. Many developmental genes were identified using polymerase chain reactions to identify closely similar sequences and then testing for homology, a technique now being used in population genetics.

Third, population geneticists focused primarily on variation in the frequency of alleles within populations or species, whereas developmental geneticists focused primarily on variations in developmental gene expression between taxa. Fourth, the population genetic approach sought to explain adaptation by natural selection ("survival of the fittest") tempered by drift, meiotic drive, and other factors, whereas the developmental genetic approach sought to explain phylogeny ("arrival of the fittest") and the production of novelties. Fifth, the population genetics approach to evolution focuses primarily on genes that affect adults and their impact on competition for reproductive success, whereas the developmental genetics approach to evolution focuses on genes expressed during development, their interactions, and their impact on the ontogeny of the organism.

These approaches, whose differences we have highlighted, are being integrated in evo-devo. Gene expression is regulated contextually and often epigenetically, so genes characterized by their effects are not seen as autonomous. At the same time, if they are to survive, developmental genes and their variants, like all genes, must have net effects that either facilitate or at least do not hinder reproductive survival within their organismal and populational contexts. Accordingly, developmental genes must be analyzable by the approaches taken by both developmental and population genetics. Which tools are appropriate depends on the problem at hand, e.g., whether we are examining embryonic development or phylogenetic histories. But in the end, we must be able to track genes and their interactions within developmental contexts and in lineages of transmission through both generational and evolutionary time.

One of the first investigators to suggest merging the approaches of developmental and population genetics, C. H. Waddington, concluded that evolutionary biology needed to study those processes that get from the genotype to the phenotype—the "epigenetics of development." Following Goldschmidt (1940), Waddington (1953b, p. 190) declared, "Changes in genotypes only have ostensible effects in evolution if they bring with them alterations in the epigenetic processes by which phenotypes come into being; the kinds of change possible in the adult form of an animal are limited to the possible alterations in the epigenetic system by which it is produced." This concept distinguishes between selection working on adults and selection working during development. Waddington claimed that both these modes of evolution work together to produce species adapted to particular environments. The evo-devo model of evolution follows from Waddington's notions of selection of alternative epigenetic processes during development.

Evolutionary developmental biology does not seek to replace the Modern Synthesis. Rather, the consensus of evolutionary developmental biologists is that their discipline complements and extends the Modern Synthesis, and that no full account of evolutionary phenomena can be achieved without both population genetics and developmental genetics. Most evolutionary developmental biologists expect that the bricolage necessary for the generation of new morphologies is consistent with the traditional microevolutionary mechanisms of mutation, recombination, and meiotic drive. In addition, they emphasize the importance of gene duplication and divergence in allowing co-option, the roles of changing the time and place of gene expression to produce variation, and the roles of gamete-binding proteins as a mechanism of reproductive isolation (Metz and Palumbi, 1996; Lyon and Vacquier, 1999). These mechanisms are now being taken seriously by population geneticists, as well.

The role of mutations in regulatory genes has long been assumed to be important in producing major phylogenetic changes, but supporting evidence was provided only recently. During the evolution of arthropods, the Ser/Thr domain of the Ubx protein in the common ancestor of crustaceans and insects was replaced by an alanine-rich region that gave it a new activity. This mutation converted the Ubx protein into a repressor of the *Distal-less* gene. This mutation is found in all insects but is absent in crustaceans, spiders, centipedes, and onychophorans. The successive replacement of the serine/threonine region by the poly-alanine repression domain is correlated with and may have facilitated the patterns of segmental diversity found in insects (Galant and Carroll, 2002; Ronshaugen et al., 2002).

Both the population genetics account of evolution and the account rooted in the genetics of development are required to produce an analysis of intraspecific and interspecific variation. Intraspecific variation provides the raw material for the population genetics account of evolution and thus is the pivot of natural selection. However, many of the genes producing such normally occurring variations are active in development, although they are not well characterized. Indeed, a major limitation of the attempt to forge a new synthesis has been the failure to identify alleles of paracrine and transcription factor genes, including their regulatory sequences, which may be responsible for variation in the amounts or efficiencies of paracrine or transcription factors, and which could then be studied at the population level. Amundson (2001) predicts (and we agree) that until developmental biology increases its attention on intraspecific variation, developmental approaches will not interest mainstream evolutionary biologists. Such steps are being taken (e.g., Kopp et al., 2000; Johnson and Porter, 2001; MacDonald and Hall, 2001), and are necessary if evolutionary developmental biology is to provide the tools for studying the ways by which microevolutionary processes relate to macroevolutionary differences.

These new tools let us employ new methods in attempting to answer old questions of phylogeny and novel morphology that provided a starting point for Darwin's theory. Moreover, we seek to synthesize the same modes of thinking that Darwin himself had to bring together: the interaction of form ("unity of type") and function (adaptation to the "conditions of existence") were integrated into "descent with modification" through "natural selection." Darwin (1859, p. 206) wrote that "natural selection acts by either now adapting the varying parts of each being to its organic and inorganic conditions of life; or by having adapted them during long-past periods of time, being in all cases subjected to the several laws of growth. Hence, in fact, the law of the Conditions of Existence is the higher law; as it includes, through the inheritance of former adaptations, that of Unity of Type."

Recent differences between population genetics and developmental genetics approaches to evolution can be seen as continuing the debate over the relative importance of function versus structural homology (Amundson, 1998). In comparison with the population genetics tradition, evo-devo elevates the importance of structural homologies, but remains within a generally Darwinian framework. Darwin wrote (1859, p. 403), "What can be more curious than that the hand of a man, formed for grasping, that of a mole, for digging, the leg of a horse, the paddle of a porpoise, and the wing of a bat should all be constructed on the same pattern and should include similar bones, and in the same relative positions?" To account for these and similar instances of descent with modification, new accounts of the generation of form, now being produced within evolutionary developmental biology, need to be completed and integrated with the findings and tools of both developmental and population genetics. This new evolutionary synthesis would be able to account not only for the mechanisms by which phenotypes are selected but also for the mechanisms by which certain phenotypes emerge.

BRUCE M. CARLSON

Developmental Mechanisms: Animal

Developmental mechanisms are fundamental to phylogenetic stability and to evolutionary change. Some mechanisms are remarkably stable over a wide range of phylogenetic groups; but others are prone to variations, though sometimes with a common underlying theme.

Although the totality of developmental processes and mechanisms is highly diverse, a very broad way of conceptualizing both the diversity and conservatism of animal development is shown in Figure 1 (Raff, 1996). Both the initiation of development and its early phases are subject to considerable diversity, not only among diverse taxa, but sometimes even between closely related groups. Somewhat later, embryos enter a period, called the *phylotypic stage,* during which they develop the fundamental features of the body plan of their major taxonomic group. This phase is characterized by relatively little variation and is phylogenetically stable over long periods of time. When vertebrates reach the phylotypic stage of development, their germ layers and body axes have been set, and they possess common features, such as a notochord, dorsal nervous and branchial arch systems, and a post-anal tail.

This period of ontogenetic and evolutionary conservatism is followed by one in which the number of developmental pathways greatly expands, and structures as diverse as fins, limbs, wings, paddles, or greatly reduced appendages form. Yet despite such diversity in form, the underlying developmental mechanisms are often remarkably faithfully conserved. Another important theme in development is the conservation of molecular mechanisms, not only across a wide spectrum of phyla but

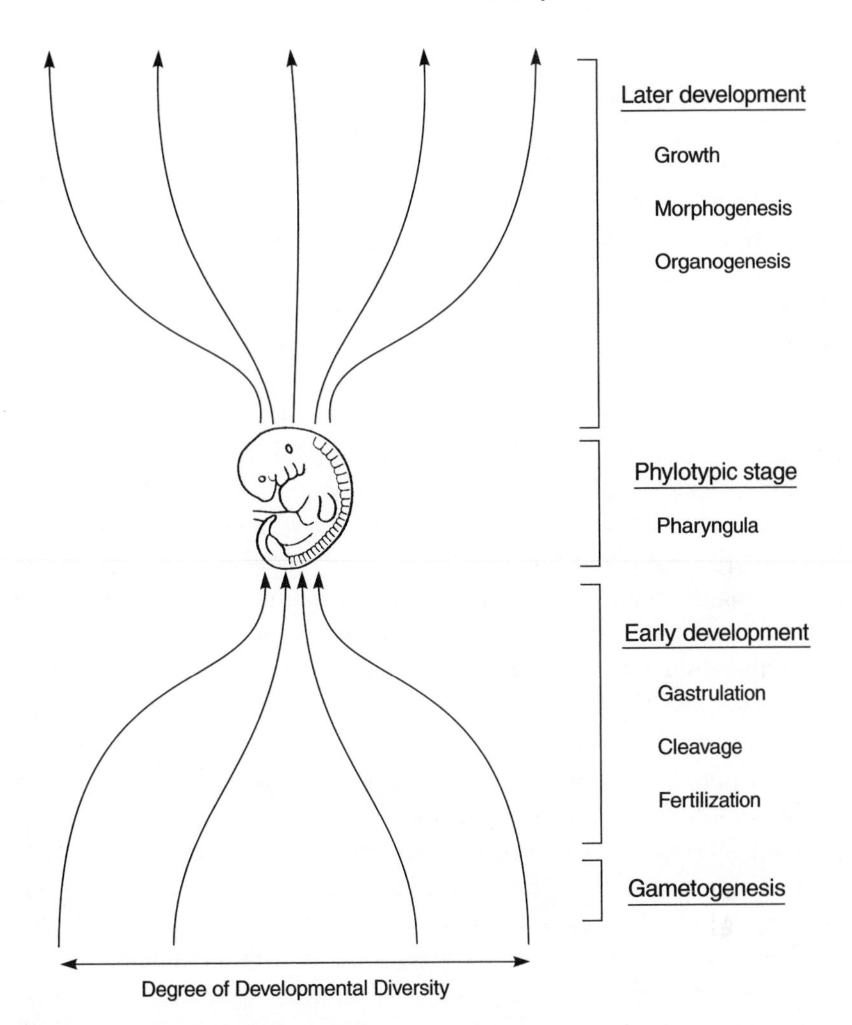

FIGURE 1. Hourglass diagram showing an expanded base with diverse patterns of initiation and early development narrowing to a highly conserved neck during the phylotypic stage. In later stages of development, diversity in developmental processes and in form increases greatly.

also in the reuse of the same molecules to mediate important developmental processes at different times and in different sites within the same embryo.

In the vast majority of species, development is initiated by the *fertilization* of an egg by a spermatozoon, with the attendant mixing of the genetic material from the two parents. Nevertheless, there are numerous examples among the invertebrates of asexual reproduction by budding, by fission, or even by evisceration in the case of some holothurians (Ivanova-Kazas, 1977). Other forms of reproduction found in vertebrates include parthenogenesis, seen in some species of fishes and birds, and polyembryony, which includes the subdivision of cleavage-stage embryos, such as those of the armadillo, into individual blastomeres, each of which goes on to develop into a complete individual. The evolutionary implications of these modes of reproduction remain poorly explored.

Once fertilization has occurred, it is inevitably followed by a period of *cleavage,* during which the number of cells (blastomeres) rapidly increases. The developmental mechanism common to all cells in cleavage is a modified form of mitosis in which blastomeres divide without any synthesis of new cytoplasm, i.e., without growth. For many species that have been investigated, another mechanism appears to be important, namely maintaining the early blastomeres in an undifferentiated condition through the actions of genes, such as members of the Oct family.

For the embryos of many species that develop outside the mother's body, products of maternal genes control much of the period of cleavage. In amphibians, for example, the embryonic genome does not become the dominant force controlling development until the end of the cleavage period. Another significant influence on cleavage in many embryos is parental imprinting, which occurs during gametogenesis through the methylation of certain bases in selected genes (Latham et al., 1995). Imprinting, which can be transmitted either maternally or paternally, appears to affect only a minority of genes (possibly 500 in humans); the effects can be significant although only felt on the immediate descendant. When gametogenesis takes place in the descendant, the genes are newly imprinted according to the sex of that individual.

Although cleavage takes on a variety of patterns, a fundamental subdivision of animals is into the protostomes and deuterostomes. The protostomes include flatworms, roundworms, arthropods and molluscs, in which cleavage occurs in a spiral pattern, leaving the blastomeres oriented at an angle to the principal axis of the embryo. Cleavage-stage embryos of protostomes are typically characterized by mosaic qualities, namely the inability of a subdivided embryo to compensate for the parts that are lost. Embryos of deuterostomes (echinoderms and chordates) undergo radial cleavage and possess regulative qualities; if a portion of a cleaving embryo is removed, the remaining cells compensate for the loss.

Because the presence and functions of *Oct* genes have been investigated most thoroughly in the chordates, it will be instructive to determine the degree of correlation between regulative properties and differentiation-suppressing gene activity in regulative embryos in comparison with the activities of similar genes in mosaic, nonregulative embryos.

By definition, a fertilized egg, or zygote, is *totipotent*—a cell that is capable of producing all of the progeny contained in the adult body. As cleavage progresses and throughout much of subsequent embryonic development, the cells become restricted in their capacity to produce different types of progeny. Restriction is defined as the progressive narrowing of developmental options available to a particular type of cell.

Of considerable developmental and evolutionary importance is the identity of factors, both internal and external, that narrow a cell's developmental options. Classical embryological studies focused on the role of unequally distributed cytoplasmic factors as determinants of the future fate of cells, especially but not exclusively in mosaic embryos (Conklin, 1905). Many biologists have considered mosaic development to operate almost exclusively on the basis of cells' responding to information inherent in the cytoplasmic determinants that they are allotted during cleavage, and they have claimed that interactions with neighboring cells during development play either minimal or no role in the cells' fate.

In other groups of embryos, primarily those with regulative development, inductive signals from neighboring tissues or groups of cells act as restriction agents and narrow the future possible developmental options of a cell. As is seen in early mammalian embryos, in which some blastomeres form trophoblast and others form the inner cell mass, even the external environment may play an important role as a restriction agent. In contrast to mosaic embryos, regulative embryos have traditionally been considered to depend on interactions with neighboring cells or tissues to guide the course of development. A more contemporary view suggests that the distinction between mosaic and regulative development is not this clear cut. For example, in amphibians, as well as in many invertebrates, cells that inherit germ cell determinants from the cytoplasm of the egg will ultimately become the germ cells.

When a cell has passed its last restriction point and its fate is fixed, that cell is said to be *determined.* As a cell passes through restriction points during development, its morphology and functional properties typically change. This series of changes, commonly called *differentiation,* reflects the actual morphological and functional expression of the portion of the genome that remains available to a cell or group of cells. The term "differentiation" is often qualified by the adjectives "high" or "low." In reality, a cell's level of differentiation is appropriate to its developmental stage, and its level of specialization may be just as complex relative to its temporal function early in development as it would be in later stages. The term *fully differentiated* typically refers to a cell that has passed the final restriction point and is thus determined.

At the point of determination, a cell can still undergo considerable differentiation. For example, when a multinucleated muscle fiber first forms, it is incapable of further mitotic division, but undergoes a long and complex series of changes in gene expression that results in biochemical and structural changes leading ultimately to a fully functional muscle fiber. This series of changes is referred to as muscle fiber differentiation. Of relevance to evolutionary studies is the degree to which variation occurs in the differentiated characteristics of cells or groups of cells and whether such variation confers any adaptive or reproductive advantage to the individual.

Toward the end of the cleavage period, many of the blastomeres develop different configurations of surface molecules and begin to undergo sweeping directed migratory movements. This period, called *gastrulation,* results in the formation of three definitive germ layers—ectoderm, mesoderm, and endoderm—in all but a few phyla. Patterns of gastrulation vary considerably, but the endpoint of the gastrulation movements is remarkably similar throughout the animal kingdom. Much of the variation in the pattern of gastrulation movements is a consequence of the amount or location of yolk in the egg. Among the vertebrates, at least, the patterns of gastrulation movements appear to be more conservative than the distribution of yolk. A classical example is the gastrulation movements of mammals, which bear a remarkable resemblance to those of birds, even though the vast majority of mammalian eggs are almost devoid of yolk.

Fundamental to the organization of the animal body is the specification and actual laying down of the major body axes. There is remarkable diversity in both the timing and mechanism of *axis specification* even within a given phylum. At one extreme, the anteroposterior axis in insects is specified well before fertilization when the oocyte is still surrounded by follicular cells. At the other extreme, the anteroposterior axis in birds is not set until the cleaving embryo contains up to 40,000 or 50,000 cells; if the embryo is surgically manipulated, this axis may be reset. It is significant that, in many forms, axis specification seems to result from self-organizing activity (Gerhart and Kirschner, 1997), a characteristic that is of considerable importance as a potential mechanism for evolutionary change.

An important consequence of gastrulation is the approximation of sheets or aggregations of different types of cells to one another. Further development often depends on interactions, called *inductions,* between adjacent groups of cells. An embryonic induction is an interaction whereby a signal emitted from an inducing tissue, often called an *inducer,* acts on a responding tissue with the result that its developmental fate is changed over what it would have been in the absence of the induction. Inductive events occur in a hierarchical fashion throughout much of early embryonic development, and the major inductive mechanisms for whole or parts of organs are strongly conserved throughout broad groups of animals.

A prominent example is induction of the central nervous system from ectoderm by the notochord, which migrates beneath the ectoderm during gastrulation. The inductive message is complex; in essence, signaling molecules emanating from the notochord block the inhibiting action of a growth factor, bone morphogenetic protein-4, on the dorsal ectoderm, thereby releasing the inherent tendency of the ectoderm to form neural tissue. In primitive chordates, such as amphioxus, the notochord underlies essentially the entire nervous system, which in this animal corresponds to the hindbrain and spinal cord of vertebrates. The inductive function of the notochord is apparently so fundamental in chordates that phylogenetically the notochord has been retained principally for that function rather than for its role in longitudinal support of the body, as is the case in amphioxus.

Formation of the head and forebrain in the vertebrates is correlated with the presence of a signaling center, called the *prechordal mesoderm,* which is located just anterior to the tip of the notochord. The hierarchy of signaling that results in the formation of the head is complex and still incompletely understood. Whether the appearance of a new signaling center is the principal phylogenetic basis for formation of the vertebrate head remains a topic of discussion. Gans and Northcutt (1983), for example, held that the formation of the head was in large part due to the evolutionary origin and expansion of the neural crest.

Neural induction is rapidly followed by molecular and morphological evidence of segmentation along the anteroposterior axis. Fundamental to this organization is the ordered expression of homeotic (Hox) genes and the subsequent appearance of segments (neuromeres) in the central nervous system and pairs of mesodermal somites and somitomeres alongside the nervous system. Patterns of Hox gene expression determine kind rather than number. For example, similar combinations would determine whether a vertebra is the first thoracic or first lumbar, rather than the twentieth from the base of the skull. Thus in snakes, which have hundreds of vertebrae, the increased numbers are considered to be an example of serial homology, whereby each one is not specified by a unique molecular code. A good example of epigenetic factors acting on anteroposterior segmentation is the influence of ambient water temperature during early development on the number of vertebrae of trout and other fish. Hox genes are also involved in the formation of other structures, such as the gut, limbs, and genitalia. The evolutionary implications of the developmental functions of homeotic genes are covered in greater detail elsewhere in this volume.

After the primary body axis has been established through patterns set up by homeotic genes and notochordal inductions have stimulated development in both the central nervous system and in the somites, a host of other organs and structures begin to take shape. In a number of cases, such as the eye and limbs, the expression of specific selector genes (e.g., *Pax6* for the eye) releases a cascade of developmental genetic events that results in the formation of the structure in question. At stages that require inductive signaling from one group of cells to another, there is a remarkable amount of reuse of signaling molecules and pathways that played a role at earlier stages.

A classic example is the gene *Sonic hedgehog (Shh)* and its signaling molecule, which is of vital importance in the induction of the motor component of the central nervous system in vertebrates. Production of Shh, again by the notochord, mediates the induction of the sclerotome in the epithelial somite. Later the same molecule is involved in early stages of development of a wide array of structures, ranging from the limbs to the gut to the lungs. According to the concept of homology of process (Gilbert et al., 1996), mechanisms and messages conserved across phylogenetic lines can stimulate the development of analogous organs, a good example being the molecular events underlying eye formation in a variety of invertebrates and vertebrates.

Another example of phylogenetic and ontogenetic conservatism is the use of specific molecules, mainly proteins, for different purposes. A striking example is the large family of crystallin proteins in vertebrate lenses (Wistow, 1993). This instructive example illustrates conservatism with respect to the use of enzyme proteins as structural components of lens fibers, while the phylogenetic distribution of the various crystallin proteins shows that quite different combinations of lens proteins are used to serve the same function.

Sex determination and sexual differentiation provide numerous examples of different mechanisms influencing phenotypic sex. Although phenotypic sex is determined genetically in most animal groups, exceptions abound. For example, in reptiles such as many turtles and crocodilians, phenotypic sex is largely determined by temperature, whereas among certain species of sea bass, functional phenotypic sex can change late in life, even in relation to the position of a specific fish in the social pecking order. The development of phenotypic sex in mammals is subject to certain chemical influences, e.g., the steroid hormonal environment; but this neither changes the gonadal phenotype nor the type of gamete that is produced.

Morphogenesis remains one of the most poorly understood aspects of development. Although the genetic blueprints underlying the formation of many organs and structures are beginning to be worked out, the mechanisms by which encoded genetic information is translated into structure remain obscure. One of the time-honored concepts in development is that of the *morphogenetic field,* a region of the body in which the cells contain the information sufficient to form a specific structure, such as a limb or an eye. Such a region has boundaries, which often shrink during the course of development; and in contemporary parlance, a field may be a region in which a selector gene or specific combination of homeobox genes is expressed. Within an active morphogenetic field, the cells in aggregate possess certain properties, such as regulation. On an individual basis, there appears to be considerable room for autonomous behavior among cells within a morphogenetic field, and this property has been viewed as one in which individual variation can be expressed.

One characteristic of many fields in embryos is a change in their proportions over time. An area that has the competence to form an eye or an ear, for example, is relatively larger in earlier stages of development than it is at later stages. For structures such as the sense organs, which require inductive stimuli for the initiation of development, timing is also an important consideration. Competence to respond to an induction typically falls off over time, and ultimately a potential responding tissue is not capable of reacting to an inductive message. Conversely, there are examples, such as the lack of teeth in birds, in which a tissue retains its competence to respond to an induction, but the inducer fails to function. Whether they have a genetic base or are due to epigenetic factors, variations in the expression of field properties can produce significant structural change.

Although a morphogenetic field sets the pattern, the ultimate shape and size of the structure that forms from it is dependent to a high degree on growth. Quantitative variations in growth patterns become manifest as allometric or differential growth. When allometric growth occurs in extremities, such as the fins of fishes, the final form can assume considerable functional significance and can confer significant selective advantages or disadvantages on the individual. Growth at a gross structural level can be a function of the number or the size of cells, or it can reflect the abundance of an extracellular matrix. Although the molecular mechanisms controlling the mitotic division of a single cell are strongly conserved, whether a given cell will enter a mitotic division depends on a large number of external influences, many of which have still not been identified. The local action of any of a number of growth factors constitutes a proximate cause of a mitotic division; but how other internal or external stimuli, such as mechanical factors, are translated into mitotic divisions is not well understood. Not only the temporal pattern but also the spatial sequence of cell division can have a significant modifying effect on the final form of a structure.

Genetically programmed cell death is a widespread and important developmental mechanism. In some structures, such as the embryonic nervous system, far more neurons are born through mitosis than can be supported by functional connections. The ultimate size and form of the spinal cord, for example, depend on the amount of peripheral tissue served by the axons emanating from the cord. Thus, the lumbosacral enlargement in the spinal cord of many dinosaurs is a reflection of the massive size of their hind limbs. Cell death in other areas can result in qualitative changes, as well. One of the best-known examples of this is the relationship between the amount of interdigital cell death and the presence or absence of webbing in the feet of birds. In humans, a number of genetic syndromes result in syndactyly, or the lateral joining of two digits by soft tissue. Cell death, often stimulated by hormonal action, also plays a prominent role in metamorphosis. Variation can occur by either the timing or degree of hormonal stimulation as well as by the degree of responsiveness of the target cells.

TSVI SACHS

Developmental Processes That Generate Plant Form

Land plants make a living in a unique way, absorbing light and CO_2 from the air and water and essential ions from the soil. They thus have more than one "mouth" and every individual concurrently exploits two different habitats. Continued development throughout the life of the plant is required for optimal use of the initial contact between the air and the soil. Organogenesis and other aspects of development are not limited to an embryonic stage; development is a major component of organization and phenotypic plasticity throughout a plant's life, even when that life lasts thousands of years (Figure 1).

Plant development is still separate from other functions, but separate in space rather than in time. Development occurs in specialized *meristems,* tissues that are distinct from those that carry out functions such as photosynthesis, transport, or defense. Organogenesis, an essential component of development, is limited to apical meristems at the tips of shoots and roots. Pattern formation thus occurs at two levels: specification of tissue organization within meristems and long-distance coordination between the various meristems of the same plant.

Unique characteristics of plant development may be due to the way plants make a living. Another reason could be the independent evolution of multicellular ontogeny, which would indicate convergent evolution of shared developmental principles (Meyerowitz, 1999).

Most recent work on plant development has dealt with genetic analysis, defining molecular components of developmental processes. Mu-

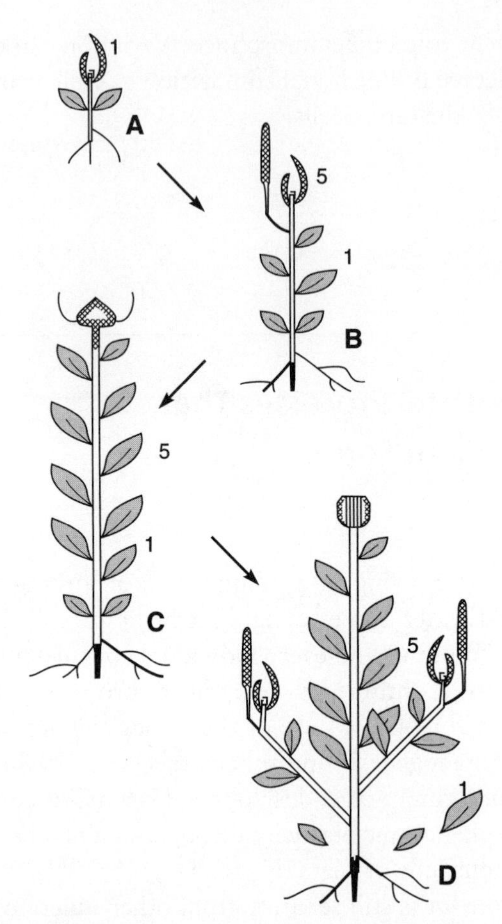

FIGURE 1. Four stages in the life of a plant, illustrating apical iteration of leaf formation and developmental changes during branching, organ death, and reproduction. Similar development occurs in the complementary poles of the root system (not illustrated). Leaves are formed only in apical meristems (A and B, and on the branches in D). The same meristem converts to form a flower (C) and fruit (D). Continued meristem activity changes the relative location of tissues and organs: leaf number 1 (the first to form above the two modified leaves present in the seed) is growing rapidly in A, is mature and functional in B and C, and has been shed in D.

tations and genetic manipulations, however, have as yet yielded only limited insights about morphology (Sussex and Kerk, 2001). Neither a system of genes "for" specific structures nor many genetic switches have been found. With the partial exception of the homeotic mutations of flower development, the results found in *Drosophila* have not had plant counterparts.

This review is based on the working hypothesis that developmental processes are an outcome of complex genetic networks. The roles of specific genes can best be understood in the context of these processes, which can also be recognized by comparative studies and by following and perturbing development. These processes can be labile or conservative in evolution and are the keys to organ homology and plant morphology (Sachs, 1982, 2001). This entry therefore summarizes examples of developmental processes at different organizational levels. Space limits the discussion to patterning, and differentiation and growth are taken for granted. Discussion is further limited to vegetative shoot development in seed-bearing land plants. For a wider coverage, see Steeves and Sussex (1989), Lyndon (1990), and Sachs (1991).

Persistent Meristems and Developmental Morphology

Land plants are bipolar branched axes that develop at their tips—the apical meristems of shoots and roots, which are the organogenic centers (see Figure 1). This localization of development is accentuated by the absence of cell movement, though localized growth does change spatial cell relations. The absence of cell movement greatly simplifies the study of development, especially of cell lineages (Szymkowiak and Sussex, 1996; Kagan et al., 1992).

Shoot apices (Lyndon, 1998) are characterized by the outgrowth of leaves, which start as meristematic bulges (Figure 2). Their development, though rapid, is finite or determinate (see Figure 1). Lineage analysis shows that leaves are initiated as multicellular structures (Poethig, 1997). One or more lateral shoot apices form in the axil of each leaf, exponentially increasing the branching potential. The actual number and location of growing branches varies, and rarely do all vegetative meristems develop.

Shoot apices iterate the formation of leaves, associated stem tissues, and lateral buds (see Figure 1). The entire apex can also differentiate to form flowers, which are modified, determinate shoots (Lyndon, 1990). Other types of apical differentiation are common and varied, and need not entail unique tissue differentiation. Thus, specialized organs, such as rhizomes, tubers, and permanently horizontal branches, are formed (Steeves and Sussex, 1989). Shoot apices also change gradually as a function of their own development—changes that are expressed by leaf shape (heteroblasty, Poethig, 1997) and the capacity to form flowers (Sachs, 1999).

Apical meristems can also form mature structures that have no apparent similarity to leafy shoots. Space limits the discussion to one example. A lemon does not resemble a group of leaves. Yet ontogeny, teratologies, and comparative work prove that it is initiated as a whorl of neighboring carpels, or leaf-like perturbances. These are said to fuse, but this fusion is congenital or evolutionary (Cusick, 1966): development starts as separate primordia and only after initiation does development shift to the joint base, below the actual perturbances. The location of comparable fusions can also be shifted by experimental treatments (Sachs, 1988).

Root apices resemble shoot apices, an outstanding difference being the absence of leaves. Roots branch by the organization of new apices within a specific internal layer, close to the tip. Their location is not predetermined, and branching is less orderly than that of shoots. Another salient character of root apices may indicate the evolutionary innovation by which they originated. Shoot apices are the source of inductive signals that influence vascular differentiation (Sachs, 1991, 2000). One such signal, auxin, is actively transported away from the shoot apex (Goldsmith, 1977), while in root apices this polarity is reversed. This criterion has been used to define an organ as a shoot or a root (Wochok and Sussex, 1974).

Continued apical growth and branching increases the canopies both above and below the ground. In most plants there is a special meristem, the *cambium,* which adds and replaces the vascular and supportive tissues connecting these canopies. The cambium is not organogenic, but forms organized tissues consisting of varied differentiated cells. Cambial orientation can be modified, even in large trunks, according to the changes of the organs of a developing plant (Sachs, 1991; Sachs et al, 1993). The cambium is missing in the Monocotyledons, whose forms (palms, for example) indicate that the contacts between the shoot and root canopies cannot increase; leaves are replaced rather than added. However, some Monocotyledons have "rediscovered" a cambium (Philipson et al., 1971), an example of convergent evolution.

A Developmental Basis for Organ Homology

Plant morphology (Bell, 1991) establishes homologies primarily on the basis of spatial relations of mature structures. There are intermediate states (Sattler, 1994), but actual morphological exceptions are rare, common only in water and parasitic plants that have evolved different ways of making a living. This brief outline points to the possibility of considering homologies in terms of meristems and shared or conservative developmental processes (Sachs, 1982, 2001). Plants also have apices that are formed in unusual locations. These, considered adventitious even when they are a regular feature of a large taxon (Bell, 1991), do not differ from other apical meristems.

Differentiation and Totipotency of Meristematic Cells

Promeristems and stem cells in shoot apices. A permanent meristem must both perpetuate itself and form mature structures. Self-perpetuation requires stable stem (or initial) cells; at least one product of each of their divisions must remain similar to the parent cell. The way in which a shoot apical meristem develops (Figure 2) identifies the location of the stem cells, in the very tip. This region, the *promeristem* (Steeves and Sussex, 1989) or *apical dome* (Lyndon, 1990), is generally a shallow cap ~0.1mm in diameter. There is no observable differentiation between promeristematic cells, and their divisions are relatively slow (Steeves and Sussex, 1989; Lyndon, 1990). A possible advantage of slow divisions is a reduced accumulation of somatic mutations. The same promeristematic region forms reproductive organs: there is no special germ line in plants, but gametes still form from embryonic cells that have not been directly modified by the environment.

Gradual maturation of meristematic tissues. The transition from promeristem to mature tissues always takes place through gradual, relatively long-term modifications (Sachs, 1991). It is never direct; an apical meristem is comparable to an embryo in which the temporal stages coexist and are spatially arranged (Figure 2).

The specialized future of the various tissues can be distinguished once the leaf primordia and their adjoining stems start developing (Steeves and Sussex, 1989, Lyndon, 1998). This is the region of the primary meristem, in which rapid yet finite development forms almost all shoot

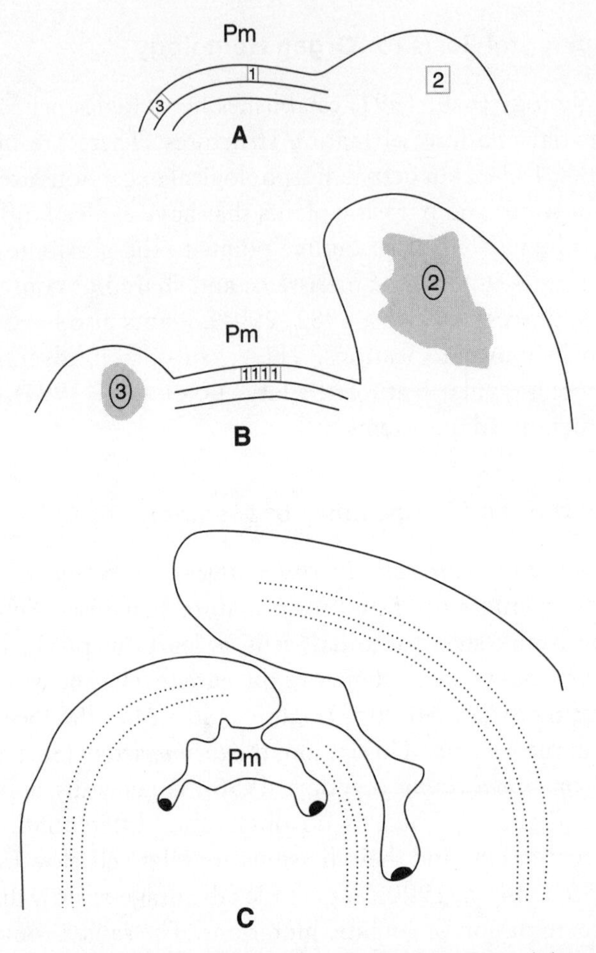

FIGURE 2. Leaf development and cell fate in longitudinal sections of shoot apical meristems. Leaves are initiated only on the flanks of the promeristematic region [Pm] and their growth is rapid but determinate. During the transition from state A to B the cell in the promeristem marked as 1 divided slowly, while cells in a leaf primordium (2) and on the flank of the promeristem (3) formed relatively large tissue patches (gray). During continued development, however, products of 1 will take over the entire meristem, 2 and 3 being left behind, outside the developing apex. A later stage of the same meristem (C), at a lower magnification, includes a number of leaves of different ages, all in direct contact with one another. It is thus comparable to a series of embryonic stages arranged in a linear sequence. Stem tissues elongate later, below maturing primordia. Vascular tissues (parallel broken lines) appear in conjunction with leaf development. Lateral or branch shoot apices (dark patches), each with its own promeristem, are formed in the axils of the developing leaves.

cells. Continued specialization of the various tissues is associated with reduced cell division, while continued growth increases cell sizes. Development ceases when growth has moved the promeristematic tip a few millimeters or centimeters (in climbers) above the maturing tissue. The cells in a developing meristem are said to undergo a transition from an undifferentiated state in the promeristems to mature differentiation. There is a nontrivial semantic problem here. Differentiation is often due to specific changes in gene expression. An undifferentiated state thus implies the patently unreal states in which either all or no genetic information is expressed. The concept of undifferentiated cells also implies that the meristem cells of the shoots, roots, and cambium are in the same state. It is therefore preferable to consider meristematic cells, including the zygote, as specialized and differentiated, and differentiation as changing rather than appearing during development.

Regeneration and totipotency. Gradual restricted changes in apical meristems are also evident in responses to microsurgery. The promeristem has the capacity of direct regeneration, replacing missing parts from a wounded surface (Sachs, 1991). In the primary meristems, only vascular continuity is fully restored. Other regeneration is partial, and becomes more so as the distance from the promeristem increases. The cells have become determined, retaining traits they acquired during development. Removed tips are still replaced, but only by indirect regeneration.

The surface of wounded mature tissues becomes covered by a callus of unspecialized cells. Callus can also be grown in culture, both as a tissue and as isolated cells (Steeves and Sussex, 1989). It is by no means uniform, and it can retain characteristics derived from the original tissue, a common expression of determination that is maintained in dividing cells (Sachs, 1991). Such calli, however, are able to divide and form apical meristems, and hence entire new plants. Isolated callus cells may develop as embryo-like structures, or embryoids (Steeves and Sussex, 1989). These are, of course, expressions of the remarkable totipotency of plant cells, which is the basis of so much genetic engineering. Callus cells are still severely limited. The callus can form an entire plant, but only indirectly, following a transition to promeristematic cells. Without divisions, the callus cells are known only to redifferentiate to become water-transporting cells. Totipotency and determination, therefore, are not mutually exclusive.

Integration of Whole Plant Development

Regeneration expresses integrating mechanisms. To introduce the relations between meristems, consider a simple experiment. A large part of a plant's shoot is removed by a person or a cow. Most plants survive this common event. Photosynthesis in any remaining leaves is enhanced and their senescence is delayed, but the long-term response is indirect regeneration. Remaining shoot apices, which may not have grown previously, develop rapidly (Figure 3), an expression of the removal of apical dominance (Sachs, 1991). Where no such lateral meristems are available, adventitious ones may form, responding to the correlative effects of other organs and to exogeneous hormones, in the same ways as preformed apices (Sachs, 1991). At the same time, root development is temporarily arrested. New vascular tissues differentiate, oriented so they connect new branches with the roots. The entire plant is restructured by varied, integrated changes.

The integration processes seen in regenerating plants also maintain undamaged plant organization (Sachs, 1991). This requires that shoot tissues, as long as they are present, be sources of information that act on the rest of the plant (see Figure 3). More generally, integrated development must mean that information is exchanged between the various parts of the plant, over minute distances in embryos and many meters in trees. This exchange chooses between the many available meristems, representing the excess of redundant alternatives (Sachs et al., 1993). This choice is an important component of developmental plasticity in plants.

Nature and roles of developmental signals. Information could be exchanged between the various organs by means of sink/source relations involving the substrates they form and consume. This possibility, however, does not account for integrated development at the expense of stored rather than synthesized and absorbed substrates, the effects of leaf primordia on root initiation and on oriented vascular differentiation before they become photosynthetic, and other phenomena (Sachs, 1991, 2000).

The alternative, which is not mutually exclusive, is hormonal signals. There is direct evidence for specific small molecules acting as hormones and integrating plant development (Steeves and Sussex, 1989), and recently also for macromolecules carrying information throughout plants (Kim et al., 2001). Hormones are often considered in terms of the spe-

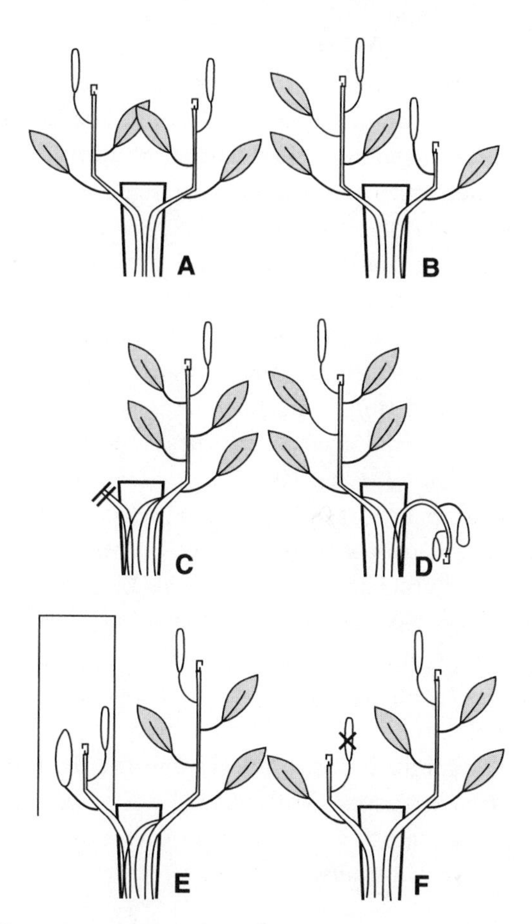

FIGURE 3. Long-distance interactions between branches and induction of oriented vascular differentiation. A model system illustrating competition between redundant alternatives, each being the source of information about its state and environment, a competition which is also found at microscopic cell and tissue levels. The figure shows a pea seedling with two branches that replaced the removed main shoot. The relative development of the branches is reflected in the differentiation of their vascular contacts, some of which are marked schematically in the cut stem. Those of a successful branch can take over the space of one that has been removed or has died (C, D). The two branches may grow equally (A) or unequally (B). The latter situation is much more common when the plants are subject to stress. Removal of one branch (C) always leads to the development of the other, demonstrating that reduced development depends on correlative inhibition. Continued presence of a successful branch can lead to the actual death of its counterpart (D). Shade (E) or the removal of a young leaf (F) reduces the status of a branch.

cific responses they elicit. Yet it could hardly be a coincidence that the application of one hormone, auxin, reproduces the remarkably varied effects of a growing shoot. Auxin is known to be produced by shoot tissues, and is therefore a signal of shoot organs, "informing" the plant of their size and state, and of the quality of their environment (Novoplansky et al., 1989). The use of one signal for many responses integrates changes throughout the plant (Sachs, 2000). A similar, though less convincing, argument can be made that another group of plant hormones, cytokinins, are major signals of the roots (Sachs, 1991). Adding only these two simple substances to wounded tissues causes the development of a partially organized tumor. Bacteria *(Agrobacterium tumefasciens)* genetically engineer such hormone synthesis to induce the tumorous growth of invaded tissue (Sachs, 1991).

The polar transport of auxin, appears to depend on localized membrane proteins in the root side of each cell (Steinmann et al., 1999). Though polarity is stable, a new polarity can be induced by auxin diffusing through competent tissue along a new orientation (Sachs, 1991, 2000). Polarization, furthermore, appears to be a gradual feedback process: the more auxin cells transport, the more polar they become. The result is a gradual canalization of auxin transport and of the vascular differentiation this auxin induces, accounting for the longitudinal patterning of the vascular tissues (Sachs, 1991, 2000). The general point is that the very same signals that correlate between organs can act in tissue organization (Berleth and Sachs, 2001).

I conclude by pointing out some characteristics of development that could contribute to understanding the genetic specification and evolution of plant morphology. The changes that occur in meristematic and other cells are always gradual and at any given stage limited. Even the totipotent cells of the promeristem don't form the various tissues of the plant without a hierarchy of intermediates. Organogenesis occurs in multicellular units rather than being focused in critical cells, and there is always an excess of cells and meristems to fulfill any given role. Development checks and corrects itself, according to integrated information from distant parts of the plant (Berleth and Sachs, 2001), not only by interactions between neighboring cells and tissues (Meyerowitz, 1996). Variable cell lineages (Kagan et al., 1992) and the distribution of tree branches (Sachs et al., 1993) demonstrate that rather than avoiding or correcting chance events, development makes use and builds upon them.

These traits suggest a view of development that is a far cry from the intuitive assumption that genetic information provides a blueprint or a program of events within single cells. A component in the generation of plant form must be the "epigenetic" or developmental selection between redundant alternatives (Figure 3; Sachs, 1991, 2002; Sachs et al., 1993). This is comparable to Darwinian selection, but is also profoundly different because the alternatives are genetically equivalent and the outcome is robust ontogeny rather than phylogenetic change. Developmental selection could reach a predictable overall result in spite of chance events if genetic information specifies not only the course of events but also their outcome, which is the relations between the components of the developing form. Such relations could depend on a balance of long-distance, integrative signals, hormones, and on rules that would enhance the development of any structure whose specific signals are limiting. The same signals act on cell differentiation (Sachs, 1991), and there are no separate controls of morphology and histogenesis (Kaplan, 2001). The specification of overall form accounts for leaf shape being independent of the orderly orientation of cell divisions (Smith et al., 1996).

What does this suggest about alternatives and constraints of plant evolution? Early stages in the development of organs, not only embryogenesis, can be conservative and form the basis of morphological categories (Sachs, 1982). This is at least partially due to the roles these early stages have in development itself, not only in mature function; a leaf primordium, for example (Figure 2), influences the entire apical meristem and is therefore essential even where no functional leaf matures. In contrast, changes in integrative controls of development that lead to many coordinated modifications can be the focus of rapid adaptations (Sachs, 1988). An essential characteristic of development that is self-correcting rather than programmed is that it is robust, and not readily perturbed by point mutations, local damage or chance events. Because many changes or "mistakes" are corrected during development, there can be considerable unexpressed or hidden information, and in fact knocking out many individual genes has no obvious phenotypic effect (Bouché and Bouchez, 2001). Point mutations will often lead to teratologies, but only rarely to novel structures adapted for a new way of life. The maize ear is a new organ that appeared in response to the unique selection of domestication (White and Doebley, 1998), but it is probably an exception rather than an example of a rule.

JASON SCOTT ROBERT

Developmental Systems Theory

Within evolutionary developmental biology (evo-devo), scientists agree that genes function within the complex context of organismal development; development is a process of material emergence; much of the variation required for evolution to occur is introduced during development and may bias evolutionary outcomes; and explaining evolution requires explaining (the evolution of) development. These ideas are common currency in Developmental Systems Theory (DST) as well, and their recognition by philosophers is due as much to scientific research as it is to the contributions of DST.

Oyama inaugurated DST in her 1985 book *The Ontogeny of Information,* though she had numerous precursors in psychology and biology (see references in Johnston, 1987; Gottlieb, 1992; Oyama, 1985, 2000a, b) who attempted to dissolve the traditional nature-nurture dichotomy, and to underscore an alternative to the gene's-eye view of evolution and development (Gray, 1992, 2001; Griffiths and Gray, 1994, 2001; Oyama et al., 2001).

DST's integration of development, genetics, and evolution challenges the widespread notion that genes are programs or blueprints for development. Genes are, rather, material resources for development, implicated in often nonlinear and nonadditive interactions between dispersed developmental resources of various kinds. According to DST, interactive developmental processes both generate the species phenotype and also introduce new variants of potential evolutionary significance.

The central and fundamental construct of DST is the *developmental system,* a fluid set of entities and influences that crosses all hierarchical levels and all levels of analysis from the molecular through the biogeographical. The developmental system thus includes the organism and those features of the extraorganismal environment that influence development. These organism-environment systems host a complex dance of reliably recurring, though nonetheless contingent, interactions. The components of this interactive resource matrix share the joint developmental and evolutionary task of reliably, though somewhat imperfectly, reproducing evolved developmental systems. The core tenets of DST are as follows (see also Robert et al., 2001, and the introduction to Oyama et al., 2001).

Contextualism. Close study of developmental processes highlights the importance of developmental context in the assembly, activation, and regulation of functional genes. As the many:many relationship between genotype and phenotype and the phenomena of epistasis, epigenetics, pleiotropy, developmental stochasticity, and phenotypic plasticity suggest, the eventual form an organism takes is not genetically predetermined. Instead, potential emerges during development, as current contexts condition possible next steps (Oyama, 1999).

Nonpreformationism. Developmental resources emerge as resources during development rather than being preformed and transmitted between generations. Developmental interactions construct these resources anew in each generation; they do not merely trigger the specific information contained putatively only in genes. Moreover, the reliable presence of developmental resources at the right times and places is explained as largely the result of their having been constrained, influenced, selected, and constructed by organisms, their conspecifics, and symbionts. Deep interrelations between developmental interactions help to explain the specificity of development, without invoking the problematic notion of preformed genetic instructions.

Expanded pool of interactants. DST rejects the blanket dichotomy of developmental resources/causes/factors into genetic and generic (usually environmental) types. There are more than just two types of interactants among the heterogeneous components of a developmental system, and there are a multiplicity of ways in which they interact in development. Within the organism, interactants include DNA sequences, mRNA, cells, extracellular matrices, hormones, enzymes, metabolites, and tissues; beyond the organism, developmental interactants include aspects of habitat (including temperature and nutritional resources), the organism's behavior and that of conspecifics and others (Johnston and Gottlieb, 1990), social structure (Keller and Ross, 1993), and depending on the system even gravity and sunlight (van der Weele, 1999; Gilbert, 2001). Like evo-devo, DST demands exploration of the specific nature of constructive interactions between all developmental resources as part of any adequate account of organismal development.

Causal interactionism and dispersion. Constructive causal interactions in development involve inducing, facilitating, maintaining, and participating in time-sensitive feedback loops at multiple levels within and beyond the developing organism—only some of which might be characterized as gene activation. The interactions comprising organismal development are complex, and their effects are not simply additive. Some aspects of development, such as cell-cell signaling, cannot be represented as simple causal pathways, but rather should be construed as networks of causal interactions (Dover, 2000). Causal power is not contained within any particular entity or class of entities; instead it resides in the systemic relations between developmental interactants. DST, like some strands of evo-devo, thus refocuses our attention away from the perspective that genes are ontogenetically (and ontologically) primary, and toward the multiplicity of factors, forces, and mechanisms operative in, and constitutive of, development.

Extended inheritance. DST emphasizes that there is more to inheritance than just DNA and cytoplasm; also inherited are all reliably present elements of the developmental context. Dover (2000, p. 1154) observes that "DNA is a far more unstable molecule, on an evolutionary scale, than is conventionally thought." But whatever stability genetic inheritance enjoys, even on a much shorter timescale, depends critically on the inheritance of those resources that are part of the expanded pool of interactants. Some of these resources are, of course, DNA and cytoplasm, but also included are chromosomal structures, nutrients, the chromatin marking system, and environments and associated selection regimes. Accordingly, DST proposes a broad interpretation of inheritance and reinterprets hereditary transmission as contingent but reliable reconstruction of resources-in-interactive-networks in the next life cycle.

Evolutionary developmental systems. Given this broad account of inheritance as construction, selection works on elements at all levels of developmental systems, at any level of complexity, and at any time in the life cycle. According to DST, then, evolution should be defined as changes in the life cycles of organisms in their co-constructed niches, tracked by differential reproduction and distribution of developmental systems.

In summary, according to a developmental systems view, genes are but one of many inherited developmental resources; these resources cannot be dichotomized into (specific) genetic and (nonspecific) generic classes; DNA sequences and other resources participate in complex, nonadditive, time-sensitive, constructive networks of interactions, such that control of development is dispersed. Accordingly, causal arrows must be drawn in multiple directions and evolution occurs through changes in organism-environment systems, reflected in their frequency and distribution. Understanding both development and evolution thus requires analyzing developmental systems.

Developmental systems "theory" may be a misnomer, if theory is interpreted in a hypothetico-deductive sense. DST has as yet generated no such theory. It has provided only verbal models in the model-theoretic sense of "theory", with Wimsatt's (2001) model of "generative entrenchment" a possible exception. DST is, however, a useful methodological heuristic in reminding us of the perils of uncautious reductionism in biological theory and practice. Moreover, in reconceiving the relations between development, genetics, and evolution, DST may serve as an integrative theoretical framework for philosophical analysis of evolutionary developmental biology.

JAMES HANKEN

Direct Development

Many animals display a complex life history in which development is *indirect:* one or more successive, free-living larval stages intervene between embryo and adult, and there is a more-or-less abrupt transition, or metamorphosis, between the last larval stage and the adult. *Direct development* is a specialized reproductive mode, in which the sequence of ontogenetic stages between embryo and reproductive adult is abbreviated in comparison to that seen in related species. In moderate instances of direct development, one or more post-hatching larval stages may be present, although these typically do not feed and are anatomically simplified in comparison to the free-living larvae of indirect developers. In extreme instances of direct development, there are no intermediate larval stage(s) or postembryonic metamorphoses of any kind. Instead embryonic development culminates in the hatching or birth of a fully formed, albeit miniature adult. The term *direct development* was first coined by the great nineteenth-century evolutionary embryologist and morphologist Francis Balfour, based on his comparative ontogenetic studies of aquatic invertebrates (cited in Hall, 2000a).

The reproductive mode in any given species may not fit easily into a simple, dichotomous classification. In many groups, variation in life history among species defines a continuum between "pure" indirect development and "pure" direct development (Altig and Johnston, 1986; McEdward and Janies, 1993; McEdward and Miner, 2001). Not surprisingly, definitions of direct development vary, with some authors reserving the term for the extreme instances described above (e.g., McEdward and Janies, 1993; Schatt and Feral, 1996) as opposed to more inclusive and less stringent definitions.

Regardless of definition, direct development is widespread in the animal kingdom. Indeed, it is the predominant reproductive mode in some taxa, ecological communities, or species assemblages. Direct development has evolved independently many times in vertebrates and invertebrates, in terrestrial and aquatic as well as freshwater and marine environments. Animal groups that have at least some direct-developing species include molluscs, priapulid and polychaete worms, crustaceans, jellyfish, ascidians, insects, arachnids, echinoderms, fishes, and amphibians. Arguably, one could regard the reproductive modes seen in amniotes (reptiles, birds, and mammals) as examples of direct development, because they include neither free-living larvae nor a discrete metamorphosis. However, the term usually is not applied to these vertebrates because there are no indirect-developing species.

Other uses of direct development should be noted. The term is frequently used for insects that seasonally bypass a period of developmental arrest (diapause) between successive larval stages, or *instars*. It also is employed in studies of plant development that involve artificial propagation from somatic embryos *in vitro,* and in reference to alternative life cycles in algae. All these phenomena are distinct from direct development as defined in this entry and are not considered further.

For many animal groups, indirect development is the plesiomorphic, or ancestral, reproductive mode, which may or may not be retained in living species. When direct development occurs in such groups, it must have evolved from indirect development and therefore represents an evolutionarily "derived" condition. However, a straightforward, unidirectional progression from indirect to direct development does not necessarily apply in all instances. For example, direct development has been offered as the ancestral reproductive mode for many invertebrate phyla (Haszprunar et al., 1995), and even for multicellular animals generally (Wolpert, 1994). If these hypotheses are correct, indirect development would itself represent a derived state, from which direct development re-evolved independently in numerous clades. Direct development also has evolved multiple times within many individual groups, ranging from clams, to sea urchins, to salamanders (Arndt et al., 1996; Wake and Hanken, 1996; Hanken, 1999; Park and Foighil, 2000).

Developmental correlates and mechanisms. Direct development offers unique opportunities to address both the nature and developmental basis of evolutionary change in embryonic, larval, and adult phenotypes, as well as the evolution of developmental mechanisms themselves. Typically, such studies begin by defining the consequences and correlates of the evolution of direct development for ancestral patterns of indirect development. Reflecting the fact that direct development has evolved independently numerous times, and that reproductive modes in particular species define a continuum between the extremes of indirect and direct development, these consequences and correlates may vary widely from group to group. Nevertheless, several generalizations are possible.

One of the most conspicuous characteristics of direct development is the loss of larval-specific features. Among the features most commonly lost are external gills and keratinized mouthparts in frogs (Elinson, 1990) and the larval skeleton and ciliary band in sea urchins (Raff, 1992). Not all larval features are routinely lost, however. Some are prominently retained and may adopt new and important functions that enhance adaptation of the novel life history. Embryos of direct-developing frogs, for example, typically retain a well-developed and highly vascularized tail. Such a tail is characteristic of larvae of indirect-developing anurans, in which it serves as an important locomotor organ. It is eventually resorbed as part of the post-hatching metamorphosis. In direct-developing frog embryos, the prominent tail is believed to function as an important respiratory organ (Townsend and Stewart, 1985). It may be one of the few larval characters retained in some species, not being rapidly resorbed until shortly before hatching.

A second characteristic of direct development is precocious formation of adult morphology. Adult-specific features that do not appear until metamorphosis in indirect-developing species—weeks, months, or even years after fertilization—may form after only a few days in direct development. In extreme cases, their formation is completed within the embryo, so that embryogenesis culminates in the hatching or birth of a fully formed, albeit miniature adult. Actual time to hatching may be extended, such that the embryonic period encompasses a larger proportion of development overall.

These and other modifications to the ancestral ontogeny offer excellent examples of *heterochrony,* or evolutionary change in the relative timing of developmental events. For this reason, heterochrony has been recognized as a dominant theme in the evolution of direct development in many groups. Formal analyses of heterochrony have been particular effective in revealing complex patterns of dissociation of developmental timing among larval and adult characters, ranging from molecules to morphology (Bisgrove and Raff, 1989; Klueg et al., 1997; Schlosser, 2001). Some instances of direct development appear to involve a relatively straightforward, but accelerated, recapitulation of the ancestral (indirect) ontogeny. Others entail a much more fundamental reorganization or repatterning of embryonic development, which yields a novel ontogeny that comprises a mosaic of both ancestral and derived traits (Wray and Raff, 1990; Hanken et al., 1997; Schlosser and Roth, 1997; Haag et al., 1999; Callery and Elinson, 2000a).

Recent analyses have begun to address specific mechanisms that underlie evolutionary changes in developmental pattern in a few model species drawn primarily from three groups—sea urchins, ascidians, and frogs. Several studies analyze the genetic and developmental bases for loss of larval structures (Jeffery, 1994; Fang and Elinson, 1996, 1999; Schlosser et al., 1999). Others investigate the mechanisms that mediate precocious (embryonic) formation of adult-specific features, which in indirect development form after hatching. In direct-developing amphibians, this invariably involves an assessment of the role of thyroid hormones in embryonic development, given the predominant role that these hormones play in regulating metamorphic changes in indirect-developing species (Jennings and Hanken, 1998; Callery and Elinson, 2000b).

Another conspicuous characteristic of direct development is greatly increased ovum size. While there is little evidence to date that directly links increased ovum size to the modifications in larval and adult features previously discussed, that increase may be related to various changes in early embryogenesis that frequently accompany the evolution of direct development. These changes include cytoplasmic rearrangements (Fang et al., 2000), morphogenetic movements during gastrulation, and germ layer formation (Schatt and Feral, 1996; Ninomiya et al., 2001).

Finally, direct-developing species offer unique opportunities to be used as model systems to investigate basic problems in developmental biology and physiology, ranging from animal-vegetal cell determination and differentiation to cardiac and metabolic physiology (Henry and Martindale, 1994; Henry and Raff, 1994; Burggren, 1999–2000; Kitazawa and Amemiya, 2001).

Evolutionary significance. In addition to its immediate and obvious effects on developmental patterns and processes, direct development may have important evolutionary consequences. Some of these follow directly from the shift in habitat requirements and preferences associated with loss of the free-living larval stage. This is especially true for groups, such as amphibians, in which aquatic larvae occupy dramatically different environments from the typically terrestrial adults. In the lungless salamanders (Family Plethodontidae), loss of both the aquatic larval stage and the associated requirement for montane stream habitats has allowed direct-developing species to expand the family's geographic range into and across broad areas and habitats that are generally unsuitable for the survival of indirect-developing species (Wake and Hanken, 1996). Loss of larvae may also affect population structure and patterns of geographic distribution and dispersal (Hoskin, 1997; Collin, 2001).

Another important consequence of the acquisition of direct development is its potential to remove constraints on the evolution of adult morphology that may be associated with a complex life history (Wake and Roth, 1989). For example, in indirect-developing species in which adult-specific structures develop from larval precursors, and in which the amount of metamorphic remodeling of these larval components is limited, the diversity of adult phenotypes may be significantly constrained by the array of larval phenotypes. Larval phenotypes, in turn, may be limited by functional or developmental relations unique to this stage of the life history or to particular clades. Abandonment of the free-living larval stage provides at least the potential for loss or relaxation of these ontogenetic or functional constraints. To the extent that diversification of adult morphology in indirect-developing species is limited by such constraints, the evolution of direct development may enable or even facilitate phenotypic diversification, including the origin of morphological novelty.

While it is an appealing concept, the *larval constraint hypothesis* has been explored in relatively few groups. Evidence comes primarily from plethodontid salamanders, in which the evolution of direct development is correlated with the origin of novel morphological and functional configurations relating to prey capture. These configurations are largely absent from metamorphosing species (Wake and Hanken, 1996). In many other groups, ranging from tropical frogs *(Eleutherodactylus)* to sea urchins *(Heliocidaris),* the evolution of direct development is not correlated with any apparent consequences for adult morphology or physiology (Hanken et al., 1992; Raff, 1992). This lack of effects is observed despite the widespread loss of larval features from each group's ontogeny, which presumably would promote the loss of any pre-existing larval constraints. Indeed, the fact that adult morphology need not be affected by dramatic changes in early ontogeny is one of the most intriguing aspects of direct development, and one that makes this reproductive mode so interesting to study from a developmental perspective. It also underscores the capacity for independent evolution among successive life history stages in many animals, which may be enabled by a complex life history (Hanken et al., 1997).

ROGER SAWYER AND LOREN KNAPP

Embryonic Induction

Formation of embryos represents one of the clearest links between development and evolution. What are the mechanisms by which seemingly homogenous assemblages of embryonic cells differentiate into complex organs and uniquely structured organisms? As early as the fourth century B.C., Aristotle established himself as a premier embryologist through his observations and insights into the development of chick embryos. He rejected the notion that an individual derived from a preformed entity and espoused what was to become our modern view of epigenesis: the progressive formation of a highly structured organism from undifferentiated primordia.

Embryonic induction is the mechanism that underlies the conversion of undifferentiated cells into the unique cells, tissues, organs, and organ systems responsible for the physiological functions of an organism. The inductive events that build an organism are ordered in time and space. The responding cells sequentially and progressively become competent, determined, and finally differentiated into the more than 200 cell types that characterize many adult vertebrates, including humans. Embryonic induction is classically defined as the process whereby one set of cells produces a substance that changes the behavior of a neighboring set of cells, causing them to change their morphology, their association with other cells, their proliferative status, and their state of cytodifferentiation.

The concept of induction originated from studies of the induction of enzymes in response to substrates in microbes (Gilbert, 2000a). In the case of the development of multicellular organisms, embryonic induction is viewed as the regulation of differential gene expression in response to external cues. It is well established through cloning studies that every nucleus of a multicellular organism contains all the genes necessary for the complete development of an organism; no genes are lost or destroyed during development, with rare exception (Gilbert, 2000a). Because all genes are not expressed in all cells, cellular differentiation must result from regulation of the expression of specific subsets of the total genomic set of genes.

Development is progressive and hierarchical; cells involved in embryonic induction are progressing along a series of one-way streets. This process, driven by complex interactions between cells within the embryo, involves intercellular cross-talk and reciprocity that lead to evermore-complicated histological and anatomical structures. Such interactions provide the opportunity to specify the fates of cells based on their position and prior inductive interactions. A cell that is primed to differentiate and awaits a signal to do so is considered to be *competent.* A cell or tissue that requires a specific set or sets of signals in order to differentiate is said to do so under the *instruction* of an inducer. Thus, there are two kinds of cells in any inductive interaction—responding cells and signaling or inducing cells—and at many points in the process their roles may be reversed. Wessells (1977) set up four principles that characterize *instructive embryonic inductions:*

In the presence of tissue A, responding tissue B develops in a certain way.
In the absence of tissue A, responding tissue B does not develop in that way.
In the absence of tissue A, but in the presence of tissue C, tissue B does not develop in that way.
In the presence of tissue A, a tissue D, which would normally develop differently, is changed to develop like B (Gilbert, 2000a).

In some cases the responding cells are already programmed to differentiate in a certain manner, but require the appropriate environment to do so. Such interactions are referred to as *permissive,* in that they provide an appropriate environment, but not specific signals.

The history of embryonic induction is a history of the means by which investigators were able to observe organisms and experiment at the genetic, histological, and molecular levels. The earliest work was at the level of the organism. For example, the study of organisms with clear genetic anomalies contributed to a better understanding of the developmental processes involved in the expression of abnormalities in limbs, eyes, teeth, and other phenotypic characters. As investigations became more sophisticated, the direct manipulation and separation of cells and tissues meant that induction could be considered at the histological level. Histotypic reorganization of cells suggested properties at the cell surface that helped to direct the differentiation of cellular assemblages (Moscona, 1974). Aggregation of cells in various combinations could be carried out over time as well as spatially. As mechanisms for cell interactions were targeted for explanation, the era of molecular biology turned the study of embryonic induction into a search for adhesion molecules, growth factors and their receptors, signal transduction mediators, and molecules involved in regulating selective gene expression.

Hierarchical Inductions

Fertilization may be viewed as the first hierarchical level of induction. The specialized sperm cell activates the egg and sets in motion the differential gene expression that will transform the zygote into an adult organism. The first experimental evidence that one embryonic tissue could directly influence the fate of another was presented by Spemann and Mangold (1924) who used techniques for transplantation of specific tissues in pre-gastrula frog embryos to demonstrate the existence of a *primary inducer* associated with the dorsal lip of the blastopore. The interactions at this early stage of development are associated with gastrulation, a time when essential inductive modules overlap as they establish the major body plan of vertebrates (Raff, 1996; Gilbert, 2000a). At the pharyngula stage, there is an evolutionary bottleneck in development in which early inductive interactions globally constrain the *Baupläne* of vertebrates (Raff, 1996, see also Gilbert, 2000a). Deviation from this basic plan would involve major heterochronic alterations in the patterns of early gene expression, which is a rare occurrence in the history of life.

Following the *primary induction* of the dorsal axis and neural tube, numerous *secondary embryonic inductions* then determine the many discrete developmental fates of specific tissues, irrespective of the fates of other regions. These inductions take the form of the classical *epithelial-mesenchymal interactions* (EMIs) and represent discrete modular inductive events, unlike the critical overlapping modular interactions that establish the pharyngula (Gilbert, 2000a). Embryonic epithelia consist of sheets of ectodermal or endodermal cells that interact in regionally specified pairings with an underlying mesenchyme to give rise to the evermore developmentally constrained grouping of cells associated with histogenesis and organogenesis. For example, in the skin, which forms epidermal appendages such as scales, feathers, and hair, the failure of appropriate EMIs may lead to the absence of these specific appendages, even if the skin from which these appendages emanate forms normally (Hall, 1996b).

Developmental and Evolutionary Aspects of Epithelial-Mesenchymal Interactions

Heterospecific EMIs in various regional or modular combinations result in unique morphogenetic patterns that share many cellular and molecular pathways (Gilbert, 2000a). One of the most thoroughly studied systems is the teeth in mice, which produces a unifying example of both the unique local and shared global mechanisms that underlie development and evolution of epithelial-mesenchymal combinations. Tooth development is regulated by a series of reciprocal inductive interactions between an overlying dental epithelium and an adjacent mesenchyme (see Peters and Balling, 1999). The early stages of tooth morphogenesis, which have been studied extensively using tissue recombinants (Kollar and Baird, 1969; Lumsden, 1988), have helped to establish a hierarchy of progressive signaling events that build on one another and are represented by an initiation phase, a bud phase, and a cap stage.

During the initial phases of this process, the epithelium induces changes in the gene expression of the dental mesenchyme, which in turn reciprocally induces bud formation in the overlying epithelium. These coordinating interactions result in an epithelium that grows into and is surrounded by mesenchyme. This process is characteristic of EMIs that regulate the development of numerous organs and structures of multicellular organisms. Modular growth specifies the type of organ or structure based on regional constraints, patterns of local influence, and heterochronic regulation of gene expression (Gilbert, 2000a).

In tooth development, as the bud cells proliferate and differentiate, they form a more complex structure known as the cap. During the cap stage, ongoing gene expression leads to differentiation of cells into an enamel knot, which influences gene expression of cells in both the proximal epithelium and the dental mesenchyme, suggesting both intra- and intertissue interactions are taking place concurrently. The enamel knot, along with secondary knots, will eventually form the cusps of mature teeth. Thus, there are at the molecular level what might be called gene networks and super networks whose products interact to drive differentiation along pathways that are specifically constrained, but whose underlying mechanisms are evolutionarily quite open and plastic (Chuong, 1998).

The dynamic nature of dental developmental processes convert a simple epithelial-mesenchymal pair into the anlagen for a tooth, which will eventually produce a number of specialized cell types, including those involved in the production of the hardest materials in vertebrates, enamel (produced by ameloblasts) and dentin (produced by odontoblasts). The system of gene regulation leading to the development of teeth has been selected for and maintained over hundreds of millions of years of evolution in untold species, both extinct and extant.

For decades, the study of tooth development provided a valuable histological and biochemical framework for understanding embryonic inductions. What was missing was a molecular basis for the regulation of these processes that would connect the development of teeth to their evolution in vertebrates and, moreover, to the regulation of EMIs driving a wide variety of specialized tissue outcomes.

As it turns out, the early signaling between dental epithelium and dental mesenchyme involves expression of a number of genes whose proteins induce the changes in morphogenesis described earlier (see Table 1 in Peters and Balling, 1999). Though the story is far from complete, sonic hedgehog (SHH), bone morphogenetic factor (BMP4), and fibroblast growth factors (FGF8) produced in the epithelial cells initially induce mesenchymal cells to differentiate. *Pax9* and *Msx1* are homeobox genes induced by BMP4 and FGF8, and have profound effects on the mesenchyme, as does the TGFβ family member Activin βA. All of these factors are essential for the mesenchyme of the tooth bud to progress to the cap stage. What of the epithelium? The HMG-box gene *Lef1* is at least transiently essential for the dental epithelium to progress to the cap stage (Peters and Balling, 1999). Evidence for this comes from studies of mutants that lack *Lef1* and as a consequence lack all teeth. With respect to the mesenchyme, mutants for *Pax9, Activin* β*A,* and *Msx1* also lack all or most of their teeth.

BMP4, produced in the mesenchyme by the inductive influences of *Msx1* and *Pax9,* signals back to the epithelium and turns on *Msx2* and *p21,* which are associated with apoptosis. Thus, morphogenesis reflects not only proliferation and differentiation of cells, but programmed cell death as well. This comes as no surprise, because programmed cell death has been shown to be an essential component of development in general, for example, markedly influencing the outcome of limb and nervous system maturation (Gilbert, 2000a). The types of genes involved and their patterns of expression reflect many common elements of control in familiar processes of morphogenesis, including cell proliferation, cell differentiation, and cell death. However, each specific morphological outcome is also influenced by the evolutionary history of the species.

While tooth morphogenesis provides a valuable histological picture of development, the underlying gene regulation of the process has important evolutionary ramifications (Sharpe, 2001). What do fish scales, teeth, and hair have in common? Are they homologous structures? In many respects it is essential to infer homology in order to understand evolution. This has most often meant morphological homologies, but might just as well be viewed as molecular homologies, especially in the context of gene networks and super networks influencing the outcome of EMIs (Chuong, 1998).

Genes of the tumor necrosis factor (TNF) signaling pathway influence vertebrate ectodermal dysplasias, dramatically reducing or eliminating both tooth and hair development. Genes such as *hedgehog, Bmp,* and *Wnt* are also involved in normal development of both hair and teeth, reflecting molecular homologies in the processes and providing a molecular connection to evolution and the underlying common mechanisms of epithelial–mesenchymal interactions.

The TNF pathway has also been implicated in the failure to form scales in the teleost fish, the Japanese medaka (Kondo et al., 2001; Sharpe, 2001). The scales in teleosts are considered to be dermal in origin and are not developmentally related to teeth or hair—or for that matter, to scales of birds and reptiles, which are all epidermally derived (Sengel, 1976). How is it that development of scales in fish are linked to morphogenetic processes occurring in the mouths and on the skin of mammals? A recently discovered mutation in the ectodysplasin-A receptor, a TNF gene product, results in the failure of scale formation, just as similar mutations in mammals lead to hair and tooth anomalies. Thus, with a shared pathway controlling the morphogenesis of such apparently disparate structures as scales, hair, and teeth, it is becoming clear that evolutionarily conserved molecular mechanisms involved in secondary embryonic inductions are, and will continue to be, common requirements for development of functionally and morphologically distinct organs. The search for more elements in these fundamental cell and tissue inductive interactions during embryogenesis is ongoing. Aristotle would have been proud.

CARL D. SCHLICHTING

Environment

Among biologists, *environment* appears to suffer the fate of receiving either scant or abundant attention. As larvae, biologists and geneticists are taught (explicitly or implicitly) that *Genotype = Phenotype*. Experimentalists strive to minimize environmental effects, and subsequently statistically partition these to an "error" term: *Genotype* = *Phenotype* ± *Error* (i.e., *Environment*). On the other hand, evolutionists give environment a quantum boost in status when it dons its cloak of Selective Agent 007 (being secret, it is referred to not as E but as S: for population genetics, $p + (q - qs) = 1$, and for quantitative genetics, $R = h^2 \times S$). And for ecologists, environmental factors are always paramount.

Does the true role of environment lie in between these extremes, or does it depend on the particular area of investigation? The answer is neither—environment always plays a fundamental role regardless of the hierarchical level of inquiry. Although we rarely think about it, even the trio of DNA, RNA, and proteins functions only within certain environmental limits. Double-stranded DNA, although notoriously stable, begins to unzip above 75°C. Proteins and enzymes of different organisms have their own temperature and pH optima: the hotsprings archaebacteria *Thermus aquaticus* has a temperature optimum of 70°C; the fish *Pagothenia borchgrevinki* swims in Antarctic waters with a mean temperature of −2°C (Brueggeman, 2001).

An organism develops "normally," within a "normally encountered" range of environmental parameters and perhaps even well beyond that range, because its proteins perform their intended functions. The wider the range of values that permits normal development, the more likely we are to forget the environment. Outside these limits, however, we are reminded of its fundamental role when organisms display developmental defects, or if development is terminated: *P. borchgrevinki,* mentioned above, dies of heat shock above 6°C (43°F)!

Both within and beyond normal ranges of conditions, environmental agents—whether temperature, light, or even mutualist symbionts—are still the ultimate arbiters of phenotypic expression. Even though we may be able to conceptualize a direct mapping of genotype to phenotype, in reality a phenotype cannot be produced outside the context of an environment. DNA may contain the information necessary to initiate construction of an organism, but every bit of that information is environment-dependent: the intended results of its instructions—an organism with fitness >0—can be achieved only within particular environmental parameters. The model of quantitative genetics, $P = G + E + G \times E$, despite the explicit recognition of interaction between genotype and environment, remains misleading because of its implication that variation in phenotypes can be parsed into variation in genotypes and variation in environment. In fact, the *G* term always contains an implicit *E* component (Lewontin, 1974).

External Environment

The external environment of a given organism consists of myriad factors, ranging from the standard abiotic features (temperature, light, water/humidity, pH, salinity) to habitat components (territory, shelter, soil type, nutrient availability) and biotic factors (predation, competition, food, parasites, mutualists, mates, and kin). For evolutionary developmental biologists, three general impacts of external environmental parameters must be considered: selection intensity, selection response, and phenotypic plasticity (i.e., responses of physiology, morphology, and behavior to changes in the environment). The impacts of each can be seen through the equation $R = S \times h^2$: the response to selection *(R)* equals the strength of selection *(S)* times the proportion of phenotypic variation that is due to genetic factors (h^2, heritability).

External environmental factors are potent forces of selection, and *S* is a measure of how much the mean value of a trait *could* be moved toward the optimum by those factors. The subjunctive "could" is required because *R* depends as well on h^2 (genetic variance divided by phenotypic variance, V_G/V_P). If $V_G = V_P$, then $h^2 = 1$, all observable variation has a genetic basis, and $R = S$. However, environmental fluctuations will increase phenotypic variance, thus making $h^2 < 1$, and $R < S$. In addition, plastic responses to environment can alter both genetic variances—via impacts on gene expression—and heritabilities, which have been found to vary substantially among different environments (Schlichting and Pigliucci, 1998; Hoffmann and Merilä, 1999).

When environmental factors vary frequently and have predictable consequences, the evolution of adaptive phenotypic plasticity is favored (e.g., Schmitt et al., 1995; Van Buskirk et al., 1997). Adaptive plastic responses move the mean value of the trait closer to the optimum, in the process diminishing *S;* if the plastic response matches the optimum, then there is no net selection in the population. Recent studies of plasticity have begun to examine the responses of groups of traits (e.g., Arnqvist and Johansson, 1998; Hammond et al., 2001). Differential plasticity of traits leads to changes in both phenotypic and genetic correlations and covariances and will result in changes in correlated responses of traits to selection (Schlichting and Pigliucci, 1998).

Despite our inclination to focus on directional forces, stabilizing selection may exert powerful resistance to change when traits are near their optima. Canalization of traits, i.e., a lack of responsiveness to environmental change (Schlichting and Pigliucci, 1998; Debat and David, 2001), may be favored when environmental change is rare, or when effective plastic responses cannot evolve (e.g., Weinig, 2000).

Internal Environment

The conventional view of environment as those factors external to the organism has increasingly expanded to encompass the internal conditions of organisms (Rachootin and Thomson, 1981; Nijhout and Emlen, 1998; Schlichting and Pigliucci, 1998; Gilroy and Trewavas, 2001). In the vertebrate gut, crypt stem cells differentiate into four cell types as they migrate along a villus. "Germ-free" mice lack intestinal microbes and do not exhibit the typical sequence of differentiation; however, "normal" development can be initiated by introducing a single bacterial species from the usual flora (Bry et al., 1996). Gut microflora appear to provide a number of signals mediating epithelial differentiation, in the process generating the spatial complexity required to maintain the extraordinarily diverse microecosystem (Gordon et al., 1997). These studies nicely illustrate the fuzziness of the boundary between external and internal environments: an external environment for the microflora, the mammalian gut is carefully controlled as an internal environment for temperature and pH, despite being open at both ends to the external world.

The role of internal signaling factors such as morphogens in determining pattern formation in organisms has long been hypothesized by developmental biologists (Wolpert, 1969). Numerous experiments have now demonstrated that animal and plant cell differentiation and patterns of gene expression are dependent on information from and interactions with their surroundings (Neumann and Cohen, 1997; Gurdon and Bourillot, 2001; Irish and Jenik, 2001; Scheres, 2001). As an example, Walter and Biggin (1996) found significant differences in DNA binding affinity *in vitro* vs. *in vivo* for the homeodomain proteins even-skipped (Eve) and fushi-tarazu (Ftz). Binding to the *rosy, Adh,* and *Ubx* genes was much lower *in vivo* than *in vitro;* conversely, binding to *ftz* and *eve* was higher *in vivo.*

Feedbacks among internal components are well known in plants.

Cucurbita texana produces separate staminate (pollen-producing) and pistillate (fruit-producing) flowers. When fruits are present, levels of the plant hormone ethylene are lowest, and relatively more staminate flowers are produced; the reverse occurs when fruits are absent (Krupnick et al., 1999). A set of elegant experiments has demonstrated such trade-offs in body part size in insects (Nijhout and Emlen, 1998). The beetle *Onthophagus taurus* is doubly dimorphic: only males are capable of producing horns, but they must achieve a certain size threshold to do so. When juvenile hormone (JH) is applied to beetles in the last larval instar, males produce smaller horns than expected, but larger compound eyes. JH-treated females however, show no changes in either horn production or eye size, indicating that eye size change in males results from a reallocation and not the effects of JH per se. Results of experimental selection on horn length supported these results: eye size responded in the opposite direction to horn length, and reallocation was local—there was no correlated size change of antennae, palps, or forelegs (Nijhout and Emlen, 1998).

Stabilizing selection for the integration of the components of functional suites of traits, e.g., the muscles, ligaments, and bones composing the vertebrate feeding apparatus, has been argued to play a significant role in the evolution of development ("internal selection": Whyte, 1965; Arthur, 1997; Schwenk, 2000; Fusco, 2001). Evidence also exists for the coordinated plasticity of elements of functional suites, in the form of compensatory adjustments when one or more of the components is altered (e.g., Smits et al., 1996; Shipley and Lechowicz, 2000). Alternatively, stabilizing selection for canalization of development may favor the evolution of genetic backup systems, e.g., maintenance of duplicate genes (Wilkins, 1997; Wagner, 1999) or the evolution of epigenetic networks in which unrelated genes can compensate for mutation or developmental errors (Wagner, 2000; Hartman et al., 2001).

Additional evidence for the impact of internal environment comes from studies that have shown that an organism's current phenotypic state cannot be fully accounted for by its present or even most recent history. There can be substantial influences of its past environmental experiences (ontogenetic contingency or state-dependent life history: Diggle, 1994; McNamara and Houston, 1996), or even the conditions that its parents encountered (e.g., maternal effects: Mousseau and Fox, 1998). In *Nicotiana sylvestris*, Baldwin and Schmelz (1996) found evidence for "immunological memory": nicotine production was significantly faster for plants previously induced to produce nicotine, compared to uninduced control plants. Sultan (1996) examined the effect of maternal environments in *Polygonum persicaria,* applying treatments of high versus low levels of light, water, and nutrients. When the growth of the offspring of these plants was assessed under normal conditions, the quality of offspring of parents from *low* treatments was equivalent to or even enhanced relative to offspring of parents from *high* treatments.

Is there a need to consider external and internal agents separately? All environmental signals, even external ones, must be received and processed at the level of individual cells. Whether the signal is light on a plant or the morphogen Decapentaplegic (Dpp) in insects, only cells with receptors can tune in, and only the subset of those cells in which the appropriate DNA sequences are accessible (e.g., unmethylated) can process the signal. Subsequent actions are mediated by the production of additional substances that serve as environmental signals for other cells—leading, for light, to the processes of photomorphogenesis (e.g., Tepperman et al., 2001) and for *Dpp,* to insect leg development (e.g., Jockusch et al., 2000).

It seems clear that the distinction between external and internal environments refers only to the source of the signal, and not to the location of its effect. In addition, the effectiveness of signal reception, transduction, and response will ultimately be measured in the currency of fitness (survival or reproduction), whether the signal was initiated within or without. We have previously suggested that development itself can be thought of as a reaction norm to internal environmental signals (Schlichting and Pigliucci, 1998; see also Wolpert, 1994; Newman and Müller, 2000). Here I conclude that our understanding of the evolution of developmental systems can benefit significantly by the application of some components of the toolkits developed for understanding organismal adaptation to changing environments.

GERD B. MÜLLER AND LENNART OLSSON

Epigenesis and Epigenetics

The concepts of epigenesis and epigenetics have deep historical roots that reach into early philosophical explanations of the organismal world. At the same time, a host of different usages exist in present day science. In using the term "epigenetic," developmental biologists wish to emphasize the context-dependence of developmental processes, geneticists refer to mechanisms of gene regulation that do not require changes of DNA sequence, evolutionary biologists imply non-DNA-based mechanisms of inheritance, and population geneticists evoke phenotypic variation in response to environmental conditions.

These usages seem vastly dissimilar. Although they address different phenomena pertaining to different fields of traditional research, these phenomena and their epigenetic components are related through the continuity of processes that link the succession of ontogenies in evolution. Here we provide an outline for a coherent discipline of epigenetic research. We begin by a brief summary of the historical background (for more detailed accounts, see Maienschein, 1986; Richards, 1992; Jahn, 2000), and then define and present examples of the four prevailing areas of epigenetic research. Finally we discuss the meaning of the epigenetic approach for evolutionary theory and evolutionary developmental biology.

An Explanatory Concept of Development

The historical origin of the epigenesis concept is not related to genetics but is rooted in a debate started by Aristotle about the nature of embryonic development. Reacting to what could be called proto-preformationist views held by the Hippocratic school, Aristotle clearly distinguished between the two possibilities of embryonic organs arising from pre-existing parts or from the formation of new parts, and decided for the latter. From here onward the history of developmental biology can be traced as alternations between these two explanatory concepts.

Contentious debate began in the seventeenth century regarding epigenesis-preformation. In *Exercitationes de Generatione Animalium* (1651, p. 121), William Harvey (1578–1657) first used "epigenesis" to argue that development proceeds as incremental formation of new entities out of nonstructured germ material. Based on the first microscopical observations of Leeuwenhook (1632–1723) and Malphigi (1628–1694), the early preformationists countered that the complete organism exists already in miniaturized form in the egg (ovist view) or sperm (spermist view) and that embryonic development only consists of the elaboration of the preexistent form. Albrecht von Haller (1708–1777) concluded categorically that there is no epigenesis, and he used the term "evolution" (unfolding) to characterize the preformationist concept of embryological development, as elaborated by Swammerdam (1637–1680) and later by Bonnet (1720–1793).

These and other authors favored the view that God had created germs containing miniature organisms, which in turn contained germs that contained miniature organisms, and so forth, with the effect that "the entire human race already existed in the loins of our first parents, Adam and Eve" (Swammerdam, 1685, p. 46). Bonnet also believed that the world had undergone several catastrophes in which the adult organisms had disappeared but their germs survived. These germs would then have developed into new forms after each catastrophe. Here Bonnet connects individual development with that of species renewal and is thus chiefly responsible for the beginning shift of the term "evolution" toward signifying the unfolding of progressively more perfect forms of "organized beings." The embryological notion of unfolding had important consequences for Darwin's understanding of species transformation and the conceptualization of evolutionary theory (Richards, 1992).

In renewed opposition against the rising preformationist explanations, Caspar Friedrich Wolff (1734–1794) developed a theory of epigenesis in *Theorie von der Generation* (1764). He noted, for example, that the heart and circulatory system of the chick form gradually from an undifferentiated mass, and considered the preformationist view that the organs were already formed, but too small to see, a fable. However, because epigenesis assumed the absence of preformed structures, it required an organizing force that produced differentiation out of uniformity, which led to vitalist notions, first expressed by Blumenbach (1752–1840) and subsequently endorsed by many empirical embryologists, such as Karl Ernst von Baer (1792–1876) and Hans Driesch (1876–

1941). In contrast, the main attraction of preformationism was that it made unnecessary the search for additional causes of the transformation from undifferentiated to differentiated form; only more simple processes, such as growth, needed to be explained.

In the nineteenth century, under the new auspices of a well-developed cell theory and a rising understanding of the factors of heredity and species evolution, the epigenesis-preformation debate famously went through another round when August Weismann´s (1834–1914) views collided with those of Oscar Hertwig (1849–1922). Weismann had originally been sympathetic to the epigenetic point of view. However, his ideas about inheritance and differentiation forced him to become "convinced that epigenetic development is an *impossibility*" (Weismann, 1893, p. xiv). Weismann envisioned that differentiation during development was regulated by "determinants" already present in the germ plasm or, more precisely, in the chromosomes that had just been detected. These determinants he thought represented the basic units of heredity, constituting a preformed entity in each new generation. His preformationism saw ontogeny as a process in which the inherited determinants became distributed in such a way that each cell receives the correct combination of determinants to make it into a specific cell type.

Oscar Hertwig criticized Weismann in *Präformation oder Epigenese* (1894). Hertwig, whose work pointed to the nucleus as the locus of heredity, argued that by allocating the causal factors that regulate development to "determinants" that could not be directly observed, Weismann had abandoned the scientific method. Despite this criticism, Hertwig agreed with Weismann that the germ plasm was already highly organized, but he emphasized that embryonic development was influenced by external factors, even in its early stages. Hertwig pointed out that complex processes, such as gastrulation, in which many cells must cooperate, are difficult to explain using Weismann´s determinants only. He also argued that the way in which embryos respond plastically to changing environmental conditions and experimental manipulations posed substantial problems for Weismann´s view.

Around the middle of the twentieth century, with genetic theory in full swing, "epigenetic" began to shift its meaning, mostly through the work and writings of Conrad Hal Waddington (1900–1975). Combining aspects of epigenesis and genetics, Waddington suggested the term *epigenetics* as an English language equivalent to *Entwicklungsmechanik*, instead of the unwieldy direct translation "developmental mechanics"

(Waddington, 1956b, p. 10). Thus, for Waddington, epigenetics generally signified the causal analysis of development and, in particular, all interactions of genes with their environment that bring the phenotype into being. Here we find the roots of the semantic shift from *-genesis* to *-genetic,* which caused much confusion in latter-day interpretations of epigenetics.

At the same time Waddington emphasized the evolutionary importance of the *epigenotype* as a historically acquired, species-specific network of developmental interactions that has further consequences for the evolvability of a phylogenetic lineage. He coined a number of concepts to address these developmental and evolutionary mechanisms, such as the "epigenetic landscape," "chreods," and "assimilation," all related to the new, causal epigenetics he advocated. Most authors in the second half of the twentieth century refer to Waddington's concepts when they speak of epigenetics, and the prevailing usage in developmental texts still evokes the context-dependency of developmental processes (e.g., Løvtrup, 1974; Hall, 1998). Less prominently, and quite independently from the developmental interpretations—although equally rooted in Waddington's views—"epigenetic" started to be used in population genetics to signify environmentally induced phenotypic variation.

The shift of meaning introduced by Waddington strengthened toward the end of the twentieth century by an increasing association of the term epigenetics with molecular mechanisms of selective gene regulation and non-DNA-based forms of mitotic and meiotic inheritance. Although a certain notion of developmental context has remained, the prevailing emphasis now is on the regulatory mechanisms of gene activity, and new definitions of epigenetics are framed in the vocabulary of genetics (e.g., Holliday, 1987, 1994; Russo et al., 1996; Chadwick and Cardew, 1998; Urnov and Wolffe, 2001). This has effectively removed the problem of embryonic form generation from the discussion of epigenetics, although the core questions of the epigenesis-preformation debate started in the seventeenth century resurface in the postgenomic era. They are cast in different terms: Is all information to build a body contained in DNA sequences? In other words, can we predict what an unknown organism would look like if its complete genome were known? And is information contained in the genome deterministically exhaustive, or does information arise through the processes of developmental interaction in which many other, nonprogrammed factors participate?

Not only does the epigenesis-preformation debate continue in new form, but different usages of epigenetics coexist at the beginning of the twenty-first century. Textbooks of developmental biology do not refer to gene silencing, paramutation, or non-DNA-based inheritance when they address epigenetic factors, and neither do population genetic texts. And molecular geneticists do not refer to reaction norms, environmental factors, or context-dependent morphogenesis when they mention epigenetics. Therefore, a closer look at the current usages and their interrelations is necessary.

Connotations of the Term Epigenetic

Four connotations of *epigenetic* can be distinguished in modern usage. Although the scientific domains to which they apply overlap, the four usages are characterized separately before their interrelationships are re-established in the final section.

Epigenetic development (epigenesis). In developmental biology epigenetic still refers primarily to epigenesis, i.e., the individual generation of embryonic form through a series of causal interactions. It emphasizes the fact that development does not consist merely of the reading out of software-like genetic programs but also depends on a species-specific and context-dependent set of regulatory exchanges, also with factors not encoded in DNA. Epigenetic factors are "all conditional, non-programmed factors that act on the materials of the zygote and its derivatives, including those specified by the genes that are necessary to generate three-dimensional biological form" (Müller and Newman, 2002). Such factors comprise all elements and processes active in the molding of tissues at various times in development. They can be internal, belonging to the embryonic system itself, or external, belonging to the environment.

Internal factors of epigenesis reside in (1) the maternal materials and non-DNA-based templates (e.g., cell membrane) passed on from the parent organisms; (2) the generic physical and self-organizational properties of cells and tissue masses; (3) epigenetic gene regulation; (4) the dynamics of interactions among and between cells, cell populations, and tissues; (5) the spatial, geometric, and biomechanical conditions of expanding cell masses; (6) the material properties of intra- and extracellular cell products; (7) activity of the tissues and of the whole embryo; and (8) the intra- and extracellular presence of foreign materials or parasitic organisms. External factors reside primarily in the physicochemical conditions of the environment in which development takes place, such as temperature, humidity, gravity, light, radiation, mechanical influences, and the chemical composition of the surroundings and nutrients. External factors can also consist of the activity and products of other organisms.

The epigenetic domain of development represents the transformational interface between molecular and macroscopic levels of organismal organization. Epigenetic factors can be specific or unspecific, permissive or instructive, and are converted by the developmental system into a specific morphological outcome. Much in the same way as the phenotype is composed of structural units, and the genotype of units of gene expression, the epigenotype can be conceived as consisting of units of developmental integration.

Epigenetic gene regulation and mitotic propagation (epigenetics). In molecular and developmental genetics, epigenetic refers primarily to modifications of gene activity that are not based on alterations of DNA sequence and that provide mitotically propagated changes in gene function. Epigenetic repression is responsible for genomic imprinting and related phenomena of allelic exclusion. Its molecular mechanisms include cytosine methylation, histone hypoacetylation, and RNA silencing, but sometimes mRNA processing and other forms of posttranscriptional modification are also termed epigenetic.

DNA methylation, primarily of cytosine, is the best-understood mechanism of gene silencing. Methylation at regulatory regions, especially within the promotor, prevents the binding of regulatory factors at these sites. There is a strong positive correlation between the extent of methylation and the degree of silencing. Already suggested in the mid-1970s, these mechanisms were experimentally vindicated in the early 1990s (reviewed in Urnov and Wolffe, 2001). DNA methylation is essential for the normal control of gene expression in development. Significantly, methylation patterns are mitotically propagated, thus maintaining differential functional states of the genome in cell lineages, also called *imprinting*. Unfortunately, this process is often subsumed under the term "epigenetic inheritance," blurring the distinction between development and inheritance.

The most striking example of imprinting is X chromosome inactivation in female placental mammals. Of the two X chromosomes, one is inactivated and heterochromatized in each cell, forming the so-called Barr body. This inactivation requires a gene on X called *XIST*. *XIST*-

RNA accumulates along the chromosome containing the active *XIST* gene and proceeds to inactivate (almost) all of the other genes on that chromosome (Lyon, 1995).

More common than the suppression of an entire chromosome is the inactivation of single genes or of small clusters of genes. When methylation goes wrong, the effects on the developing organism can be dramatic. In humans, a number of genetic diseases are caused by loss of imprinting, probably triggered by defective methylation (Chadwick and Cardew, 1998). Loss of imprinting is also an important factor in tumor formation (Jones and Laird, 1999), possibly because the effect of a mutation on an imprinted allele will not be rescued by the other, nonmutated allele. The study of epigenetic gene regulation has become an important field of biomedicine including cancer and stem cell research, somatic gene therapy, transgenic technologies, cloning, teratogenesis, and so forth.

Epigenetic Inheritance in Sexual Reproduction

In evolutionary biology, epigenetic inheritance refers to the transmission of epigenetic states from one generation to the next, via the germ line, without a change in DNA sequence. This is viewed as a second inheritance system, based on the same mechanisms as the passing on of gene deactivation patterns in cell lineage propagation. Although differential methylation states are generally erased during sexual reproduction through reprogramming in germ cell and pre-implantation development (Reik et al., 2001), certain epigenetic marks seem to be able to escape erasure.

Parental imprinting is the best known phenomenon of this kind. It confers to the offspring an asymmetry between the activation states of parental genes. Only one allele is normally active, either from the father's or from the mother's pronucleus; the other allele is permanently silenced. In general, the silent allele is hypermethylated, and the active allele is often hypomethylated. Among vertebrates, parental imprinting is known to occur only in mammals, such as in mules and hinnies, in which it can make an important difference whether a gene comes from the mother or the father. In the early 1980s it was first shown that male and female genomes cannot substitute for each other (Surani et al., 1984). The exchange of the pronuclei of mouse eggs produced zygotes with two male or two female haploid genomes, but both failed to develop normally. These effects are now explained by the absence of differently imprinted genes.

Paramutation is another case in which epigentic inheritance is implicated. It is a type of gene silencing that involves, in a heterozygous condition, the inactivation of one allele by a mutated allele, and this condition is retained after the alleles have segregated in the offspring. The mechanism is possibly related to cosuppression, involving homologous pairing of genes with direct effects on transcription. Such meiotic inheritance of epigenetic states was first described in plants (Brink, 1960), but was recently also observed in fission yeast, insects, and mammals.

Epigenetic inheritance reflects the functional history of a gene and thus raises the controversial issue of the transmission of individually acquired, functional states from one generation to the next. Because it is known that methylation and other forms of epigenetic chromatin marking can depend on environmental influences, epigenetic inheritance has been argued to represent a kind of Lamarkian mechanism in evolution (Jablonka and Lamb, 1995).

Epigenetic variation. In population genetics, the variation of a phenotypic trait caused by environmental or behavioral factors is sometimes called "epigenetic variation." Epigenetic variation takes place within the range of phenotypic plasticity for a given trait, depending on its genetic reaction norm (Suzuki et al., 1986). A reaction norm is defined at the population level, representing the complete range of phenotypes that a genotype can express in interactions with the environment. Epigenetic variation thus reflects the influence of the environment on individual development and represents one of the modes through which the same genotype can give rise to different phenotypes (Gilbert, 2001).

Although reaction norms are usually defined in genetic terms alone, the factors of epigenesis, epigenetic gene regulation, and epigenetic inheritance are equally important for epigenetic variation. Individuals that are genetically identical (i.e., sufficiently similar that the small genetic differences play no role) can differ epigenetically and therefore differ in their phenotypes. Hence epigenetic variation is a term that encompasses the phenotypic consequences of the epigenetic factors discussed above.

Epigenesis and Epigenetics: A Common Agenda

The results of genome sequencing projects indicate that organismal evolution has proceeded not so much by an increase in the number of genes but rather through regulatory modification. Part of this modification is epigenetic. Epigenetic modulation acts at different levels of the evolutionary process, which can be distinguished as the generative level, the integrative level, and the inheritance level. The different types of epigenetic mechanisms interact in a contingent, lineage-specific way and have different roles at each of the levels of evolution. Brief characterizations follow.

The generative level of evolution is the level from which variation and innovation take their origin. Whereas neo-Darwinian theory has focused on variation, the mechanisms of phenotypic innovation are much less explored (Müller and Wagner, 1991; Müller and Newman, 2002). It has been argued that epigenetic factors governing the behaviors of cells and tissues had an important influence on morphological evolution, both at the early origination of multicellularity and metazoan body plans, and during later innovations in advanced organisms. These generic forms would have been based on the physical processes characteristic of condensed, chemically active materials, such as cells and cell masses, and on the conditional, inductive interactions that became established in evolved developmental systems (Newman and Müller, 2000). This concept—that epigenetic mechanisms are the generative agents of morphological innovation—has been suggested to contribute to explanations of complex phenomena in evolution, such as the origins of novelty and homology and rapid morphological change.

The evolution of epigenetic gene regulation might have played an important role both during these generative events and also at the integrative level of evolution, at which generative changes become consolidated and made routine by molecular control mechanisms. Epigenetic regulation probably arose from cellular defense mechanisms against viruses and parasitic DNA and was later recruited for differential gene regulation (Matzke et al., 1999). One important evolutionary aspect is that these mechanisms might have contributed significantly to genome evolution, via the silencing of duplicated gene loci, and thus might have facilitated the evolution of developmental differentiation and determination processes. On the other hand, these epigenetic regulation mechanisms might have assisted in the developmental and genetic integration of successful innovations that arose from the epigenesis-type mechanisms listed earlier. This picture is supported by the finding that gene-silencing mechanisms seem to have gained their major influence only in higher plants and vertebrates, whereas in other organisms, including most invertebrates, methylation seems to be a relatively unimportant event. In vertebrate evolution, a moderate increase of DNA content was accompanied by a substantial increase of methylation (Bird, 1995), indicating new roles for gene silencing.

At the inheritance level, the transgenerational transmission of methylation patterns and other epigenetic states represents one of the most challenging aspects. It infers that a second, autonomous, non-DNA-based inheritance system is at work, and it includes the possibility that epigenetic states acquired in one generation can be transmitted to the next (Jablonka and Lamb, 1995). Environmental stimuli may thus have a more direct influence on adaptive modification than is generally assumed under the neo-Darwinian model. This type of epigenetic inheritance is likely to have its greatest impact on the evolution of organisms that lack a distinct segregation of soma and germ line, such as fungi and plants. It could also assist in reproductive isolation of higher plants and animals and could indirectly affect DNA base changes by influencing the frequency of mutations at certain loci. Furthermore, there is the clear possibility that epigenetic states become fixed by random mutation and natural selection (Maynard Smith, 1990). This transition from an epigenetic to a genetic state corresponds to the process of "genetic assimilation" proposed by Waddington (1953a).

In each traditional research area—molecular genetics, development, heredity, evolution—"epigenetic" refers to a different issue. The molecular mechanisms of gene regulation, the embryonic generation of form, the transmission of information from one generation to the next, and environment-induced variation do not represent the same biological problem. However, these issues are linked through an integrative, evolutionary perspective. In this view, the epigenetic agenda emerges as the science focusing on the role of the nonprogrammed, regulatory, and modulating factors in biological processes. Whereas evolution was traditionally studied either from a genetic or from a phenotypic perspective, the common epigenetic agenda represents a new level of analysis, focusing on the causal interactions between genes, phenotypes, and the environment. This epigenetic agenda will be central for the formulation of integrative models in evolutionary developmental biology.

KARL J. NIKLAS

Evolution of Plant Body Plans and Allometry

Much attention is devoted to the evolution of animal body plans (e.g., Carroll et al., 2001; Davidson, 2001). Much less has been written about the evolution of plant body plans, and most of the literature treating this topic has been largely confined to the organography and development of the stereotyped vascular plant (see, however, Hagemann, 1999; Niklas, 2000). Despite their complexity and ecological importance, the vascular plants provide a limited view of the totality of land plant diversity. Moreover, a number of distinct, predominantly aquatic lineages (the "algae"), evolved early in the history of life (Schlegel, 1994; Graham and Wilcox, 2000). Although these lineages are virtually impossible to typify other than on the basis of molecular or cytological criteria, no synoptic treatment of plant body plan evolution is possible without reference to the algae.

Despite their tremendous diversity, all plants manifest nearly identical fundamental size-dependent (allometric) relationships. Thus, across twenty-two orders of body size magnitude, across unicellular and multicellular plant species, growth scales as the 3/4-power of body mass (Niklas and Enquist, 2001). Likewise, across eight orders of magnitude of vascular plant size, annual leaf, stem, and root growth in biomass scale isometrically each to one another. These seemingly invariant "rules" for proportional growth and biomass relationships cut across all lineages and are remarkably indifferent to species habitat preference.

This entry reviews the different plant body plans, enumerates their occurrence among the various lineages, and briefly describes how each achieves its organized growth and development. I adopt a nontraditional approach to defining plant body plans because each body plan type is theoretically capable of manifesting a broad range of morphology or anatomy. I also briefly review plant allometry because each major plant body plan has the potential to grow in size indefinitely, which necessitates correlated changes in shape or geometry (Niklas, 1994).

Body Plan and Unity of Type Concepts

The body plan concept comes from zoology. Long ago, comparative zoologists recognized that the species of monophyletic metazoan groups share adult phenotypic features whose spatial arrangement or juxtaposition typifies each major lineage. These body part layouts are called body plans or *Baupläne*. Thus, extinct or extant mammals have two sets of limbs, three middle ear bones, and a lower jaw composed of one bone, whereas adult insects have three sets of limbs, a dorsal heart with paired ostia, and abdominal tracheae. These features distinguish insects from mammals and each from all other major metazoan lineages.

The evolutionary persistence of the body plan characterizing each major animal lineage is called the "unity of type." That the organization of each metazoan clade is highly conserved may appear logically circular. If organisms are grouped based on what they look like, each grouping must manifest a unity of type. However, the unity of type of each animal lineage reflects a shared developmental and genomic repertoire stemming from a last common ancestor. Although each repertoire may be modified by mutation, it cannot be radically modified or abandoned altogether without producing what would probably be considered a new lineage. Thus, the limb buds of mammals can develop to produce legs, arms, or flippers in different species. But these different appendages are recognizable as "limbs" by virtue of their many shared features. As such, their modification neither obscures the mammal body plan nor diminishes its unity of type.

Two explanations are traditionally offered for unity of type. One emphasizes natural selection and the other, developmental constraint. If each body plan is a highly integrated arrangement of parts, any significant mutation would probably be deleterious. Alternatively, if each body plan reflects only what is possible developmentally, no other body plan could evolve in a particular lineage. These two explanations are not as highly polarized as they seem, because the meaning of "constraint" can be ambiguous. Traditionally, a constraint refers to the impossibility of development achieving anything other than what it does. Development is seen as intrinsically self-limited and the external environment (and thus natural selection) is of little importance.

However, a constraint may also reflect the persistent elimination of all developmentally maladaptive variants. A variety of developmental results is thus possible, but only those that are fit survive. Here, development is not self-limited and natural selection plays an important role. In the absence of fastidious developmental studies, the unity of type of a particular lineage logically provides neither evidence for nor against the role of developmental constraints as traditionally defined. In contrast, convergent evolution on the same body plan by taxa with very different ancestries provides circumstantial evidence for adaptive evolution, just as the divergence of body plans in a lineage may provide evidence for adaptive evolution.

Plant "Plasticity"

Plants, defined here as all photoautotrophic eukaryotes, have evolved independently many times (Figure 1). Some lineages originated in deep evolutionary time, ostensibly as a result of primary endosymbiotic events involving heterotrophic and photoautotrophic prokaryotes.

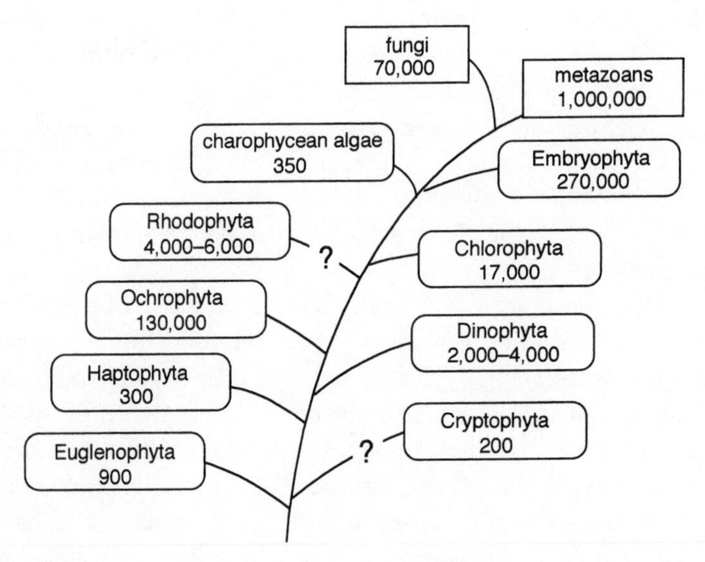

FIGURE 1. Plant and animal phylogenetic relationships based on nuclear-encoded small subunit ribosomal RNA sequence (and other) studies. Uncertain relationships indicated by question marks. Approximate number of extant species indicated below formal or informal taxonomic grouping. (Relationships adopted from Schlegel, 1994.)

Other more recent, yet still old lineages evolved presumably as the result of secondary endosymbiotic events among heterotrophic and photoautotrophic eukaryotes. Collectively, these lineages are called the "algae," a polyphyletic constellation of primarily aquatic clades (Graham and Wilcox, 2000). The most recently evolved plant lineage, the embryophytes, is monophyletic, descending from an ancestral plexus of algae sharing a last common ancestor with the charophycean algae (Graham, 1983; Graham and Wilcox, 2000). The embryophytes colonized and subsequently radiated on land.

Typifying any of these lineages based on the size or general appearance of their representative species is virtually impossible. Unicellular, colonial, and multicellular pelagic and attached species are found in all but a few algal lineages. The size and appearance of some algal species approach those of some embryophytes, while some algae possess conducting tissues that are functionally analogous to those found in land plants. Within the embryophytes, we find diminutive, essentially filamentous organisms that look much like some algae. Moreover, a single species may manifest two or more very different morphologies during its reproductive cycle—some fern gametophytes are filamentous in appearance, whereas their corresponding sporophyte generation may be treelike in size or shape.

The tremendous range in size, shape, and anatomy seen for most plant lineages is a probable evolutionary outcome of the contingency of plant growth and development on external environmental cues. In general, plant development is not self-limited or constrained, and natural selection can act directly on the ability of an individual to alter its phenotype adaptively in response to environmental conditions, which can change over the course of a plant's lifetime. The ecological advantage of this *phenotypic plasticity* is that an individual can survive in diverse environments. The evolutionary advantage is that a species or higher taxon can survive and reproduce in potentially very different habitats.

That plants differ profoundly from animals is evident. All eukaryotic photoautotrophs rely on the chloroplast to convert radiant into chemical energy and require the material resources (water, carbon dioxide, oxygen, and a variety of minerals) to operate metabolically. With only a few rare exceptions, each plant protoplast produces and is surrounded by a "wall" of extracellular materials whose chemical composition varies across lineages. These materials are typically permeated with enzymes that regulate the removal, addition, or chemical relaxation of specific constituents, which are essential for growth in cell size, shape, geometry, or number. The lifestyle of plants also differs profoundly from that of animals. Plants do not capture, ingest, or digest food, nor do they excrete wastes. Their metabolism and metabolic requirements change little over the course of ontogeny. Moreover, plants lack any semblance of a neurological or muscular system, and so many begin and end their existence in much the same location. Finally, plant dispersal is governed initially by predominantly random events. Because plant development and growth rely on physical external cues, we see why phenotypic plasticity is likely to be far more pervasive among plant than among animal lineages.

Plant Body Plans

Plant body plans are profitably classified on the basis of how each achieves organized growth and development. The simplest scheme identifies only three basic body plans: unicellular, colonial, and multicellular (Figure 2). Each of these achieves its organized growth by a comparatively few, albeit basic, developmental features. Variations of each basic body plan can result from the superposition of additional developmental features (e.g., the multinucleate variant of the unicellular, colonial, or multicellular body plan is achieved by delayed or repressed cytokinesis; indeterminant growth of the multinucleate body plan produces the siphonous growth habit). Some of these variants may appear so different from their basic body plan type as to warrant recognition as a distinct category. However, these variants are rare and not fundamentally distinct (Table 1).

Divergent and convergent body plan evolution is apparent when the occurrence of each is enumerated across and within the plant lineages (Table 1). Among the algae, the unicellular body plan—which is presumed ancestral for each algal lineage—occurs in all algal lineages, reflecting its numerous selective advantages in an aquatic habitat (e.g., cell mobility, high surface-area-to-volume ratio, efficient light harvesting capacity, rapid growth). The colonial body plan occurs in all but one algal lineage. The loss of protoplasmic continuity after cell division and the aggregation of derivative cells by a shared extracellular matrix, loricas,

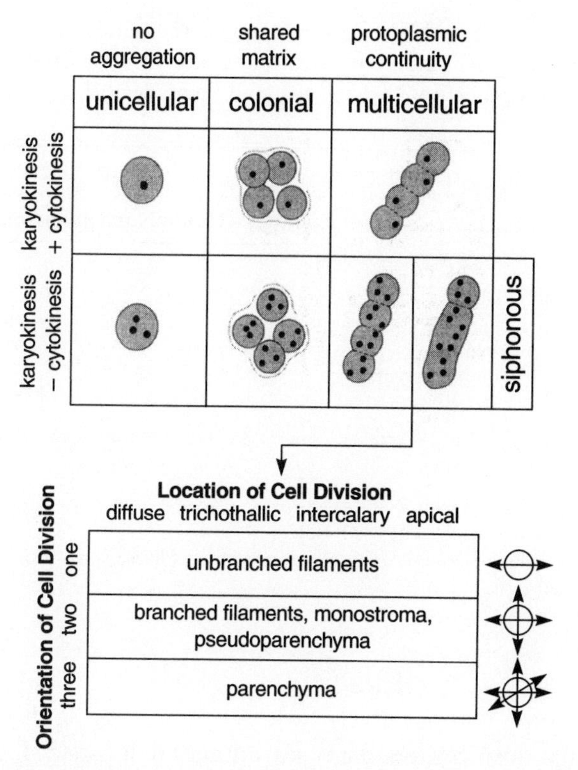

FIGURE 2. The three basic plant body plans (unicellular, colonial, and multicellular) are distinguished on the basis of a few developmental features. The unicellular body plan is achieved by karyokinesis followed by the separation of derivative cells; the colonial body plan requires the adhesion of derivative cells by means of a shared matrix of extracellular materials; and the multicellular body plan is achieved by plasmodesmatal protoplasmic continuity among adjoining cells. The multicellular body plan permits additional developmental features. Regulation of metabolic and cellular activity by phytohormones determines the location of cell division as generalized (diffuse) or confined to basal, proximal, or apical regions of the plant body (tricothallic, intercalary, or apical cell division, respectively). Depending on the orientation of new cell walls with respect to the principal axes of cells or the body as a whole, unbranched or branched filaments can be formed and, in concert with persistent plasmodesmata, division patterns may produce monostromatic (one cell thick), pseudoparenchymatous, or parenchymatous tissue constructions. (Adapted from Niklas, 2000.)

TABLE 1. Occurrence of the three basic plant body plans (unicellular, colonial, and multicellular) and multicellular variants (with filamentous, pseudoparenchymatous, and parenchymatous tissue organization) among the land plants (embryophytes) and the major algal lineages

Taxon	Unicellular	Colonial	Multicellular		
			Filamentous	Pseudoparenchymatous	Parenchymatous
Embryophytes	−	−	+	+	+
Chlorophyta	+	+	+	+	+
Rhodophyta	+	+	+	+	+
Ochromphyta[a]	+	+	+	+	+
Dinophyta	+	+	−	−	−
Haptophyta[b]	+	+	−	−	−
Euglenophyta	+	+	−	−	−
Cryptophyta	+	+	−	−	−

Taxonomy based on Graham and Wilcox (2000).
[a] The stramenophiles include diatoms, raphidophyceans, chrysophyceans, synurophyceans, phaeophyceans, oomycetes, etc.
[b] The prymnesiophytes.

or stalks requires comparatively little developmental sophistication. Yet this body plan also confers many advantages (e.g., specialization of cell types, shared arsenals of defensive or regulatory metabolites, the ability to alter water flow patterns favorably by adjusting colony appearance or size). Siphonous algae are rare, perhaps because they are susceptible to systemic viral or microbial infection as a result of a single wall breach. Yet some siphonous algae are remarkably invasive and ecologically successful.

Multicellularity is absent in four algal lineages, each of which is presumed to have evolved through secondary endosymbiotic events. However, the advantages of multicellularity are numerous. The ability to form and sever plasmodesmata among adjoining cells can establish physiological lines of communication, which becomes increasingly important with increasing body size. Cytoplasmic domains of intercellular communication are also requisite for tissue differentiation, specialization, and phytohormonal coordination of meristematic, vegetative, and reproductive effort. In this respect, all embryophytes are multicellular, develop parenchymatous tissues, typically manifest an open indeterminate ontogeny, and collectively achieve very complex and specialized tissues and organs. These highly conserved features are as functionally important in an aerial habitat as they are testimony to the monophyletic origin of the land plants (Graham, 1993).

Allometric Trends Dependent on Body Plan Size

Despite their many differences, each plant body plan appears to abide by the same size-dependent *(allometric)* scaling rules, which also seem to hold true for animal life. For example, across the various lineages, growth G scales as the 3/4-power of body mass M and likewise scales isometrically with respect to the ability to harvest sunlight (Figures 3A, 3B). These "invariant" properties suggest that biophysical constraints have shaped evolution at a very fundamental level.

Many theories have been advanced to explain these and other invariant scaling relationships. None has escaped well-reasoned criticism. However, each theory shares a common starting point—the size-dependent relationship between body surface area S and volume V. Most organisms use their surface area to exchange mass and energy with their external environment. Thus, S is a crude measure of the capacity for growth. Likewise, body volume is a convenient, albeit crude measure of the demand for the resources required for growth. Thus V reflects body mass. As such, natural selection is expected to favor body plan development, maximizing S with respect to V as an organism increases in size.

Geometry shows that the minimum scaling exponent for S versus V is $2/3 \sim 0.667$ (i.e., S increases as the square of a linear reference dimension, whereas V increases as the cube of that dimension). Growth is thus expected to scale *de minimus* as the 2/3 power of body mass. Indeed, empirically, we find that $S \propto V^{3/4}$ for most unicellular plants (Figure 3C). This larger-than-expected exponent indicates that even the unicellular body plan can alter shape and geometry independently with increasing size such that its development maximizes S with respect to V and thus presumably G with respect to M. Similar trends are seen for the colonial and multicellular body plans, suggesting that plant evolution has been influenced as much by physical phenomena as strictly biological ones.

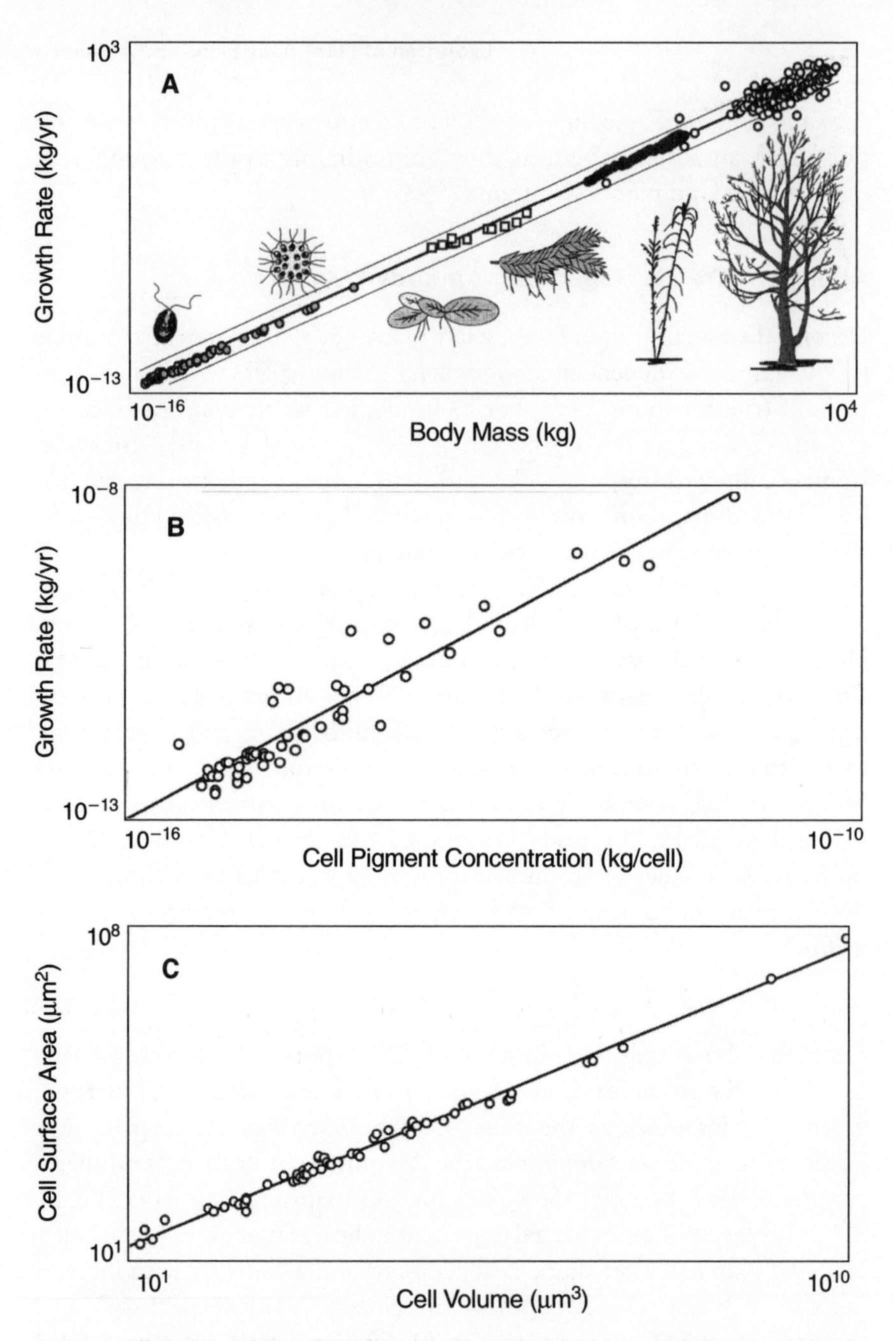

FIGURE 3. Allometric relationships for plant growth, body size, surface area, and volume. A. Plant growth rates (biomass annual production per plant) plotted against body mass (measured as dry mass). B. Plant growth rates plotted against the capacity to harvest sunlight (measured as photosynthetic algal cell pigment concentration or as foliage biomass). C. Algal cell surface area plotted against algal cell volume. (Adapted from Niklas, 1994; Niklas and Enquist, 2001.)

JOHN GERHART AND MARC KIRSCHNER

Evolvability

The means by which organisms generate heritable phenotypic variation is a central issue of evolutionary biology and evolutionary developmental biology (evo-devo). Random mutation is widely accepted as necessary for such variation. Less accepted is the role of *evolvability,* which is defined as the organism's capacity to facilitate the generation of nonlethal selectable phenotypic variation from random mutation. We summarize proposals that show that this capacity is an aspect of the phenotype that has itself evolved, and that random mutation serves more as a trigger of phenotypic change in an evolvable organism than as a sufficient cause.

Over the past century, various authors have proposed kinds of evolvability. Baldwin, Osborn, and Lloyd-Morgan (Baldwin, 1902), and later Schmalhausen (1949) and Waddington (1975) discussed physiological and behavioral adaptability of the individual, including its adaptive norm of reaction, as a source of variation. By these means the organism would initially alter its phenotype under conditions of stress (selective conditions) without genetic change, thereby surviving. Some individuals would subsequently accumulate heritable changes that stabilize aspects of the adaptive response. Random mutation, mostly present as standing mutation, would not have to create variation *de novo,* just stabilize and exaggerate alternatives latent in the phenotype. More recently, Liem, Wake, Alberch, Gould, Oster, Dawkins (who may have first used the term evolvability in 1986), and Wagner have discussed general properties of the phenotype that might facilitate the generation of variation. One is the existence of modules of developmental function that could readily dissociate and reassociate in new combinations, or duplicate and be used independently (see Wake and Roth, 1989). The present authors (1997) discussed properties of various cell biological and developmental processes that might facilitate the generation of phenotypic change by random mutation—namely, modularity, robustness, flexibility, weak regulatory linkage, and exploratory behavior.

Evolvability is an explication of descent with modification by tinkering. Heritable genetic change sets off small regulatory alterations that act on conserved modular cell biological and developmental processes, leading to their use in different combinations at different times and places to generate various selectable traits. Due to their suitability for multipurpose use, the conserved processes facilitate the generation of phenotypic variation by random regulatory mutation. This account of evolvability can therefore be called a hypothesis of "facilitated variation." Organisms with high evolvability would (1) have less lethality from random mutation, and (2) need fewer independent mutational changes to generate a particular phenotypic change, compared to a less evolvable organism.

What are these conserved processes as exemplified in Metazoa? They are the subject matter of biochemistry, molecular biology, genetics, cell biology, and development. They are common to all animals despite great differences of morphology and lifestyle. Recent DNA comparisons have revealed extensive sequence conservation for protein components of these processes. Recent advances of developmental biology have identified conserved processes of intercellular signaling, controlled cell proliferation, morphogenesis, and many genetic regulatory circuits. These processes are used repeatedly at different times and places within the same organism during its development. The processes are used to generate many different traits within the same individual and, we suggest, differences between individuals as well. Whenever a particular trait is favorably selected upon, all processes generating that trait are also selected.

These multipurpose generative processes have been called the organism's "toolkit," implying that evolutionary change has mostly concerned regulatory changes of the time, place, amount, and combinations of use of these processes (see Carroll et al., 2001). Regulation is exerted at any of many levels: transcription, splicing, mRNA stability, translation, post-translational modifications of proteins, protein breakdown, availability of other components of the process, and substrate availability. Regulatory processes are themselves subject to regulation. Regulation is a vast target for mutational change. We surmise that all regulatory means have been used in evolution to effect different combinations of use of conserved core processes, although transcriptional regulation is most often cited as a prime target.

The toolkit analogy also implies the tool-like properties that make processes regulatable, versatile, and robust, befitting use in combinations to different ends. Three kinds of properties are discussed here as facilitating the generation of variation. Although each introduces constraint into a process (lethality of some mutations), each provides a trade-off, namely, de-constraint in forming regulatory connections:

1. *Compartmentation or modularity.* Complex multifunctional processes are subdivided into smaller independent subprocesses, thus reducing pleiotropic damage from mutation. Fewer components and conditions must be satisfied by a change. Compartments are independent of one another in their variation and selection. The compartment has properties of flexibility (its range of outputs), robustness (its self-regulation to operate over a wide range of external and internal conditions), and a responsiveness to regulatory inputs external to it. Examples include: numerous Hox domains subdividing the anteroposterior dimension of the embryo, each expressing a different Hox transcription factor or combination of factors; different genetic regulatory cascades of numerous cytodifferentiated cell types of the multicellular animal; and different organellar units of function within the individual eukaryotic cell.

2. *Weak regulatory linkage* of modular processes to each other and to various signaling conditions (times and places) is a second property. Two features of modules favor weak linkage. First, they already contain regulatable elements, loci at which regulatory inputs can be registered to affect the module's function. Second, the module is built with an inherent capacity to be active or inactive. With these characteristics, a new regulatory input arising by variation and selection needs only play on features already present rather than produce them *de novo*. Examples include numerous allosteric (two-state) proteins of the cell; various on-off transcriptional control circuits such as those controlling bristle specification/differentiation in insects; and the excitable two-state nerve cell. When a module with these properties is compared to one without, presumably fewer mutational changes would be required to achieve a new regulatory connection and more solutions would be available.

3. *Exploratory behavior* of some processes allows a great reduction of information needed to achieve a specific output or a new output. These versatile processes generate variation (varied outputs) and are sensitive to selection of specific outputs by physiological agents. Examples include numerous embryonic inductions, developing nerve connections, chromosome attachment to spindle microtubules, the adaptive immune system, and neural crest cells of vertebrate embryos. To take one example, axons extend and overconnect randomly to targets in a locale within the embryo. Connections are tentative, though; a specific subset is stabilized

by local agents reflecting functional use of those connections. Other connections disappear. A correct connection is thereby made with relatively little information. Individual axons are not directed accurately to a target from long distances. In evolution, a new connection could be made if just the location of the selective stabilizing agent changes. It is simpler to effect a change than in a nonexploratory process in which both partners must change accurately in concert.

This list of three properties may or may not be exhaustive; no general theory of stable-yet-changeable complex systems exists against which to examine it. Evolvability has been called the organism's greatest adaptation because it contributes to the generation of all particular variant traits on which incidental selections act. Coselection of the properties of these processes might have occurred during the selection of the individual traits they participated in generating. The selection for, or of, evolvability is difficult to discuss because proposals evoke selection for future as opposed to present benefit and altruistic group selection, benefiting others rather than the individual. Four possibilities can nonetheless be raised for its selection.

First, robust, flexible, regulatable, modular core processes are plausibly the kinds most suited to generate complex organisms under inevitably variable external and internal conditions. These characteristics would be directly selected in individuals, and evolvability would be a byproduct. Properties good in the short run for the development of the organism might also facilitate evolutionary change. The byproduct idea does not demean the evolvability concept. Perhaps organisms have evolved processes in which phenotypic stability complements evolutionary change and in which conservation complements variation.

Second, processes with these properties may be best suited for descent with modification by regulatory tinkering. They are repeatedly used in the development and physiology of many traits within an individual, that is, at many times, places, and levels, in many combinations. Modifiability might have been directly selected or coselected with new traits. Processes and components might have gradually gained tool-like properties. Descent with modification, entailing repeated exaptation, may have increased in efficacy over the course of evolution as processes were put to second, third, and more uses. These processes may be the ones most adapted to such evolution.

Third, processes with these properties would reduce lethality of random genetic change. They can adjust to and compensate for variable conditions and levels of other components, including ones introduced by random mutation. A population would carry more genetic variation and concomitant nonlethal phenotypic differences. Clade properties of emergent fitness for the individual may enter this discussion, just as they do when sexual reproduction is proposed to be maintained selectively because it increases the genetic variation of individuals of a population. These properties, now conserved, also increase genetic variation.

Fourth, evolvability of this kind may be selected in the evolution of groups of organisms distinguished by their diversity and species richness. Selection may be direct or indirect, acting on the traits of the radiating group. Organisms generating more variation might preemptively occupy niches under the special ecological conditions of radiations. Many lines of animals have undergone repeated radiations in their evolution, such as birds from reptiles from amphibians from jawed fish, from the earliest jawless vertebrates.

We have emphasized the robustness, versatility, and modifiability of conserved generative processes of the organism. Mutationally triggered regulatory changes would drive the use of these processes to different ends. From this viewpoint, the major events of evolution may not have been the regulatory refinements themselves, but the rare occasions on which effector processes arose and acquired the capacity for use in different times, places, amounts, and combinations, and were then conserved.

ARMAND DE RICQLÈS AND KEVIN PADIAN

Fossils and Paleobiology

Fossils include any remains of past life, whether they preserve body parts (skeletons and soft tissues) or records of behavior (tracks and traces, coprolites, gastroliths). These remains may include original material, or the material may be altered physically or chemically, or even replaced by other materials. *Paleobiology* deals with all aspects of biology involved in reconstructing the history of life and its evolution; it may also be called historical biology, and it does not rely only on fossil evidence. Here we focus on the relationship between development and evolution—that is, between ontogeny and phylogeny, two terms coined by Haeckel (1866). How can fossils inform us about both issues?

Reconstructing Phylogeny

Since Darwin (1859), evolutionary paleontology has tried to reconstruct phylogeny by more or less directly reading the changes in fossil lineages through time (the stratigraphic column, or geologic record). This trend has historically taken two main pathways. On the one hand, minute changes among fossils, seen in short but uninterrupted local stratigraphic columns, were viewed as direct evidence of evolution and even of its processes at the populational level (Hilgendorf, 1867; Waagen, 1869; Neumayr, 1873; Depéret, 1907). This approach essentially became the paleontological contribution to the New Systematics, which purported to reconstruct both patterns and processes of evolution within the paradigm of the Synthetic Theory. It is currently represented by stratophenetics (Gingerich, 1979), which purports to read actual evolutionary changes in the fossil record (usually at the level of fossil metapopulations and species) directly from fine-scaled stratigraphic sequences.

On the other hand, we have the construction of phylogenetic trees. Larger-scale evolutionary relationships are constructed in space and time by recognizing progressively evolving fossil lineages that link supraspecific taxa. Starting with Gaudry (1866), Haeckel (1866) and Cope (e.g., 1889), this trend later grew within the framework of the Synthetic Theory to reconstruct broad evolutionary patterns rather than processes. In practice, the Synthetic Theory focuses on evolutionary processes almost exclusively at the levels of populations and species. Little theoretical attention was paid to macroevolution, although supraspecific taxa were often recognized or used as metaphoric ancestors (e.g., Simpson, 1944). In the New Systematics there were no set rules or methods for the assignment of taxonomic rank or even for membership in a monophyletic lineage (Mayr, 1942). Indeed, neo-Darwinian systematics explicitly permitted evolutionary grades of organization that subtly retained some spirit of the time-honored *scala naturae*. From the 1970s onward, this approach was replaced by phylogenetic analysis (cladistics). This approach does not recognize grades and requires monophyletic groups (the last common ancestor and all its descendants) and the exclusive use of synapomorphies (shared-derived character states) to diagnose them.

Stratophenetics and cladistics are both valid and useful, but for different approaches to reconstructing evolutionary patterns and processes from the fossil record. Stratophenetics and related approaches are most useful at perispecific levels; cladistics is more useful at supraspecific levels. In both cases, evolutionary patterns can be deciphered, but the claim to reconstruct or test mechanisms that produce evolutionary patterns directly from fossil evidence remains much more debatable (Fitzhugh, 1997). From these fossil sequences we are able to discuss the evolution of ontogenies in paleontological terms.

Reconstructing Ontogeny

How can fossils tell us about ontogeny? In practice, some organisms are more conducive than others to the analysis of development—namely, those whose fossilized parts offer a direct record of the individual's growth trajectory. Many invertebrate phyla produce a mineralized skeleton that is essentially accretionary and not remodeled during life. The American neo-Lamarckian paleontologist Alpheus Hyatt (1838–1902), a student of Agassiz, was one of the first evolutionary paleontologists to make extensive use of the fossil cephalopods called ammonites to study the relationship between ontogeny and phylogeny (Hyatt, 1866, 1894). Ammonites are admirably suited for such approaches because their ontogeny is fully recorded in the structure of the early coils of the shell, and because the marine sequences of the Mesozoic produced abundant specimens that allowed statistical analyses of morphological changes through finely grained stratigraphic columns (e.g., Brinkmann, 1929).

Ammonites have remained pre-eminent paleontological models for the study of evolutionary patterns and processes, both within the framework of the orthodox Synthetic Theory and of heterochronic interpretations (e.g., Donovan, 1973; Wiedmann and Kullmann, 1988; Olóriz and Rodríguez-Tovar, 1999). Ontogenetic history is more difficult to decipher when skeletons are either molted or remodeled extensively during growth. Nevertheless, even in these cases such difficulties may be partly overcome.

Extinct forms can provide insights into developmental possibilities that are not clear or revealed by living forms, especially in phylogenetic context. For example, according to Morse's Law, tetrapods should reduce digits from alternate sides. Theropod dinosaurs followed this pattern in their pedes (I and V are reduced); in their manus digit V was lost first, followed by IV. Most theropods, including birds, retain digits I-III, but tyrannosaurs lost III, and the single finger of alvarezsaurids is the first (Padian and Chiappe, 1997).

Early Conceptual Advances

The introduction of fossils into the relationships among what we now call ontogeny, phylogeny, and recapitulation began during the late eighteenth century and flourished early in the nineteenth century, attendant upon the availability of precise concepts and data in the relevant fields, e.g., stratigraphic geology, systematics, comparative anatomy, and embryology. This period started with the "Romantic" German *Naturphilosophen,* including Kielmeyer (1765–1844), Oken (1779–1851), Meckel (1781–1833), and Goethe (1749–1832), who emphasized recapitulationist interpretations of development in the context of pseudotransformist or transcendentalist views of nature (Sloan, 1992). According to Gould (1977), many of their insights originated with the renowned English physiologist John Hunter (1728–1793; see Sloan, 1992).

The French "founding fathers" of scientific paleontology, Lamarck (1744–1829) and Cuvier (1769–1832), offer an interesting contrast to the *Naturphilosophen* and to each other. In spite of the influence on the young Cuvier of his mentor Kielmeyer, Cuvier generally refrained from broad hypothetical and theoretical constructs to build a science firmly based on generalized empirical data. Lamarck, on the other hand, remains famous for his synthetic approach: his was probably the first well-articulated general theory of the evolution of life as an historical reality. The two men had different opinions about development, but these do not seem to have been of great influence on their contrasting views about evolution. In general, perhaps oversimplified terms, Cuvier put forth a strongly empirical, discontinuist, and sometimes catastrophist (and probably creationist) view of the history of life. Lamarck, on the other hand, insisted on a theoretical, continuist, explicitly transformist (and materialist) interpretation. This fundamental difference was apparently rooted in the patterns they found in the taxa they worked on: terrestrial tetrapods for Cuvier (1812) and marine invertebrates for Lamarck (1802–1806). Of course, preconceived philosophical or ideological considerations formed the hidden substrate of their open technical disagreement.

In this context it should be recognized that the "laws" of Von Baer (1729–1876), which were based on his meticulous embryological works (Von Baer, 1828), were first set in a fixist, Cuvierian pattern. Simultaneously, the great paleontologist Louis Agassiz (1807–1873) offered a different fixist vision of recapitulation. Based on his famous research on fossil fishes, which formed the foundation of paleoichthyology, Agassiz (1833–1843) proposed a theory of recapitulation that detailed a "threefold parallelism" among (1) the stratigraphic succession of fossils, (2) the data from comparative anatomy and systematics, and (3) the evidence from embryology. Thus, extant organisms, during their development, recapitulate the adult stages of successively less remote ancestors. It is stunning, in retrospect, that Agassiz's threefold parallelism was set in an almost "neo-Cuvierian" fixist intellectual framework, because it later became a favorite argument in teaching the evolutionary canon.

Étienne Geoffroy Saint-Hilaire (1772–1844), who proposed the term "embryology" in 1830, cogently applied the transformist views implicit in his concept of "unity of design" or "unity of type" to the evolution of terrestrial vertebrates from aquatic ancestors (Appel, 1987). He emphasized both ontogenetic and phylogenetic aspects of the problem and perceived how transformations experienced during the course of individual development could explain some unusual features of extant adults and could even suggest evolutionary changes in remote ancestors. He emphasized the need to look among ancient fossil forms to discover the actual ancestors of extant organisms. Along with the embryologist Serres (1786–1868), Geoffroy set a trend of interpreting Von Baer's embryological laws in a more dynamic, nearly transformist context. Serres (1860) explicitly compared ontogeny and evolution: "A natural classification is thus no more than a table of development that expresses the step-by-step march of improvement" (p. 352). Although today we may tend to see many such comments of that era as clear evidence that embryology was being integrated for good into generalized evolutionary thought more or less based on paleontological evidence, the situation in fact remained much more ambiguous (Tort, 1996).

Meanwhile, across the Channel, Richard Owen was forging his own synthesis of paleontology, development, and evolutionary theory (Russell, 1916; Desmond, 1979, 1982; Sloan, 1992; Padian, 1995, 1997). Owen accepted evolution but not naturalistic (i.e., strictly materialistic) transformation of species. He saw evolution as mediated by development and influenced by environmental habitus. He did not see fossils as directly ancestral to living forms, but he did accept that their parts were homologous (in several senses: general, special, and serial). Furthermore, he saw that many extinct animals such as dinosaurs presented forms and developmental possibilities that are not seen in the living world.

In synthesizing the ultimate pre-Darwinian view of homology, Russell (1916) stressed the importance of the criteria of position, histological structure, and development in establishing homology of tissues and organs, notably in his works *On the Archetype and Homologies of the Vertebrate Skeleton* (1846) and *On the Nature of Limbs* (1849). Thus, his sense of the continuity of life was based on the transcendentalist concept of *Bedeutung* (roughly, "deep meaning") of biological structures. For Owen, evolution occurred as environmental habitus influenced development of archetypal forms and produced new forms through time that still retain fundamental typological signatures of their true affinities. Although Owen insisted that the mechanisms of these transitions were primarily naturalistic, he definitely accepted a prime mover behind all such change, though he was opposed to intelligent design (Paley, 1803). His obscure prose and shadowy philosophical principles have continued to defy clear interpretation, but in his own way he regarded development as just as central to morphogenetic change (in the present as well as the past) as Darwin did.

Shortly after the middle of the nineteenth century, the stage was set for a renewal of the relationship between ontogeny and phylogeny, stimulated by Darwin (1859) and Haeckel (1866). However, general schemes of relationships were often conflicting and based on different formulas, depending on one's philosophical and methodological training, as well as the evidence in question (e.g., Bowler, 1983). For Haeckel, ontogeny was a condensed recapitulation of phylogeny. For Garstang (1922), on the contrary, ontogeny did not recapitulate phylogeny: it created it (Mengal, 1993). The recognition of the confluence of development and evolutionary history culminated in the concept of *heterochronies* (de Beer, 1930). For various reasons, these approaches declined markedly during the hegemony of the Synthetic Theory in its most orthodox form (circa 1940–1970). Only after the publication of Gould's seminal book *Ontogeny and Phylogeny* (1977) did an explosive renewal of interest occur, in connection with breakthroughs in developmental biology and genetics. Current evo-devo research integrates evidence from the formerly disjunct but more anciently united fields of paleontology and developmental biology, which are both reacquiring their well-deserved places at the core of evolutionary biology.

The Use of Paleohistology in Illuminating Ancient Ontogenies and Their Evolution

Many fossil structures, including eggs with embryos, provide information about ontogenetic stages of extinct animals. As an example we discuss *paleohistology,* the study of preserved microstructures or tissues of fossils. It uses standard techniques of thin sectioning and observations under ordinary or polarized light. Fossilization often preserves surprisingly precise microscopic details; in some cases the actual mineral composition of the fossil may be completely preserved. Paleohistology is thus useful for several purposes. Many plant and invertebrate skeletal microstructural characters have considerable taxonomic value, allowing determinations even from minute or scrappy remains (Carter, 1990). For fossil vertebrates, similar determinations are often possible from dental tissues (enamel, dentine, and related tissues), but the situation is more complex for bone tissues. Although some indirect taxonomic conclusions may be sometimes extracted from structures of bone tissue, a much more promising field is to use fossil bone as a record of vertebrate ontogeny, including growth rates and growth strategies. This is where paleohistology intersects with evolution and development.

The "primary" (i.e., not remodeled) bone tissues are laid down accretionally and thus provide a record of the skeleton's ontogeny and growth. However, this record is never complete bone because bone is remodeled throughout growth: pre-existing bone is locally eroded, and new secondary bone is later deposited *in situ,* often in several erosion and deposition cycles. This process erases records of earlier bone growth and creates local secondary records of later bone history.

Comparative observations and now experimental data show that many characters of primary bone tissue are functionally linked to its rate of deposition (Amprino, 1947; Castanet et al., 1993). If a given bone tissue type expresses a given range of growth rates in extant animals, we infer the same rates of deposition for similar fossil tissues. As an extended example, we summarize how this generalization has been used to infer that most dinosaurs had a very active growth, much more similar to ostriches and large extant mammals than to any large extant reptiles (Padian et al., 2001).

Primary bone also records interruptions and cycles of growth (Castanet et al., 1996), as do very generally the mineralized skeletons of invertebrates and growth rings of trees. The recorded cycles are annual in many vertebrate bones, although the record can be complex and its yearly significance must be assessed on a case-by-case basis rather than assumed. When we can use this approach to age fossil organisms, we can independently test assessments of radial growth based on Amprino's work (Horner et al., 1999, 2000).

Embryonic material and growth series (Carpenter, 2001) supplement our understanding of the dynamics of endochondral ossification in important ways. It is well known that growth is generally more rapid at younger stages and tapers through life, and this has been determined for extinct organisms as well (Erickson et al., 2001). The record of early tissues that are destroyed or remodeled later in life provides the basis for estimating growth rates. As a result, we can reconstruct growth trajectories and time to maturity of extinct animals (e.g., Horner et al., 2000). In addition, details of epiphyseal growth can reveal the relative and absolute contributions of each epiphyseal growth plate to the growth dynamics of the bone and to the overall longitudinal growth in a given limb (e.g., Horner et al., 2001).

With sufficient phylogenetic and ontogenetic control, it is becoming possible to see how growth strategies change throughout the evolution of an extinct clade, and how they allow animals to exploit new evolutionary opportunities. For example, the ontohistogenetic patterns in fossil archosaurs (archosaurs include birds and crocodiles and all descendants of their most recent common ancestor) reveal quite general (though not unexceptional) patterns of tissue deposition and reflected growth strategies (Padian et al., 2001). On the crocodilian side, we found a conservative pattern of bone deposition. The bones were not extensively remodeled; preserved tree-like growth with distinct, probably annual growth lines; and retained most records of early growth because their marrow cavities were relatively small. On the avian side, which includes other dinosaurs and pterosaurs, we found tissues that reflect much more rapid growth, fewer growth lines, a larger marrow cavity that obscured most early stages of growth, and extensive secondary reworking of especially the deep cortex, beginning in subadult and even juvenile stages in some bones. Growth tapered far more sharply in adult stages than it did in crocodiles and their relatives, which grew more slowly and steadily. The tissues of smaller dinosaur and pterosaur taxa reflected patterns of slower growth, and this allowed us to infer that birds achieved their smaller adult size compared to their theropod relatives by shortening the rapid phase of early growth while retaining the trajectory of shape change, essentially becoming miniaturized adults similar to other theropods (see also Ricqlès et al., 2001).

Altogether, the possibilities offered by paleohistology and related approaches in evolutionary development and paleobiology now allow us to extend some kinds of morphogenetic analyses of developmental biology to a spectrum of fossil taxa. This provides a potentially fascinating research program linking developmental biology and paleontology, set in a common phylogenetic context (Ricqlès, 1993). In other words, we have the potential to use ontogenetic data from extinct organisms in phylogenetic context, which opens further possibilities of understanding how evolution works through heterochronic changes in development.

DAVID STERN

Gene Regulation

Gene regulation is the control of the concentration and spatiotemporal distribution of a gene product at the transcriptional and post-transcriptional levels. Control of transcription is by far the most common mode of gene regulation, and alterations in transcriptional regulation appear to predominate developmental evolution. There are theoretical reasons for thinking that transcriptional control should be most amenable to evolutionary change. However, other modes of regulation are less well studied, and examples of evolutionary change in these mechanisms do exist. I describe the molecular nature of each of these mechanisms of gene regulation and, when available, provide examples of how they contribute to developmental evolution.

Regulation of Transcription

Gene transcription is performed by RNA polymerases, which bind to a region of DNA, called a *promoter,* located immediately upstream of the transcription start site (for general reviews of transcriptional regulation, see Latchman, 1995, 1998; White, 2001; Davidson, 2001). Three RNA polymerases in eukaryotes transcribe different classes of genes. RNA polymerase I transcribes most of the rRNA, RNA polymerase III transcribes various small RNA molecules, while most genes are transcribed by RNA polymerase II; most gene regulation therefore involves differential activation of RNA polymerase II, which is part of a large complex of at least forty proteins that are recruited to the promoter. This complex is sometimes referred to as the *basal transcription apparatus.*

The basal transcription apparatus itself exhibits low levels of endogenous activity. Many of the proteins found in this complex are thought to regulate transcription by their interactions with other proteins. Hence, high levels of transcription normally require the presence of transcription factors bound to adjacent regions of DNA variously called upstream promoter elements, enhancers, or *cis*-regulatory regions. The more generic term *cis*-regulatory regions is used here because these regions often have repressive as well as activating effects on transcription and because these regions may be located upstream, downstream, or within the transcription unit.

Cis-regulatory regions are stretches of DNA, usually several hundred bases long, that contain binding sites for transcription factors. Individual transcription factors bind to short specific stretches of DNA, on the order of 6 to 15bp. Multiple binding sites for each of several transcription factors are often found within a single regulatory region. These transcription factors may be enhancers or repressors of transcription, and a single region normally contains a combination of both. This combination of binding sites forms a module driving transcription in a spatially and temporally distinct pattern during development. The output from a single module (up- or down-regulation of transcription) is a function of the combination of transcription factors expressed in a cell at a particular time. For example, if transcription factors that promote transcription and recognize binding sites within a module are present, and if repressors are absent, then transcription of this gene is promoted in the cell. Although in most cases *cis*-regulatory regions are located within several hundred or thousand bases of the promoter, there are many examples of *cis*-regulatory regions located tens of thousands of bases from the promoter. Transcription factors bound to regions distant from the promoter are thought to influence transcription by looping the DNA to allow contact between transcription factors and the basal transcription apparatus.

A single gene involved in development may contain multiple *cis*-regulatory modules directing transcription. These modules often act independently of each other. The complex patterns of expression that are often observed for genes involved in patterning embryos are the result of output from multiple *cis*-regulatory modules.

How is all of the information that is encoded as the binding of transcription factors to regulatory regions integrated to generate a particular level of gene expression in a single cell? The best-studied example of the operation of a *cis*-regulatory region is the *Endo16* gene of the sea urchin *Strongylocentrotus purpuratus* (Yuh et al., 1998). Several independent modules integrate inputs (the presence or absence of transcription factors) and output a quantitative response to a module located adjacent to the promoter. Remarkably, the operation of this regulatory region can be modeled as if it were a microprocessor integrating Boolean operations to generate a quantitative output (Yuh et al., 1998).

Transcriptional regulation is therefore the product of the binding of a combinatorial code of transcription factors present in a cell at a particular time to specific *cis*-regulatory modules. The required level of transcription is essentially computed by these modules and presented as an output to the basal transcription apparatus.

Evolution of transcriptional regulation. There is now abundant evidence that evolution of transcriptional regulation has played a major role in the evolution of development (Carroll et al., 2001; Davidson, 2001). Although most of the evidence comes from comparisons of gene expression patterns between distantly related taxa, evidence is slowly accumulating for a significant role of changes in transcriptional regulation at the microevolutionary level (Kopp et al., 2000; Stern, 1998, 2000; Sucena and Stern, 2000).

It is not obvious, *a priori,* why metazoan genomes would have evolved with a modular regulatory structure. Organisms with modular regulatory circuitry may be more evolvable (Gerhart and Kirschner, 1997). For example, the extreme modularity of *cis*-regulatory regions has several consequences for evolution. Perhaps most importantly, modular *cis*-regulatory structure allows a single gene to be recruited for multiple developmental functions without generating a conflict with its original role(s). Thus mutations in a single module can alter this specific gene function with little or no pleiotropic consequences (Stern, 2000; Sucena and Stern, 2000).

A second evolutionary advantage is that regulatory modularity allows transcription factors to be easily co-opted for patterning different stages of the same tissue (Davidson, 2001). For example, the *Pax6* gene is required for specification of eyes in divergent metazoans, even in taxa that are believed to have evolved complex eyes independently (Gehring and Ikeo, 1999). However, *Pax6* is also required to specify various elements of the complex eye, including the photoreceptors. One attractive model to explain the predominant role of *Pax6* in specifying eyes across the metazoans involves the progressive co-option by *Pax6* of additional regulatory roles associated with eye development (Davidson, 2001). The use of *Pax6* for co-option would have been favored simply because this transcription factor is already expressed in the right place and time, originally simply to specify the production of photoreceptors in the head. In general, this model allows the evolution of increasing complexity without the invention of new transcription factors.

An evolutionary consequence of the redundancy of transcription factor–binding sites within *cis*-regulatory modules is that this structure allows the accumulation of compensatory mutations and turnover of binding sites within a module. That is, selection probably acts mainly at the level of the entire module, and there is probably weak selection on at least some individual binding sites (Ludwig et al., 2000). The long-term consequences of such binding site turnover is that modules may retain their function without retaining any sequence similarity (Takahashi et al., 1999b). The ability of regulatory regions to gain and lose individual binding sites is probably a function of the redundancy and cooperativity of binding sites within a module. This fact complicates the comparative study of *cis*-regulatory regions. Although stretches of conserved noncoding DNA are often found adjacent to genes, not all binding sites—nor perhaps some entire modules—will be identified by comparative approaches.

Epigenetic control of transcription. Transcription can also be influenced by methylation of the cytosine residues of DNA. Most such methylated C residues reside adjacent to a G residue on their 3′ sides. Methylation can be propagated throughout mitosis because the CG sequence is palindromic and there is a specific enzyme that recognizes double-stranded DNA carrying a methylated C from a CG couplet on one strand and methylates the C residue on the opposite strand. Methylation typically results in a reduction in gene expression and, because methylation can be tissue specific, this is an additional mode of tissue-specific gene regulation (Latchman, 1995). Methylation is also an important mechanism of transcriptional control of imprinted genes (Ohlsson et al., 1995).

Cubas et al. (1999) demonstrated that a natural polymorphism for floral structure (bilateral versus radial) is caused by variation in methylation of a gene controlling floral dorsal-ventral asymmetry. Methylation causes elimination of gene expression in the floral meristem and the methylation status of the gene is inherited in a Mendelian fashion. Thus, epigenetic regulation of transcription controls a dramatic floral polymorphism, and it will be well worth examining the generality of this finding.

Post-Transcriptional Regulation

After, and in some cases while, the mRNA transcript is generated, the mRNA is subject to modifications or other controls that influence the type, distribution, or amount of protein produced. The major phenomena are briefly reviewed here, together with evolutionary examples where relevant.

Alternative splicing. While being transcribed, mRNA is modified in several ways. First, the 5′ end is modified by the addition of a 7-methylguanosine cap. In addition, the introns are spliced out and in many cases some exons are excluded or included in a process called alternative splicing. Alternative splicing is regulated by splicing factors that can be cell type specific and recognize *cis*-regulatory information. Alternative splicing can generate an enormous diversity of final mRNAs from a more limited number of genes (Graveley, 2001), and therefore may provide extensive opportunities for evolutionary modification. Alternative splicing plays a key role in several developmental processes, notably sex determination in *Drosophila*. In addition, the molecular pathways of sex determination have evolved within flies (Schütt and Nöthiger, 2000). However, there is not yet clear evidence of how the evolution of alternative splicing has contributed to developmental variation.

In the last stages of mRNA processing, the 3′ end of the mRNA is polyadenylated. The length of the poly-A tail can have a dramatic effect on translation efficiency and polyadenylation of specific mRNAs can be controlled in a cell type–specific manner. Therefore, polyadenylation is another potentially important source of variation in gene regulation.

RNA editing. RNA editing involves changing the sequence of a RNA molecule from that encoded in the DNA after transcription. This editing can transform the amino acid sequence encoded by a mRNA. RNA editing has been found in a wide variety of organisms, from protozoa and slime molds to mammals, plants, and *Drosophila* (Bass, 2001). Notably, in mammals and *Drosophila,* RNA editing is highly developmentally regulated and appears to be important for the generation of receptor protein diversity in the nervous system (Keegan et al., 2000; Palladino et al., 2000a, b; Reenan, 2001).

RNA transport. After the mature mRNA is generated, it must be exported to the cytoplasm for translation. This is an active process involving specific proteins in the nuclear envelope and is susceptible to regulation (Latchman, 1995).

RNA stability. An important mechanism of post-transcriptional gene regulation is the rate at which specific mRNAs are degraded. Obviously, mRNAs that are quickly degraded will be available to produce less protein. However, rapid mRNA turnover is an important mechanism when the production of protein product must be quickly extinguished. There are many examples of cell type–specific regulation mediated by changes in RNA stability (Latchman, 1995).

Translational gene regulation. The rate of translation is also subject to tight regulation. Regulation can be either global, as represented by the global repression and activation of translation in reticulocytes and sea urchin eggs, or specific, in which the translation of only certain mRNA molecules is regulated in a cell type–specific manner (Mathews et al., 2000). Although global control is clearly important in certain contexts, only specific regulation is discussed here because this regulation is dependent on information encoded within the mRNA molecule itself and is therefore most likely to be accessible to specific evolutionary changes. Although control elements have been found in both 5′ and 3′ untranslated regions (UTRs), the 3′ UTR appears to be the most common location of control elements.

Specific subcellular RNA localization and translational control are common modes of gene regulation during the early stages of embryogenesis of a wide variety of animals. This is probably because when maternal mRNAs are deposited within the egg, the spatial and temporal distribution of gene function must be controlled post-transcriptionally. Therefore, many important molecular events prior to the onset of zygotic transcription are controlled at the translational level (Wickens et al., 2000).

Because the molecular pathways involved in early patterning have clearly evolved, these translational control mechanisms have probably also evolved. We do not currently have any examples of the evolution of translation control in early embryos. However, one example of the evolution of translation control later in development is discussed in summary.

The diversity of appendages along the body axis of crustaceans has proven to be a useful model to study evolutionary changes in Hox gene regulation. For example, Averof and Patel (1997) demonstrated that the segmental boundary between regular legs and the more anterior maxillipeds has shifted between different crustaceans; changes in the expression pattern of *Ubx/abd-A* mirrors these shifts. Abzhanov and Kaufman (1999) showed that this morphological transformation is recapitulated in the first thoracic segment during the ontogeny of the isopod *Porcellio scaber.* Remarkably, the Hox gene *Scr* is transcribed in this limb throughout ontogeny, but Scr protein is only observed temporally coincident with the morphological transformation. *Scr* is probably playing an instructive role in this morphological transformation. Therefore, post-transcriptional regulation of *Scr* is crucial to defining the specific limb morphology expressed throughout ontogeny.

Variation in development is thought to be caused largely by variation in the regulation of gene expression. This in turn has focused attention on the considerable variation in the expression patterns of transcription factors and other genes (such as cell-cell signaling molecules) involved in pattern formation. This variation is most likely to be due to changes in *cis*-regulatory regions that influence transcription, although our understanding of how these regions evolve to alter development is rudimentary. In addition to the important role of *cis*-regulation of transcription, post-transcriptional mechanisms also influence gene regulation, although they tend to be less well studied and we have fewer examples of their evolutionary import.

ELIZABETH L. JOCKUSCH

Genome Size

Genome size, also known as C-value because it is generally constant within a species, is the haploid DNA content of a cell. Genome size varies greatly across organisms, ranging from 5.9×10^{-4} to 0.013 pg in Eubacteria, 1.6×10^{-3} to 4.2×10^{-3} pg in Archaea, and 0.009 to 700 pg in Eukaryota (Grauer and Li, 2000). The variation in eukaryotes has attracted attention because of the so-called C-value paradox, the observation that the complexity of organisms is not correlated with their DNA content. Genome sizes vary greatly in most groups of eukaryotes, including angiosperms (range 0.05 to 127.4 pg; Bennett and Leitch, 2001). The variation in protists, a nonmonophyletic group that includes *Amoeba dubia,* which has the largest known eukaryote genome, is particularly striking.

In contrast, genome sizes are relatively low and constant in most vertebrate clades. Notable exceptions include amphibians (0.95–120.6 pg), with the largest genomes found in salamanders and lungfishes (40.1–123.9 pg) (Gregory, 2001a). Model organisms used in developmental studies typically have small genome sizes. The yeast *Saccharomyces cerevisiae* has one of the smallest known eukaryotic genomes (0.009 pg). The genome sizes of other model organisms are as follows: 0.09 pg in the nematode *Caenorhabditis elegans,* 0.18 pg (the eighth smallest known angiosperm genome size) in the thale cress *Arabidopsis thaliana,* 0.18 pg in the fruit fly *Drosophila melanogaster,* 1.8 pg in the zebrafish *Danio rerio,* and 3.25 pg in the lab mouse *Mus musculus* (Bennett and Leitch, 2001; Gregory, 2001a).

In Eubacteria and Archaea, the vast majority of DNA encodes proteins or RNA genes. Nongenic DNA, which usually comprises 5% or less of the genome, includes gene regulatory regions, spacers, and a limited number of repeated sequences. In these groups, genome size is directly correlated with the number of genes (Grauer and Li, 2000). In contrast, in many eukaryotes, most DNA is nongenic. Repetitive sequence elements ranging from a single base pair to thousands of base pairs in length usually comprise the bulk of the nongenic DNA, while regulatory regions, pseudogenes, and introns typically contribute a smaller fraction (Grauer and Li, 2000).

Many hypotheses have been proposed to explain the C-value paradox. They can be classified into two major groups. One group proposes that most nongenic DNA is parasitic or junk DNA and thus either deleterious or neutral to organismal fitness (Ohno, 1970; Doolittle and Sapienza, 1980; Orgel and Crick, 1980). The other proposes that DNA has a "nucleotypic" function along with its genotypic one (e.g., Bennett, 1972; Cavalier-Smith, 1985). A nucleotypic function is one that is carried out by the bulk quantity of DNA rather than by the encoded information. In this view, variation in genome size results from natural selection on the nucleotypic function. More recently, variation in genome size has been viewed in a classic mutation-selection balance framework. These views make it clear that consideration of the mechanisms that produce variation in genome size as well as of the fitness consequences of differences in genome size is needed to resolve the C-value paradox (Petrov, 2001). Interest in the C-value paradox has led to exploration of phenotypic and molecular correlates of genome size and, more recently, to a search for the molecular mechanisms generating the variation.

Genome size correlates with a wide variety of phenotypic characters, many of which are directly relevant to organismal development. At the cellular level, genome size is tightly correlated with cell and nuclear size and with cell division rate (Gregory, 2001b). These traits clearly can have important consequences for organismal fitness—for example, by influencing the size of unicellular organisms (which may in part explain the extreme range of C-values in protists), number of cells in structures in multicellular organisms, and lengths of various stages in the life cycle. Life history parameters that have been shown to correlate with genome size include embryonic development time in salamanders and crustaceans (Jockusch, 1997; White and McLaren, 2000), time to metamorphosis in frogs (Goin et al., 1968), longevity in birds (Monaghan and Metcalfe, 2000), and minimum generation time in plants (Bennett, 1972).

It is also notable that among salamanders those with the largest genome sizes are either paedomorphic or direct developing and thus lack metamorphosis, suggesting that large genomes may place a constraint on metamorphosis, or that metamorphosis may limit genome size. Model organisms for developmental studies are selected in part because of their rapid development and short generation times, which may explain their low genome sizes. Other phenotypic features correlated with genome size, at least in some groups, are metabolic rate (shown in several vertebrate groups; e.g., Vinogradov, 1995, 1997), body size (e.g., in some copepods; White and McClaren, 2000), and complexity of brain organization in amphibians (Roth et al., 1994).

The significance of these correlations is not resolved. While causal relationships between genome size and some of these features are plausible (e.g., Gregory, 2001b), causation has generally not been demonstrated. For example, in the case of the tight correlation between cell and genome size, this relationship applies only when a single cell type is compared. Within an individual, different types of cells can be of very different sizes but usually contain equal amounts of DNA, indicating that other factors play a large role in determining cell size. Caution is also warranted because many of the correlations across taxa have only been tested using nonphylogenetic methods and thus may reflect inheritance of similar trait values from a common ancestor rather than evolutionary correlations.

More recently, molecular correlates of genome size have also been identified. Progress in this direction should accelerate with the tremendous amount of data being generated in genome sequencing projects. Genome size is correlated with repetitiveness of simple sequences across model organisms (Hancock, 1996) and with intron size across eukaryotes (Vinogradov, 1999). Molecular processes that generate variation in genome size on which selection and drift act have also been compared in species with large and small genomes. A high rate of deletions relative to insertions is found in nongenic DNA in *Drosophila* (Petrov and Hartl, 1997), and higher rates of deletions or larger deletions in nongenic DNA occur in *Drosophila* than in species with larger genomes (Petrov and Hartl, 2000; Petrov et al., 2000; Bensasson et al., 2001).

Although a causal link to variation in genome size is clear in some cases, the interpretation of these data is not straightforward. First, these comparisons are generally restricted to a few taxa and, as with phenotypic correlates, analyses do not always take phylogeny into account. Second, the variation in these molecular processes might have been shaped by selection on a nucleotypic function, and thus it may reflect an effect of selection on genome size rather than an evolutionary cause. Ultimately, explaining the distribution of genome sizes in eukaryotes requires integrating data on both the factors generating variation and on the selective consequences.

MARY LOU KING

Germ Cells and Germ Plasm

In animals, sexual reproduction is accomplished through the fusion of two haploid gametes, the egg and sperm. As the agents of heredity, germ cells play essential roles in maintaining diversity and adaptability as well as genetic continuity in organisms. Germ cells are distinguished from somatic cells by their ability to give rise to all cell types, including themselves. A key question in developmental biology concerns how germ cells retain their developmental totipotency while their somatic cell neighbors do not. Given the competence of somatic cell nuclei studies to support complete development of all cell types after transplantation into enucleated ova, this mechanism must reside within the egg cytoplasm.

The processes by which gametes are formed are very similar across the animal kingdom. Future germ cells arise outside the somatic gonad early in development as a small population of primordial germ cells (PGCs). All PGCs, except in *C. elegans*, migrate to the future gonadal tissue (genital ridges in vertebrates) before the gonads have formed, where they invade the somatic cells and begin the proliferation and differentiation steps that will produce gametes. PGCs are sexually determined after they enter the gonad.

Although extragonadal PGC formation and migration are similar in many organisms, the specifics of how, when, and where PGCs are formed differ greatly. PGCs can form either as a result of interactions between tissues through inductive mechanisms, or in a cell-autonomous fashion by inheritance of cytoplasmic determinants (reviewed in Nieuwkoop and Sutasurya, 1979, 1981). PGC specification by induction necessarily occurs later in development and is common in, but not exclusive to, organisms that have embryos with extraembryonic tissues. For example, mammals are in this category, but so are urodele amphibians.

Autonomous specification, which can occur early (often during oogenesis), is characterized by the asymmetric inheritance of maternal germ cell determinants, localized in a cytologically distinct cytoplasm, the *germ plasm,* which is distinguished by dense clusters of mitochondria and unique electron-dense organelles called *germinal granules.* These granules, which are composed of specific RNAs and proteins, are important to germ cell formation (reviewed in Saffman and Lasko, 1999). This material, inherited from one generation to the next, underlies the continuity of the germ line. Many invertebrate and vertebrate species use this mode of germ cell formation. The ultrastructural similarity of germ plasm among species suggests a common mechanism in specifying the germ cell lineage for invertebrates and vertebrates. Although it is not clear what embryonic constraints determine the mode that an organism uses to specify PGCs, some hypotheses claim that PGCs form in places that are the least influenced by adult body plan specification (Dixon, 1994; Johnson et al., 2001).

Autonomous Specification of PGCs

Drosophila. The mitochondria-rich germ plasm or pole plasm forms at the posterior pole of developing oocytes. Germinal (polar) granules are first evident in the pole plasm during vitellogenesis. The posterior pole plasm undergoes cytokinesis first in the embryo, and these pole cells become PGCs (reviewed in Williamson and Lehmann, 1996). Germ band extension moves the pole cells dorsoanteriorly and deposits them inside the embryo. Subsequently, the polar granules fragment and associate with the nuclear envelope, and pole cells migrate through the midgut to the gonadal mesoderm, which surrounds the pole cells forming the definitive gonads. Germ plasm or germ plasm–like material *(nuage)* has been observed throughout the life cycle of pole cells, establishing the continuity of the germ plasm.

A number of maternal-effect mutations affecting pole plasm formation have been identified in *Drosophila* (reviewed in Rongo et al., 1997). Mutant embryos lack pole cells and polar granules and have defects in abdomen formation. Many of these genes are involved in the localization of *oskar* RNA and protein to the posterior pole. Oskar appears to control the amount of pole plasm, but genes related to *oskar* have not been found in the germ plasm of other species. In contrast, *vasa*-like and *nanos*-like genes have been found in virtually every germ line examined, including those of mice (Saffman and Lasko, 1999). In *Drosophila,* Vasa is required for pole plasm assembly as well as the recruitment of the genes required for pole cell development, such as *nanos (nos), germ cell-less (gcl),* the 16S mitochondrial large rRNA (*mtlr*RNA), and a non-translated RNA, *polar granule component.* Vasa, an initiation factor–like RNA helicase, is involved directly in regulating translation within the pole plasm. Nanos appears to have two activities, one as a repressor of translation (*hunchback* and *cyclin B* RNA) and one regulating transcription. Pole cells mutant for maternal *nanos* do not display proper migration or gene expression patterns (reviewed in Rongo et al., 1997; Saffman and Lasko, 1999; Houston and King, 2000).

One mechanism used by *Drosophila* and *C. elegans* germ lines to protect PGCs from somatic differentiation involves the delay of new transcriptional activity in PGCs until after somatic cells are committed to their fates (Saffman and Lasko, 1999; Seydoux and Strome, 1999). At present, it is not known whether transcriptional repression occurs in the *Xenopus* germ line, and there is evidence against it operating in zebrafish PGCs (Knaut et al., 2000). It remains to be seen whether this mechanism has been generally used in other organisms with germ plasm.

Caenorhabditis elegans. The complete embryonic cell lineage of *C. elegans* has been determined and the precursors to the germ line identified (Figure 1). Typical germinal (P) granules have been traced throughout the germ cell life cycle by immunofluorescent staining of a granule component (Seydoux and Strome, 1999). Initially P granules are found throughout the egg cytoplasm but then become asymmetrically distributed exclusively into the P4 blastomere that yields the primordial germ cells Z2 and Z3. All the progeny of the P4 blastomere give rise to germ cells. The P granules associate with the nuclear envelope from P4 throughout germ cell proliferation and into the meiotic stages. Unlike other organisms, the PGCs in *C. elegans* develop next to the gonads; they do not migrate. Z2 and Z3 appear to have contact with endoderm, as do the PGCs in *Drosophila* and *Xenopus.* The nature of this interaction, whether simply permissive or instructive, is not known.

A number of genes that affect *C. elegans*' early germ cell development or P granule segregation have been isolated (reviewed in Seydoux and Strome, 1999). As in *Drosophila,* many of these genes potentially encode RNA-binding proteins and appear to have roles in somatic and germ cell

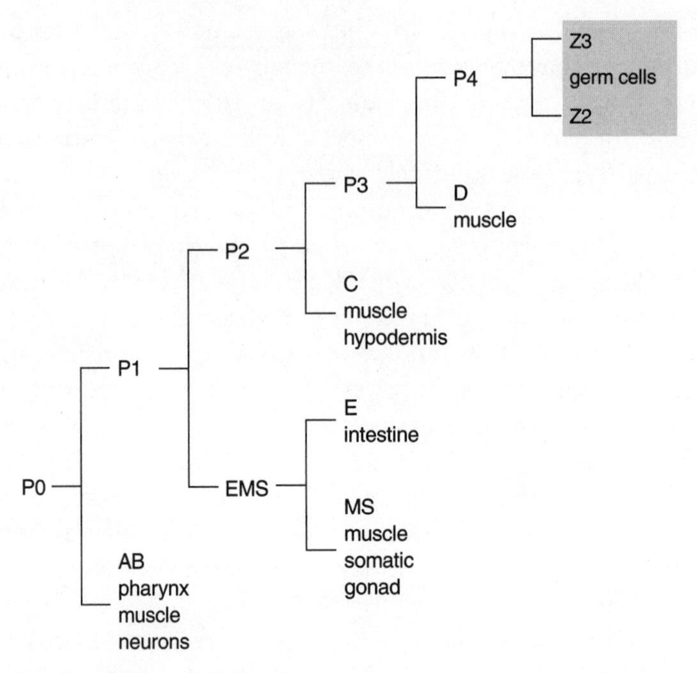

FIGURE 1. Early embryonic cell lineage of *C. elegans* showing clonal derivation of primordial germ cells.

development. They fall into the zinc finger (PIE-1, MEX-1, POS-1, *nos-2*), DEAD-box helicase (GLH-1 and GLH-2), and RGG-box RNA-binding *(pgl-1)* protein families, suggesting that they are involved in regulating RNA metabolism or translation. In *C. elegans,* germ line–specific transcriptional repression is mediated by the zinc finger protein PIE-1. A more permanent repression may be mediated by the *mes* (maternal effect sterile) gene products that encode homologues of the *Drosophila* Polycomb group, well known for their role in the long-term repression of homeotic genes during development. Homologues for PIE-1, MEX-1, POS-1, and PGL-1 have not been identified in other animals.

Additional genes related to *Drosophila* germ plasm components have also been found in *C. elegans.* The *germ line helicase* genes (*glh-1* and *glh-2*) were identified by homology to *Drosophila vasa,* which is known to regulate translation. Mutant analysis of *glh-1* is consistent with it playing a similar role to *vasa* in the localization and translation of germinal granule components. Recently, several homologues of *Drosophila nanos* have also been identified in *C. elegans.* Like *Drosophila nos, nos-2*

RNA is localized to the P granules, is required for incorporation of PGCs into the somatic gonad, and requires the activity of *pumilio* homologues (Subramaniam and Seydoux, 1999), suggesting that the mechanisms for controlling gonad entry are conserved between *Drosophila* and *C. elegans*. The Nos/Pum partnership in translational repression may very well be conserved in general, and it is important to see whether it is operating in the *Xenopus* germ line.

Xenopus and Danio. Germ plasm has been found in a variety of frogs, but not in salamanders (Nieuwkoop and Sutasurya, 1979). In *Xenopus*, material required for the germ cell lineage is present in the developing oocyte, where the germ plasm is concentrated within the mitochondrial cloud or Balbiani body (Wylie, 2000). Germ plasm is segregated unequally during cleavage stages into only four cells—the presumptive PGCs—which become distributed within the endodermal mass. Subsequent divisions result in the germ plasm being equally distributed between daughter cells. A stable population of 20–30 PGCs is eventually established. By the late tail bud stage, the PGCs begin to migrate dorsally through the lateral endoderm and later become incorporated into the dorsal mesentery. In later stages, PGC migration continues to the dorsal body wall and then laterally to the forming genital ridges, where the PGCs repeatedly divide, enter meiosis and, soon after, sexually differentiate. These processes are very similar to those that occur in zebrafish *(Danio)*, while the later migration steps are similar in mice and *Drosophila* (Wylie, 2000).

Ultrastructural analysis of germ plasm in frogs revealed a similar structure and life history as in *Drosophila* (Kloc et al., 2002). Whether the germinal granule material is present continuously in the germ line of *Xenopus* is uncertain. However, germinal granule material does appear in germ cells of all types: in oogonia as perinuclear nuage; in previtellogenic oocytes as mitochondrial cloud GFM (granulofibrillar material); and in mature oocytes, eggs, and embryos as germinal granules (Houston and King, 2000; Kloc et al., 2002).

Strikingly, three *Xenopus* genes (*Xcat2, DEADSouth,* and *Xdazl*), selected in a screen for maternally localized RNAs, are related to germ cell components in *Drosophila* (Houston and King, 2000). *Xdazl* encodes an RNA-binding protein homologous to the human Deleted in Azoospermia (DAZ/DAZL) family of meiosis and gametogenesis regulators. *Xdazl* was recently shown to be required for early PGC differentiation and capable of rescuing the meiotic entry phenotype in testes of *boule*-

mutant fruit flies. The latter result is a direct demonstration of the conservation of *dazl* function in the germ line of invertebrates and vertebrates. *Xcat2* RNA, a component of the germinal granules (Kloc et al., 2002), encodes a protein with Nanos-like zinc finger homology (Houston and King, 2000). Given the conservation of function among *nos* homologues in *Drosophila* and *C. elegans,* it will be interesting to determine whether *Xcat2* has a similar role in *Xenopus*. *DEADSouth* encodes a DEAD-box helicase closely related to eIF4A and similar to *vasa*. It may function in PGC-specific translation initiation as does *Drosophila vasa*. *Xenopus* germ plasm also contains *mtlr*RNA localized to the germinal granules as in *Drosophila,* fueling speculation that it plays a role in translation in the germ line (Saffman and Lasko, 1999). *Xlsirts* encodes an untranslated RNA (Kloc et al., 2001). With the exception of *Xdazl,* no functions are known for any other germ plasm RNAs.

Most recently, genetic and molecular studies in zebrafish have identified *vasa*- and *nanos*-related genes within the germ plasm of PGCs (Knaut et al., 2000; Koprunner et al., 2001). *nanos-1* was found to have an essential role in the migration and survival of PGCs—a role strikingly similar to the *nanos* function in *Drosophila* and *C. elegans*. Key steps in the life cycle of PGCs in both vertebrates and invertebrates appear to use genes in the same families.

Inductive Formation of Germ Cells

All the cells of the early mammalian embryo are capable of entering the germ line prior to gastrulation (reviewed in Pesce et al., 1998; Wylie, 2000). During gastrulation, mouse PGCs form from proximal epiblast cells that are adjacent to the extraembryonic ectoderm. These cells mostly give rise to extraembryonic mesoderm in addition to PGCs. Location appears to be critical here as PGC precursors transplanted to a different location in the epiblast develop as somatic cells, while distal epiblast cells transplanted proximally can become PGCs (Tam and Zhou, 1996). These results suggest that PGC precursors are induced by extracellular factors or cell interactions present locally. BMP4, a secreted product of the extraembryonic ectoderm, has been identified as a key player in this process. In mice, inducers act on a population of pluripotent cells to specify PGC fate. Oct-4, a transcription factor whose expression becomes restricted to the germ line, may function to keep the PGCs totipotent (Pesce et al., 1998). Interestingly, the mouse and axolotl germ lines share related genes (Johnson et al., 2001). PGCs in this amphibian arise in the posterior-lateral plate mesoderm. They are thought to be induced, along with other mesodermal derivatives, by unknown secreted factors from vegetal cells (Sutasurja and Nieuwkoop, 1974).

Comparative analyses of the germ line in invertebrates and vertebrates reveal several interesting common features. Perinuclear nuage has been observed in germ cells of animals from at least eight phyla (Eddy, 1975). Germ cells are specified early in embryogenesis, by or during gastrulation. Almost all PGCs undergo a programmed active migration to the developing somatic gonad, where they achieve sexual identity and proliferate. Germ plasm invariably contains a variety of RNA-binding proteins and noncoding RNAs, but not transcription factors. Germ plasm also invariably associates with the nuclear envelope, suggesting a role in PGC-specific regulation of nuclear transport. Regulation of translation appears to be an important general mechanism in PGC formation, as may be transcriptional repression. Translational repression in the germ plasm may serve to limit expression of specific proteins until the PGC lineage is established. It remains to be seen to what extent conserved germ line genes such as *Oct-4, nanos, vasa, pumilio,* and *dazl* represent conserved mechanisms for specifying the germ line.

JOAN T. RICHTSMEIER

Growth

Bertalanffy (1960, p. 252) was not overstating the importance of growth to the biological sciences when he said: "It is hardly an exaggeration to state that the problem of growth is a focusing point where the most fundamental problems of biology, from the biophysical bases of the phenomena of life up to the questions of human behavior converge and intersect." Growth is essential for the production of adult forms and for the evolution of morphology. Because growth produces anatomical structure, growth can establish limits and provide opportunities for the development of function, thereby causing changes in behavior. Differences in growth patterns produce phenotypic variation, providing a source for the evolution of phenotypes. Growth is defined here as *change in the arrangement of component parts with increase in spatial dimensions occurring over time.*

Growth is rarely mentioned except in juxtaposition with "and development." Development includes pattern formation, cell differentiation, and change in form (Wolpert, 1992). Growth is an aspect of development, but development can occur without growth. For example, early embryonic sequences are defined by changes in organization due to cell differentiation and pattern formation, while the size of the embryo remains essentially stable.

The importance of growth for the evolution of phenotypes is based on the unique and intricate nature of the growth process. Growth is hierarchical, highly integrated, and sequential—characteristics that contribute to the evolution of form through individual diversity in growth patterns.

Growth is hierarchical because each stage and process is dependent on every other stage and process. Cells contribute to growth by expanding in size, dividing, or forming condensations that, in turn, form tissues. These tissues coalesce to form organs and systems that combine into organisms. Acceleration, deceleration, and/or maintenance of localized growth rates enable parts at each level to grow simultaneously following similar or dissimilar rates, or experience quiescent periods while other parts grow. There are many levels to this hierarchy and many controlling factors that affect growth either at one level or at several levels simultaneously. For example, limits on cell size or on the number of cells within a condensation, changes in the rate of cell division, and the nature of cell migration may cause changes in growth patterns, and ultimately changes in morphological structure.

Growth is highly integrated because it is the product of cascades of interactions between genes and controlling factors acting directly or indirectly. Growth of any given structure is driven by a large number of separate loci. Many genes and gene products are shared among developmental processes that operate at different levels of the hierarchy and that occur at different times during growth. In a coordinated progression, the genetic instructions and controlling factors make the required morphological adjustments to allow change in the dimensions of a functioning organism.

Growth is sequential (Atchley, 1990) and centered on temporal parameters such as rate, duration, onset, and offset (Raff, 1996). Modifications of the ontogenetic trajectory (in the sense of Alberch et al., 1979) can result from interruptions, changes, or rearrangements in the sequence of growth events. Evolutionary change in the timing of developmental events is called *heterochrony.* A quantitative analysis of growth patterns may contribute to an understanding of the morphological effects of heterochronic change as well as of other changes in growth that contribute to the evolution of morphological diversity.

Size-Related Issues

Growth plays an important role in evolution by generating diversity during ontogeny. After embryonic patterning is complete, slight deviations in growth patterns can produce notable differences in phenotypes. These differences in growth can affect the entire organism or be limited to specific cell populations, tissues, anatomical locations, or functional matrices (in the sense of Moss and Salentijn, 1969). Evidence of differences in growth suggests possible relationships between observed growth patterns and mechanisms of morphogenesis.

When studying growth quantitatively at the gross morphological level, scientists need comparable data from varying stages during ontogeny. For example, among primates, a chimpanzee of two chronological years of age is not developmentally comparable to a gibbon of two chronological years.

Two categories of data are used to study growth. The first includes longitudinal data for individuals that have been measured repeatedly over time. The second category comprises cross-sectional data with a single measure across age groups for each individual, so that only one measure exists for each individual. Not surprisingly, analyses of longitudinal and cross-sectional data do not always give the same results, even if the data intervals are comparable. Longitudinal data record the continual lags and spurts of individual growth, whereas cross-sectional data can mask the irregular trajectories of individual growth because measures are averaged over many individuals.

While the same scientific questions can be addressed using either type of data, each data category can uncover particular patterns in the data. Longitudinal data can separate cohort and age effects (Diggle et al., 1994), for example, while cross-sectional data cannot. Moreover, longitudinal data require particular statistical models and methods that recognize the relationships within the set of measures taken on one subject. Because longitudinal data from one subject are intercorrelated, a measure obtained at time ($t+1$) is dependent on the corresponding measure obtained from the same individual at time t. The research question posed in a comparative growth analysis dictates which types of data are more appropriate.

Growth is often considered at the gross anatomical level. Growth curve analysis has been used to define differences in growth among species, ethnic groups, and sexes, and to compare and contrast the timing of specific developmental events between groups (Zeger and Harlow, 1987; Diggle et al., 1994; West et al., 2001). Standard growth curves provide models of change in a given dimension over time; velocity curves represent the average intensity and tempo of growth; crown-rump length or standing height can be used to study the growth of specific anatomical features. Growth curve analysis allows comparison of the curves themselves, as well as comparison of the parameters (e.g., rate of growth, asymptote of growth) estimated from the data (West et al., 2001). Because this method can pinpoint changes in the relative timing of events, it has been used most profitably in studies of heterochrony (for reviews, see McNamara, 1995; Klingenberg, 1998; Hall, 1999b; McKinney, 1999; Gould, 2000; Smith, 2001).

Change in form due to growth is often thought of as change in size and in shape. The quantitative study of size and shape change in populations is most closely linked with allometric studies. *Allometry* is size-dependent shape change (Gould, 1977). Studies of growth allometry as hypotheses of underlying developmental programs focus on the nature of the association between changes in shape and in size that occur over ontogenetic time (Chernoff and Magwene, 1999). In general, metric data are collected and size and shape are defined through the application of various quantitative approaches that fit a given model (usually bivariate) to the data (Olson and Miller, 1958; Gould, 1977; Alberch et al., 1979; Shea, 1983; Chernoff and Magwene, 1999). Growth allometries are often used to identify evolutionary changes in the rate and timing of development *(heterochrony)* and to study the effects of these changes on morphology (Shea, 1983; McKinney, 1988; 1999; McNamara, 1995; 1997).

Size is plotted on the x axis, while the variable plotted along the y axis typically measures some aspect of ontogenetic change (Godfrey and Sutherland, 1995). Studies of growth allometry use a measure of size as a surrogate for age to enable comparison across disparate groups. However, if the relationship between size and chronological or developmental age varies in the populations being compared, between-group growth comparisons, especially related to timing, become problematic. Discussion of this issue has filled the literature (see Smith, 2001). Another issue is perhaps just as important though not as thoroughly discussed: the choice of the surrogate for age (i.e., the size measure) requires empirical results to be framed in terms of shape change *with respect to a specified size variable*. The choice of size variable has important consequences because it affects empirical results (Kowalski, 1972; Mosimann, 1979; Mosimann and James, 1979), which in turn affect scientific inference. Godfrey and Sutherland (1995) present additional considerations of allometry as a tool for comparing growth and for identifying heterochrony, including a chronology of definitions of shape as operationalized in these studies. Smith (2001) has proposed an approach that is not dependent upon size and shape parameters.

Comparisons Using Morphometric Methods

Beginning in the 1980s several morphometric methods for the study of shape change were proposed that enabled consideration of the geometry of an organism and the changes in geometry that occur during growth (Bookstein, 1978; Cheverud et al., 1983; Bookstein et al., 1985; Goodall, 1991; Zelditch et al., 1992). Inspired by the transformational grids of Sir D'Arcy Thompson (1917), these methods define patterns of growth in the two- or three-dimensional geometric spaces in which they occur. Most morphometric methods are landmark-based, using the coordinates of biological landmarks as input data, though methods for the analysis of outlines have also been developed (Lestrel, 1982, 1989; Lohmann, 1983; Lohmann and Schweitzer, 1990; MacLeod and Rose, 1993). Landmark-based morphometric methods can be classified into three categories: deformation, superimposition, and linear distance–based methods.

Deformation methods include finite-element scaling analysis (Lewis et al., 1980; Cheverud et al., 1983; Richtsmeier and Cheverud, 1986; Vogl, 1993) and thin-plate splines (e.g., Bookstein, 1989; Rohlf, 1993). The basic idea is to use the younger form as the reference object and to deform it so that, after deformation, the reference object matches the older specimen (the target object). The deformation needed to go from reference to target object provides a summary of the changes in growth pattern required to produce the older from the younger form. Changes are defined by the application of an interpolant function that uses the observed locations of landmarks to predict the locations of unobserved intermediate points. In this case, the interpolant function estimates the deformation local to landmarks that are required to go from reference to target form and generalizes the deformation to those regions between landmarks.

Landmark-based *superimposition methods* include traditional comparative techniques like roentgenographic cephalometry (Broadbent et al., 1975) and Procrustean approaches (e.g., Sneath, 1967; Siegel and Benson, 1982; Goodall, 1991; see also Cole, 1996, for an informative historical discussion of these methods). The younger form is fixed as the reference object. An older form is rotated, translated, and scaled so that the relative locations of landmarks match those of the reference object as closely as possible according to some criterion. For each landmark, the magnitude and direction of a vector depicts the difference between the forms.

Euclidean *distance matrix* analysis (EDMA) has been used to study growth (Lele and Richtsmeier, 2001; see Strauss and Bookstein, 1982, for an alternate distance-based method used in paleontology). The first step is to rewrite the landmark coordinate matrix as a matrix of the linear distances between all unique pairs of landmarks. A ratio of like linear distances is then computed in the younger (reference) and older (target) forms. By convention, the linear distances from the target form are the numerator and those from the reference form are the denominator. The matrix of ratios of all possible linear distances is studied to determine the local changes that occur during growth. In contrast to deformation and superimposition methods, EDMA does not use a coordinate system and changes are localized to linear distances, not to landmarks. Inspection of results identifies those landmarks involved in the observed form change.

To determine the contribution of growth to evolutionary developmental mechanisms, scientists use these methods to compare growth patterns between groups in a phylogenetic context. Simply noting whether growth patterns are similar or different, however, does not explain *how* growth might have contributed to evolution. To achieve that level of understanding, scientists must be able to determine the location in space (anatomical, morphological) and time (developmental, chronological) of those differences. The ability to identify the ontogenetic timing of growth differences lies primarily in the informed collection of appropriately timed data. The ability to localize differences in growth to anatomical structure depends on the judicious choice of anatomically relevant data and the properties of the morphometrics method used.

Consider a comparative analysis of mammalian mandibular growth in two species. An analysis that cannot localize differences simply states that the mandible of one species grows "bigger" than the other species. A method that localizes the increase in growth to the angular process and the corpus enables the researcher to formulate hypotheses about mechanisms that could be responsible for localized increases or decreases in growth. Possible mechanisms include an increase in the number of neural crest cells in a specific migrating population, the size of a particular condensation at an exact time in development, or the rate of localized cell division. A morphometric method that localizes incorrectly will misidentify the loci of the growth difference, resulting in the proposal of hypotheses based on invalid results.

A method that enables valid localization of differences in growth allows the researcher to formulate testable hypotheses. Depending on the identified loci of growth differences, the researcher can plan experiments to determine the lability of varying processes in the development of those structures and their response to selection, molecular engineering, or environmental conditions. Observations at the molecular, cellular, or tissue level can be checked against what was determined morphologically.

Even if landmark data are appropriate to the scientific question, analytical results cannot provide more information than that provided by the data. If the relative location of points is all that is available for analysis, the researcher should not place much emphasis on results that describe changes in form that occur on surfaces between the landmarks. Too often, the results of interpolant functions generalized across surfaces are used to make oversimplified generalizations regarding the evolutionary or developmental basis for change in form. Just as fossils leave us a static fingerprint, data collection for growth analysis gives us a single example of a form during a developmental trajectory. Closer sampling gives us more information regarding details of the trajectory, but we can never know the exact path without infinite data. We can make assumptions about the most probable path of growth based on the interpolant function or the superimposition criteria, but, like parsimony in evolutionary analysis, what is convenient for analysis may not always be biologically sensible or valid.

The choice of a coordinate system affects the description of growth provided by any particular superimposition or deformation method. Superimposition and deformation techniques need to specify an orientation (i.e., a coordinate system) for comparison (Lele and Richtsmeier, 2001; Richtsmeier et al., 2002). If the reference and target forms in two groups are different, their growth trajectories will be described in two different, arbitrary coordinate systems. There is no unique way to place the growth trajectories of the two groups under study into the same coordinate system (Richtsmeier and Lele, 1993), so how can these growth trajectories be compared? Several approaches have been suggested (e.g., Cheverud and Richtsmeier, 1986; Goodall and Green, 1986; O'Higgins, 2000), but there is no way to know the effects of the choice of the coordinate system for each analysis. One solution is to avoid coordinate systems altogether and use a linear distance-based method (Richtsmeier and Lele, 1993). These methods have been criticized as being less biologically useful because they do not provide vectors local to landmarks that describe change in form (O'Higgins, 2000), but applications of the methods suggest otherwise (e.g., Richtsmeier et al., 1993, 1998; Cole et al., 2002; DeLeon et al., 2001).

These cautionary notes on the use of quantitative methods for studying the contribution of growth to evolutionary developmental biology are not meant to discourage these sorts of growth analyses. The key elements to remember are:

Landmark coordinate data are limited, and results that go beyond the information they comprise should be accepted only cautiously.

All geometric morphometric techniques have limitations for analysis of growth.

Results of any growth analysis conducted at the gross morphological

> level can only suggest hypotheses about mechanisms responsible for the differences found.

The process of growth offers the potential for evolutionary change through the range of diversity in, and interaction between, elements of timing, direction, and magnitude of local change in growth patterns of individuals. An astute application of quantitative methods to the growth of organisms can define testable hypotheses to be evaluated with subsequent data from alternate levels of the hierarchical system.

STUART A. NEWMAN

Hierarchy

Hierarchy has several senses in evolution and development. Some of these are unexceptionable. Organisms are made of cells, cells of molecules, molecules of atoms, and so forth; it would be hard to argue that the existence of entities at "higher" (larger scale, composite) levels did not depend on entities at "lower" (smaller scale, basic) levels. Furthermore, atoms, molecules, organelles, cells, and cell aggregates appeared on earth in this precise order. This *synchronic* "constitutive" (Mayr, 1982) hierarchy is thus also *diachronic* or evolutionary. As an example, eukaryotic cells are not only partially composed of DNA and protein, but they also evolved subsequently to DNA and protein. A synchronic hierarchy is not inevitably diachronic: modern houses, for example, can be built from steel and concrete, but houses appeared on earth prior to these materials. The same pattern can be seen in organisms, whose descendants' molecular composition and mode of development can change over the course of evolution without overt changes in outward form, ecological role, or taxonomic relationships to other lineages.

The concept of hierarchy becomes problematic when applied to a complex system such as an organism if the vector of causation of the system's properties and behaviors is considered to point in just one direction. In pre-Darwinian views of biological organization, the most fundamental causal level was the *essence* of the organism, imagined as an idea in the mind of God and manifested in special creation. The organism's physical matter, at the distal end of the causal chain, was held to be organized in conformity to this essence.

Georges Cuvier (1769–1832) was the champion of the most systematic version of the essentialist view, the principle of the *correlation of parts*. He held that the various functions and parts of an organism are mutually dependent, subject to laws "which possess a necessity equal to that of metaphysical or mathematical laws, since it is evident that the seemly harmony between organs which interact is a necessary condition of existence of the creature to which they belong" (Coleman, 1964, p. 67–68). Interestingly, modern-day fossil reconstruction draws on the general validity of this pre-Darwinian assumption that the organization of even previously unknown organisms is discernible from a set of principles that have little to do with regularities at the molecular level.

However useful such "rational morphology" may be in paleontology, contemporary cellular and developmental biology is informed by the Darwinian rejection of the notion that organisms have fixed essences or regulative principles to which smaller-scale processes are subordinated. Descriptions of experimental results in these fields typically point the arrow of causation in the opposite direction, with the origin being the set of molecular mechanisms on the basis of which all form and function must be explained. The old idea of nonmolecular organizing principles setting the terms for molecular-level processes is so far from the current discourse in experimental biology that its direct inverse has been considered: the notion that taxa (body plans, "zootypes") are defined by characteristic patterns of expression of certain genes (Slack et al., 1993).

But genes do not act autonomously in evolution or development. "The consequences of mutation for phenotypic change are conditioned by the properties of the cellular, developmental, and physiological processes of the organism, namely, by many aspects of the phenotype itself" (Kirschner and Gerhart, 1998, p. 8420), and the same holds for the consequences of changes in gene expression in individual ontogeny. With respect to the evolution of form, homologous structures are not always built from the same gene products even in closely related taxa (Felix et al., 2000). Moreover, when a given set of gene products participate in building a similar structure in different organisms, they may do so using apparently different mechanisms (Lowe and Wray, 1997; Wray, 1999).

Computer simulations suggest that genetic networks that are sufficiently versatile to generate complex structures early in evolution will eventually be replaced by less versatile, but more reliable, networks if there is a premium on maintaining the structure (Salazar-Ciudad et al., 2001a,b). Here the organism's form itself provides a regulative principle to which changes at the molecular level must adhere if such changes are to persist (Müller and Newman, 1999).

With respect to development, while control by relatively lock-step hierarchies of gene expression are known *(Drosophila),* in other forms (e.g., vertebrates) the wide disparity between the number of genes and the number of proteins specified ensures that coordination of developmental pathways is distributed across multiple levels of the organism's biochemical hierarchy (Graveley, 2001). In still other cases the external environment is actually incorporated into the developmental process as a source of cues and signals (e.g., *Dictyostelium*).

The decoupling between genotypic and phenotypic change in both evolution and development implies that causality runs in both directions, not that these levels are causally independent of each other. And because phenotype is itself a multileveled concept with morphological and biochemical aspects, determination is actually multifarious. The integrity of the organism's overt phenotype influences which genetic changes are retained during evolution; this concept has been discussed in the scientific literature under the rubric of "stabilizing" (Schmalhausen, 1949) or "canalizing" (Waddington, 1962) selection. It corresponds to a premium on breeding true: because survival of an organism that is well integrated with its environment and biologically compatible with other members of its species would typically be undermined by developmental aberrations, selection for the maintenance of phenotype in the face of environmental or even further genetic change is to be expected.

Attempts to explain evolutionary stability of the morphological phenotype ("stasis") have had the constraint of also aiming to steer clear of Cuvierian essentialism. The assumption that genes are the foundation of the organism's organizational hierarchy and that, all things being equal, genetic change should lead to morphological change—particularly in the face of the episodes of extensive environmental change through which many static morphotypes have nonetheless persevered—has led to the suggestion that stasis is the result of the "cohesion of the genotype" (Mayr, 1982). It is not, however, the genotype that exhibits cohesion; it is the morphological phenotype itself. Molecular composition and generative mechanisms can change over the course of evolution while forms remain constant. This concept has been referred to as the "autonomy" of units of construction, or homologues (Müller and Newman, 1999). The notion that an organism's morphological phenotype has an integrity that is not simply an expression of its developmentally active genes has gained confirmation from results of "knockouts" of such genes, many of which have no phenotypic effect (Shastry, 1995).

If not a coherent genotype, what is stabilized during stabilizing evolution? The same question in a different context is, "What does the phenocopy copy?" Here Oyama (1981) was referring to classic experiments in which morphological phenotypes in *Drosophila* that were originally identified as genetic variants were "copied" in wild type strains by exposing embryos or larvae to heat shock or ether. Why certain forms seem to have a "privileged" or all-but-inevitable status is inextricably connected to the issue of hierarchy in evolution and development. The connection is not only to how the different levels relate to one another in generating a functional organism phylogenetically and ontogenetically, but also to the basis on which the current distribution of taxa have come to be organized hierarchically (into the phyla, class, order, genus, species of the Linnean system of classification). Mayr (1982) referred to the Linnaean system as an "aggregational" hierarchy because, unlike "constitutive" hierarchies, the units at the lower levels are not compounded by any interaction into higher-level units. He described this hierarchy as "strictly an arrangement of convenience," a view that entails a different concept of the relationship between taxonomic levels than that discussed next.

One aspect of the organism's hierarchical organization that is usually ignored in discussions of evolution and development is its identity (particularly at early embryonic stages) as "soft, excitable matter." Soft matter is a category of material studied by condensed matter physicists that may exhibit viscous or viscoelastic properties and can undergo phase transitions (de Gennes, 1992). Excitable media are complex materials that exhibit active mechanical, chemical, or electrical responses to their environments (Mikhailov, 1990). Both soft matter and excitable media exhibit characteristic morphologies and spatiotemporal patterns, and cell aggregates are unusual in that they simultaneously fall into both physical categories. Many of the properties that characterize the body plans and organ forms of the various metazoan phyla—tissue multilayering, lumen formation, spatially periodic arrangements of cells—as well as the features that distinguish them from one another may be the result of generic mechanical and dynamical behaviors of soft excitable matter employed in various combinations (Newman, 1993; Newman, 1994; Newman and Müller, 2000).

Related to its account of recurrent themes in the evolution of morphology, this viewpoint also provides the basis for a scientific approach to what has been termed "emergent" properties in the hierarchy of biological organization. For example, all single-celled organisms, undoubtedly including those ancestral to the Metazoa, display or displayed proteins on their surfaces. If those proteins become sticky through a mutational or environmental change, the result is a population of multicellular aggregates. This macroevolutionary step, thought to have occurred several times in the history of life, may thus have a simple, "generic" physical explanation. Similarly, many single-celled organisms such as yeast exhibit biochemical oscillations—temporal periodicities in the concentration of gene products or metabolites. On the level of the individual cell, these oscillations have no morphological consequence. But temporal periodicities can readily be converted into spatial periodicities in a multicellular aggregate (Newman, 1993), and something of this sort has evidently occurred in the evolution of vertebrate segmentation (Maroto and Pourquié, 2001).

The emergent but predictable patterns and forms assumed by excitable soft matter are only a starting point for understanding the regulative role played by organismal morphology during evolution. Modern-day embryos are the product of more than 500 million years of natural selection. If genetic circuitry with canalizing functions has accumulated around original, physically dictated "generic" forms over time by the process of stabilizing selection, the morphologies of more extensively evolved organisms, as well as the developmental processes that bring them about, must be more resistant to external influences than their ancient counterparts. While the morphotypes themselves may approach stasis, undergoing less and less change over time, the generative processes themselves continue to evolve. Starting out as extensively physically determined, interactive with the external environment, and self-organizing, development eventually becomes genetically integrated, programmatic, and indeed hierarchically organized.

These considerations compel a revised notion of the taxonomic hierarchy because they imply that the large-scale, macroevolutionary steps in the history of the Metazoa occurred early on, employing uncanalized mechanisms of "large effect" in response to both genetic and environmental change. Over time, increasing canalization must have caused both environmental and genetic changes to be of smaller effect, bringing us into the Darwinian era of gradual microevolution. In this view, the taxonomic hierarchy was thus generated first by a broad brush in which the major phyla were established and then by a progressively constrained evolutionary exploration of the residual morphological possibilities. This hypothesis contrasts with the neo-Darwinian view that the large-scale differences among the top categories in the taxonomic hierarchy are just the accidental end products of the gradual accumulation of small differences. It also provides an account for the otherwise-puzzling temporal structure of evolution of metazoan forms: an early burst of morphological proliferation followed by phylogenetic stasis (Gould, 1989a).

By relinquishing an organismal hierarchy in which the genetic level stands at the causal apex, both individual development and evolution of morphology can then be treated as a branch of the science of materials. While attributing to both developmental and taxonomic transitions a certain physical necessity, this picture also recognizes a tendency toward increasing correlation between genotype and morphological phenotype as forms that were originally physically determined become genetically assimilated in the course of evolution. To achieve a tenable concept of hierarchy, then, evolutionary developmental biology must divest itself of the uniformitarianism that Darwin inherited from Lyell. That uniformitarianism was both the most revolutionary and the most restrictive aspect of his theory of biological change.

ANNE C. BURKE AND SUSAN BROWN

Homeotic Genes in Animals

William Bateson (1894) originally used the term *homeotic* to describe naturally occurring variants in both vertebrates and insects. In these mutants, one segment had transformed morphologically into a different segment in the same organism. When Lewis (1978) described genes that harbor these mutations in *Drosophila,* he coined the name *homeotic genes.* The structural similarity of these genes was elucidated in the 1980s when the *homeobox,* a highly conserved 180-nucleotide sequence that encodes a homeodomain of 60 amino acids was described (McGinnis et al., 1984; Scott and Weiner, 1984). Proteins that contain this helix-turn-helix DNA-binding domain act as transcription factors. In *Drosophila,* the homeotic genes are located in two distantly spaced clusters on Linkage Group 3, the Bithorax (BXC: *Ultrabithorax/Ubx, abdominalA/abdA, AbdominalB/AbdB*) and Antennapedia complexes (ANTC: *labial/lab, proboscipedia/pb, Deformed/Dfd; Sex combs reduced/Scr, Antennapedia/Antp*). The term *HOMC* was originally used for the single cluster of homeotic genes found in the red flour beetle, *Tribolium castaneum* (Beeman, 1987). The term *Hox genes* was created by consensus to name all genes homologous to the *Drosophila* homeotic cluster genes (Martin, 1987). The terms *Hox cluster* or *Hox gene cluster* have eclipsed HOMC.

In addition to the remarkable conservation of the homeobox sequence, these genes display a phenomenon known as *colinearity* during embryonic development. First reported by Lewis (1978) as a parallel between the order of mutations along the chromosome and the body region affected, colinearity actually entails a parallel between the physical position of the individual genes along the chromosome and the pattern of their expression in embryos. The most 3′ gene *(labial)* is expressed most anteriorly in embryos. Moving in a 5′ direction along the chromosome, each gene has a progressively more posterior expression pattern. Lewis suggested that colinearity allowed for the translation of information at the genome level into regional identity in the embryo. Indeed, the most dramatic role of the Hox genes is in the establishment of patterning along the anteroposterior (AP) axis of the embryo.

In the last 20 years, Hox genes have become the subject of a research industry with two very distinct approaches. On the level of gene function, studies range from models of chromatin structure and the possible mechanism of colinearity (e.g., Duboule, 1994; Deschamps et al., 1999) to gene behavior, *cis* regulation, and interaction with downstream targets (e.g., Maconochie et al., 1996; Mann and Affolter, 1998; Scott, 1999; Li and McGinnis, 1999; Chauvet et al., 2000). At the other extreme are studies that address the evolution of the gene family, or focus on sequence variation within the gene family as characters for reconstructing phylogenies of animals (Amores et al., 1998; Finnerty and

Martindale, 1998; Cook et al., 2001). Other studies focus on the role of Hox genes in the evolution of morphology (e.g., Burke et al., 1995; Sordino, et al., 1995; Levine, 2002). All these directions of inquiry continue to yield tremendous amounts of data, though many questions remain unanswered.

Genes in the Hox cluster are thought to have arisen as tandem duplications of a single gene. In cases in which genes undergo duplication within a genome, the resultant related genes are called *paralogues.* In this sense, the Hox genes in *Drosophila* represent a group of paralogous genes. The names of the individual Hox genes in *Drosophila* have been adopted to name classes of genes in other organisms based on orthology relationships. *Orthology* refers to specific homology between individual genes in different taxa, i.e., those cases in which direct common ancestry can be reliably assumed. Orthologous relationships are not always easy to differentiate from tandem duplications, so the status of genes in certain taxa remains ambiguous.

Each class of Hox gene is also given a number that indicates position within the cluster. In bilaterians, the hypothetical ancestral cluster is currently estimated to include a minimum of seven genes, but this number may well change as data from new taxa and new genes accumulate. Within the current model of the ancestral cluster, five are straightforward orthologues of *Drosophila* genes and are designated *Hox1 (lab), Hox2 (pb), Hox3 (zen), Hox4 (Dfd),* and *Hox5 (Scr).* In a recent phylogenetic analysis of Hox genes, *Hox1* and *Hox2* were designated anterior class genes and *Hox4* and *Hox5* central class genes, while the derivation of *Hox3* was not clear (Kourakis and Martindale, 2000). The function of *zen (zerknüllt)* in insects has diverged (Falciani et al., 1996), and it is not considered a homeotic gene. However, it is clearly a Hox gene. A single central ancestral gene underwent independent tandem duplications in different bilaterian lineages to produce *Hox6, Hox7,* and *Hox8* or *Antp, Ubx,* and *abdA.* Finally, the ancestral cluster is thought to have had a single posterior gene represented by *AbdB* or *Hox9* through *Hox14.*

ProtoHox, ParaHox, and MetaHox

Genes containing the diagnostic homeobox sequence form a very large multigene family. The Hox genes are included in the ANTP superclass, defined by relation to the *Antp* homeobox of *Drosophila,* and are apparently restricted to animals (Holland, 2001). In addition to the Hox genes, a ParaHox cluster consisting of three members *(cdx, Xlox, gsx)* was discovered in the cephalochordate amphioxus and in mammals (Brooke et al., 1998). It is thought to represent a very early "sister" to the Hox cluster, resulting from a duplication of a ProtoHox cluster, which is hypothesized to have consisted of an *evx* ancestor along with one anterior, one middle, and two posterior "Hox" members. Cnidarians have recently been shown to have both Hox and ParaHox genes (Finnerty and Martindale, 1999; Gauchat et al., 2000). Holland and Brooke have proposed that the ParaHox genes might have originally been restricted to the endoderm and the Hox to the neurectoderm (see references in Holland, 2001).

The clustered *Drosophila* NK genes (Kim and Nirenberg, 1989) also belong to the ANTP superclass. Their vertebrate homologues, the NK-like (NKL) genes, are also clustered (Pollard and Holland, 2000). Additional members of the ANTP superclass, the EHGbox genes, are clustered in vertebrates but not in flies. The complex of NKL genes has been termed the MetaHox cluster (Coulier et al., 2000). These clusters, along with Hox and ParaHox, have all been duplicated in the vertebrate lineage.

Distribution of Hox Genes

The presence of a Hox cluster has now been confirmed in all metazoan taxa examined except sponges. Full sequence and linkage maps are not available for many phyla, so there are still difficulties assigning true orthology.

Among the most basal diploblasts, Placozoa are reported to have a gene that could represent either Hox or ParaHox (Schierwater and Kuhn, 1998). Homeobox genes have been found in ctenophores, but Hox representatives have been illusive. Cnidarians have representatives of both anterior and posterior Hox and ParaHox genes, though no central genes have been found (reviewed in Finnerty and Martindale, 1999). Extensive analysis of Hox distribution and sequence data suggests that the Hox genes underwent tandem duplications to produce a cluster of at least seven genes in the common ancestor of all bilaterian groups (Finnerty and Martindale, 1998; de Rosa et al., 1999). Current knowledge of the number of Hox representatives in various bilaterian taxa is summarized in Table 1.

TABLE 1. Variation in numbers of Hox genes in the Bilateria

Protostome Phyla	Hox genes	Deuterostome Phyla	Hox genes
Arthropoda[1]		Chordata[2]	
Hexapoda	10–11	Cephalochordate	14
Crustacea	6–9	Teleost	48
Chelicerata	7–28	Amniote	39
Myriopoda	5–13	Urochordata[3]	
Nematoda[2]		Ascidian	6
C. elegans	6	Echinodermata[2]	10
Priapulida[2]	10		
Brachiopoda[2]	8–9		
Mollusca[2]			
Gastropoda	6		
Annelida[2]			
Hirudinea	6		
Polychaeta	11		
Nemertea[2]	6		
Platyhelminthes[2]	10		

[1] Sources summarized in Cook et al. (2001); see text for details.
[2] Sources summarized in de Rosa et al. (1999); see text for details.
[3] Gionti et al. (1998).

Hox Genes in Protostomes

Within protostomes, the Hox genes of arthropods have been most widely studied. Analyses of both gene sequence and expression have been used to address long-standing questions of phylogeny. There are ten Hox genes in the *Drosophila* ANTC and BXC: eight homeotic genes and *zen,* which have bona fide vertebrate orthologues, and *fushi-tarazu (ftz),* which functions during segmentation. These ten classes of Hox genes have been identified in other arthropods, with sequence analysis supporting the hypothesis that the Crustacea and Hexapoda are sister groups (Cook et al., 2001). The *zen/Hox3* orthologues in chelicerates are expressed along the AP axis in register with the other Hox genes, suggesting that they may function as homeotic genes in these organisms. *ftz* and *zen* orthologues are also found in onychophorans, indicating that the protoarthropod Hox cluster contained all ten classes of Hox genes (Grenier et al., 1997).

Anterior Hox genes are expressed in segment-specific patterns along the AP axis of Crustacea and insects. In contrast, chelicerate Hox genes are expressed in extended overlapping domains with colinear anterior borders, which more closely resemble the expression of vertebrate Hox genes. In mites, the *ftz* and *zen* orthologues are expressed similarly to the homeotic genes, suggesting they too had ancestral roles in patterning the AP axis (Telford, 2000). Interestingly, ectopic expression of *Tribolium ftz* in *Drosophila* embryos results in the common homeotic transformation of antenna to leg, suggesting *Tribolium ftz* retains some properties of a homeotic gene (Lohr, et al., 2001). In addition, sequence comparison suggests that the arthropod *ftz* gene is related to *Hox5* in lophotrophozoans.

Ten classes of Hox genes have been identified in various lophotrochozoan phyla (deRosa et al., 1999, and references therein), but all ten have not been reported for any one species. Also, strict orthologies do not always apply because one of the central genes duplicated independently in lophtrochozoan and ecdysozoan lineages (Finnerty and Martindale, 1998, and references therein).

In *Drosophila,* the best-studied member of its phylum, Hox genes function at the end of the segmentation cascade. At least six known levels of regulation including *cis-* and *trans-*acting elements are involved in Hox regulation (reviewed in Carroll et al., 2001). Gap, pair-rule, and segment polarity proteins all affect Hox transcription. In addition, there are interactions between individual Hox genes as some Hox proteins affect transcription of other Hox genes. Holcomb and Trithorax proteins serve to maintain Hox boundaries indirectly through their association with chromatin. The extent to which these regulatory interactions are conserved in other arthropods is an area of intense investigation. Studies have revealed differences in segmentation gene expression or function between various insects, indicating differences in the regulation of homeotic genes (for a recent review, see Davis and Patel, 2002, and references therein). On the other hand, Hox gene functions appear to be largely conserved, as orthologues perform many of the same functions as the native gene when expressed in *Drosophila* (Galant and Carroll, 2002). These comparative data contain significant evolutionary information, and differences in the expression pattern (Averof and Patel, 1997; Abzhanov and Kaufman, 1999b) or function of certain Hox genes have been correlated with the morphological evolution of arthropod limbs (Galant and Carroll, 2002; Ronshaugen et al., 2002; Levine, 2002).

The Hox complex in the nematode *C. elegans* contains one anterior, two central, and three posterior group genes (Burglin and Ruvkun, 1993; Van Auken et al., 2000). In addition to the loss of several genes, an inversion within the complex places one central gene 3′ of the anterior group gene). Only the anterior and two of the posterior class Hox genes pattern the AP axis during embryogenesis (Brunschwig et al., 1999; Van Auken et al., 2000), while the rest apparently evolved alternative functions in postembryonic development (e.g., Salser and Kenyon, 1996, Ferriera et al., 1999).

Hox Genes in Deuterostomes

In the four deuterostome phyla, only the hemichordates have no published account of Hox genes. The echinoid echinoderms have been fairly extensively studied and consistently show a single cluster of ten genes representing eight classes. There are single members of the Hox groups 1, 2, 3, (4 and 5), 6, 7, 8, and three from groups 9 to 13; expression studies indicate they are predominantly expressed in adult rather than larval tissues.

Six Hox genes representing four classes have been found in the ascidian urochordates, though linkage has not been demonstrated for all of them (Di Gregorio et al., 1995, Gionti et al., 1998). Studies on a gene identified as a *Hox1* homologue in *Halocynthia* show that expression is exclusive to the ectoderm, and response to retinoic acid indicates a characteristic role in AP patterning.

Homologues of the *Drosophila* Hox genes were discovered in vertebrates in 1984 (*Xenopus,* Muller et al., 1984; mouse, McGinnis et al., 1984; human, Levine et al., 1984). These genes also exhibit temporal and spatial colinearity during embryonic patterning of the AP axis (Krumlauf, 1994, and references therein).

An initial duplication of the single cluster apparently occurred between cephalochordates and craniates (Pollard and Holland, 2000). The lamprey genome has proved particularly difficult to work with, though three clusters have been tentatively identified (Carr et al., 1998; Sharman and Holland, 1998). Two clusters have been definitively mapped in a primitive shark (Kim et al., 2000), and between four and seven clusters have been identified in teleosts (Prince et al., 1998; Amores et al., 1998; Aparicio, 2000). The definitive phylogenetic pattern of duplication within the vertebrate crown groups is still a matter of debate (Ruddle et al., 1999).

Four clusters are present in the best-studied amniote taxa (mice, humans, chickens). These clusters have been designated A, B, C, and D. The duplicates of individual genes in each cluster form paralogue groups numbered 1 to 13 running from 3′ to 5′ through the clusters (Scott, 1992). A fourteenth paralogue group gene has been identified in amphioxus (Ferrier et al., 2000).

The vertebrate Hox genes are expressed in colinear order in many embryonic tissues. There is colinear AP expression in the neural tube, the migrating neural crest (Trainor and Krumlauf, 2000) and the paraxial and lateral plate mesoderm (e.g., Cohn et al., 1997; Burke, 2000) and the endoderm (reviewed in Grapin-Botton and Melton, 2000). The expression boundaries in different tissues are generally offset from each other. For instance, the anterior borders of expression of individual Hox genes are usually more anterior in the nervous system than in the mesoderm.

Regulation of Hox expression is thought to occur in three phases in the early vertebrate embryo (Lemaire and Kessel, 1997; Deschamps et al., 1999). An initial phase begins during formation of the posterior primitive streak and extends from posterior to anterior. A second phase, established during somitogenesis, involves the refinement of an anterior border of expression (Gaunt and Strachan, 1994). This model has been recently elaborated by data that tie a "segmentation clock" to Hox expression in the presomitic mesoderm (e.g., Dubrulle et al., 2001; reviewed by Vasiliauskas and Stern, 2001). A third phase is involved with maintenance of the expression borders.

Vertebrate Hox Gene Function

During embryonic development in the fly, Hox genes play a critical role in fly segment identity by controlling the expression of multiple downstream targets. Loss-of-function phenotypes generally produce anterior transformations of segment identity *(Ubx)*, while gain-of-function phenotypes produce posterior transformations *(Antp)*. This pattern led to a model called "posterior prevalence," in which more posterior acting genes are dominant over more anterior genes (Duboule and Morata, 1994).

The structure of the Hox gene is strongly conserved, and mouse sequence substituted in flies performs many of the same functions as the native gene. For example, ectopic expression of *Antp* or the human

HoxB6 gene produces the classic Antennapedia phenotype (Malicki et al., 1990). Functional studies with vertebrate Hox genes, predominantly in the mouse, have demonstrated a role similar to that found in insects: the apparent control of AP patterning and limb/segment identity (reviewed by Carroll et al., 2001).

In 1991, Kessel and Gruss proposed a model in which a combinatorial "Hox code" specified the morphology of vertebral segments. This was originally based on the observation that timed doses of retinoic acid resulted in homeotic transformations of vertebra and concomitant changes in Hox expression borders in the paraxial mesoderm.

Perspective on Hox function in mammals has been strongly influenced by loss-of-function phenotypes displayed by transgenic mice (reviewed by Capecchi, 1997). As of this writing, thirty-eight of the thirty-nine Hox genes have been knocked out with transgenic technology. The number and complexity of these studies cannot be done justice here, but several examples are presented.

A majority of Hox knockout mice show homeotic transformations of vertebrae. Many show additional mutant phenotypes in a wide variety of structures, from sphincters (Kondo et al., 1996), to hair (Godwin and Capecchi, 1999), digits, genitalia (Favier and Dollé, 1997), sternum and body wall (Manley et al., 2001), and mammary glands (Chen and Capecchi, 1999). Straightforward interpretation of the vertebral phenotypes, arguably the most simple, has not generally been possible. For instance, posterior prevalence does not hold because both anterior and posterior transformations of vertebrae occur. In addition, while phenotypic effects may be centered on areas at the anterior border of expression, this is not consistent (e.g., Kostic and Capecchi, 1994).

Several double and triple knockouts have now been successfully generated. Double mutants have shown functional redundancy as well as synergism between paralogue members (e.g., Condie and Capecchi, 1994, Horan et al., 1995). All members of paralogous group 8 have been disrupted sequentially, and the axial effects accumulate differently in anterior versus posterior ranges of expression and in different body regions (van den Akker et al., 2001). Extremely large deletions of Hox genes also give conflicting indications of overall interdependence. Almost the entire B cluster has been targeted for deletion (Medina-Martínez et al., 2000). These mice show a sum of the effects observed in individual *HoxB* gene deletions, but no synergistic effects. When the entire *HoxC* cluster is deleted, the mice die at birth due to respiratory failure, but the skeletal phenotype is considerably milder than when only *HoxC9* is deleted (Suemori and Noguchi, 2000). This has been interpreted as indicative of functional interactions within the cluster.

The vertebrate limb is one of the best-studied systems of Hox patterning. The *AbdB* members of the A and D clusters (paralogues 9 to 13) are redeployed in a colinear fashion along two independent axes in the limb bud. The *HoxA* genes are expressed in nested sets along the proximodistal axis and *HoxD* genes along the anterior posterior axis of the limb (Nelson et al., 1996). Targeted deletions in many of these genes produce limb phenotypes as well as vertebral transformations. The double mutant of *HoxA11* and *HoxD11* demonstrates both synergy and redundancy in the Hox system. In these mutants, the axial transformations are more severe than in single mutations, and the radius and ulna of the forelimb are almost entirely lacking (Davis et al., 1995). The corresponding elements in the hindlimb, the tibia and fibula, are not normal but are present. Their presence is attributed to the possible redundant function of *HoxC11,* which is expressed in the hindlimb, but not the forelimb.

Identification or estimates of the number of genes that act as downstream targets of Hox genes have varied widely. Some studies have suggested that the majority of genes in the *Drosophila* genome respond either directly or indirectly to homeotic genes (reviewed by Mannervik, 1999). Another review reported just nineteen known downstream targets, representing a wide range of functional classes from cell adhesion to transcription factors (Graba et al., 1997). The search for candidate downstream targets of Hox genes has been augmented by new microarray and gene-chip technologies. For example, quantitative transcript imaging identified 6.3 percent of the arrayed genes as targets of *labial* (Leemans et al., 2001).

Regardless of the exact number of targets, there is consensus that the specificity of Hox action is extremely complex and is probably mediated through cofactors to impact multiple direct and indirect targets. Indeed, the overall action of Hox genes to orchestrate axial patterning may reside in myriad mundane and generic targets, finely tuned to generate global patterns through the regulatory underpinnings that manifest as colinearity. The strong conservation of this regulation through all the Bilateria studied has prompted the idea of a Zootype, a stage on which AP Hox expression and the establishment of the phylotypic stage coincide. Though the generality of this model has been challenged (e.g.,

Schierwater and Kuhn, 1998: Aranda-Anzaldo, 2000), the conservation of such a seemingly complex phenomena across such deep phylogenetic time demands an explanation. The need for this explanation has recentered the comparative method as a fundamental tool in biology. Hox genes provide a fascinating subject for study on multiple fronts and a powerful tool for prying evolutionary information from biological diversity.

HONG MA

Homeotic Genes in Flowering Plants

Plants possess several organ types, including vegetative (e.g., leaves) and reproductive organs. In flowering plants, reproduction occurs in the flower, which has characteristic types of organs. Abnormal development of plant organs, particularly homeotic changes in floral organs, has been recognized for over two thousand years (see Meyerowitz et al., 1989). In fact, when defining the term "homeosis" as the conversion of one member of a series to another, Bateson (1894) indicated that in plants homeotic change is "one of the most familiar forms of abnormalities." Moreover, similar homeotic changes occur in a variety of plants, suggesting that they affect conserved gene functions. In the last fifteen years, genetic and molecular studies of floral homeotic mutants and their corresponding genes in two species, *Arabidopsis thaliana* (thale cress) and *Antirrhinum majus* (snap dragon), have provided much of our current understanding of the molecular machinery controlling flower development. For more extensive treatment of floral regulatory genes, their regulation, and evolution, see Ng and Yanofsky (2001), Theissen (2001), and Zhao et al. (2001a).

Diversity in Flowering Plants

Flowering plants, or angiosperms, number more than 300,000 by some estimates. The largest monophyletic group, the eudicots, includes both

Arabidopsis and *Antirrhinum* and is defined based on pollen morphology, strongly supported by molecular phylogenetic studies (e.g., Bowe et al., 2000; Ma and dePamphilis, 2000; Soltis and Soltis, 2000). Eudicots produce seeds containing two cotyledons or embryonic leaves. Many familiar and economically important plants are eudicots, such as rose, apple, pea, tomato, sunflower, oak, and orange. Another important group of flowering plants is the monocots (one embryonic leaf), which includes maize, rice, orchid, asparagus, and coconut. The remaining flowering plants, which have seeds with two embryonic leaves, have traditionally been classified as dicots. However, they are phylogenetically separate from the eudicots and form a number of smaller groups; currently these species, such as magnolia and water lilies, are generally referred to as basal angiosperms.

The enormous diversity of flowering plants is reflected in the wide spectrum of variation in floral form. In fact, floral morphology was an important criterion in traditional systematics. In general, flowers of eudicots and monocots have a relatively small and fixed number of floral organs arranged in concentric rings or whorls, whereas basal angiosperm species often have numerous floral organs that are spirally arranged. In eudicots such as *Arabidopsis* and *Antirrhinum,* flowers typically have four types of floral organs, each occupying a whorl. The outer or first whorl consists of sepals that are usually green and leaf-like; they protect the internal floral organs during flower development. Inside the sepals are petals, which are often prominent and colorful. Further inside are male reproductive organs called stamens. At the center is the pistil, which usually is composed of two or more fused carpels. Floral organ primordia are produced from cell divisions in the floral meristem, which initially contains a group of undifferentiated cells and is subsequently consumed by the formation of floral organ primordia.

Floral Organ Identity Genes

Several *Arabidopsis* and *Antirrhinum* mutants display homeotic changes of one type of floral organ for another (Coen and Meyerowitz, 1991; Ma, 1994; Ma and dePamphilis, 2000). Because these mutations affect the type or identity of floral organs, the corresponding genes are referred to as *floral organ identity genes.* In *Arabidopsis,* the *APETALA1 (AP1)* and *APETALA2 (AP2)* genes are required for defining sepal and petal identities. Mutations in these genes cause homeotic changes or abnormalities, e.g., the *ap2-2* flowers produce carpels in place of sepals and stamens instead of petals. Two other genes, *APETALA3 (AP3)* and *PISTILLATA (PI),* are required for specifying petal and stamen identities. Mutations in these genes cause homeotic replacement of petals and stamens by sepals and carpels, respectively. A fifth gene, *AGAMOUS (AG),* is needed for stamen and carpel identities. Plants that are defective in *AG* produce flowers with petals occupying the positions of stamens and sepals taking the place of carpels, with additional floral organs interior to the abnormally placed sepals. In *Antirrhinum,* mutations in the *DEFICIENS (DEFA)* and *GLOBOSA (GLO)* genes result in homeotic floral organ abnormalities similar to those of *ap3* and *pi* mutants, and *plena* mutants have flowers resembling *ag* mutant flowers.

Analyses of these mutants led to the proposal of the ABC model for controlling floral organ identity, which states that the floral organ identity is determined by three separate functions, designated as A, B, and C (Figure 1; Coen and Meyerowitz, 1991; Ma, 1994; Ma and dePamphilis, 2000). In addition, each function is active in two adjacent whorls of the flower: A in whorls 1 and 2, B in whorls 2 and 3, C in whorls 3 and 4. These three functions, either alone or in combination, control the four distinct floral organ identities. In other words, A function specifies sepal identity, A and B functions control petal identity, B and C functions determine stamen identity, and C function alone directs carpel identity. Another feature of the ABC model is that the domains of A and C functions are mutually exclusive and one expands in the absence of the other.

The ABC model was tested using genetic and molecular analyses. *Arabidopsis* plants carrying mutations affecting two or three functions were generated, and they exhibited the expected floral phenotypes. For example, because of the absence of A and B functions and the expansion of the C function, *ap2 ap3* double mutants produce flowers with only carpels. When *AP3* and *AG* are both mutated, the flowers have only sepals; A function is the only remaining function.

Further tests for the model came from expression studies. The mRNAs of the *Arabidopsis AP1, AP3, PI,* and *AG* genes and the *Antirrhinum DEFA, GLO,* and *PLENA* genes are present in whorls as expected for their functions defined in the ABC model. Additional support for the ABC model came from studies using transgenic *Arabidopsis* plants that ectopically express these genes from a heterologous promoter in whorls outside the normal functional domains. Specifically, ectopic *AG* causes the formation of carpels and stamens in the outer two whorls, respectively, and ectopic expression of both *AP3* and *PI* causes the production

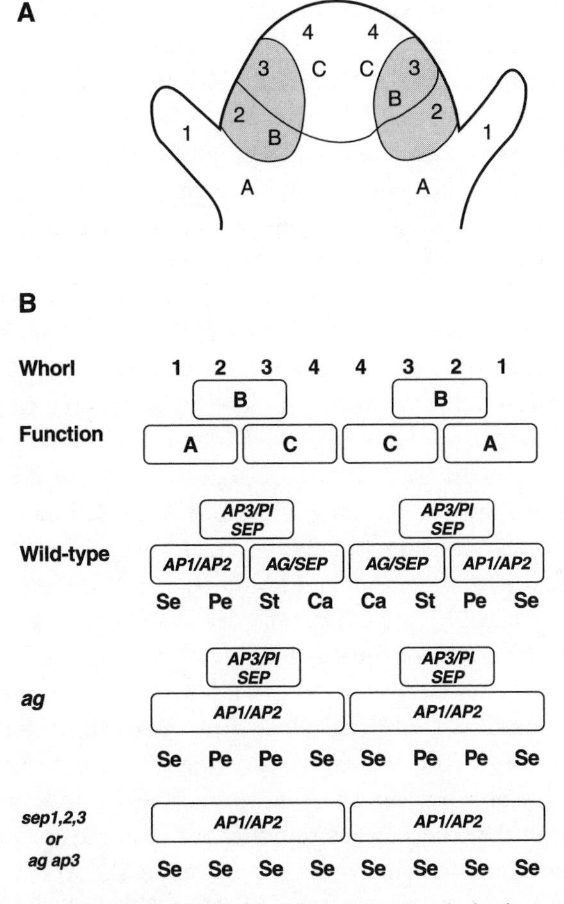

FIGURE 1. A. A longitudinal section of a floral meristem, with the four whorls and regions of ABC functions indicated. B. A representation of ABC function in *Arabidopsis* wild type and mutants. Ca = carpel; Pe = petal; Se = sepal; St = stamen.

of petals and stamens in the first and fourth whorls, respectively (Mizukami and Ma, 1992; Krizek and Meyerowitz, 1996).

Most ABC genes are members of the MADS-box gene family (Schwarz-Sommer et al., 1990; Ma, 1994; Riechmann and Meyerowitz, 1997; Ma and dePamphilis, 2000), which includes members in all eukaryotic kingdoms and is named after the founding members: *MCM1, AG, DEFA,* and *SRF.* The yeast MCM1 protein regulates DNA replication and transcription. The human SRF protein activates transcription of the c-*fos* proto-oncogene and other genes. *AG* is homologous to the *An-*

tirrhinum PLENA gene; similarly, *AP3* and *PI* are homologous to *DEFA* and *GLO,* respectively. DNA-binding activity has been demonstrated for the *Arabidopsis* AP1, AP3, PI, and AG as well as the *Antirrhinum* DEFA and GLO proteins. Furthermore, AG and AP1 were found to form homodimers, whereas AP3 and PI (and DEFA and GLO) can only bind to DNA as a heterodimer with each other. Unlike other floral homeotic genes, the *AP2* gene encodes a member of a plant-specific family of DNA-binding proteins (Jofuku et al., 1994).

Although the ABC genes are both necessary and sufficient for specifying floral organ identities, the fact that ectopic expression of these genes cannot convert leaves to floral organs indicates that other flower-specific factors are required for floral organ identity. Several recent studies have revealed the nature of these factors. Three genes—*AGL2, AGL4,* and *AGL9,* isolated as homologues of *AG*—are all expressed in early floral primordia in regions overlapping with both B and C functions; they encode proteins that can interact with AG, AP3/PI, and AP1 (Ma et al., 1991; Fan et al., 1997; Mandel and Yanofsky, 1998; Honma and Goto, 2001; Pelaz et al., 2001a). However, forward genetics did not identify mutations in these genes by using phenotypic screens. Indeed, when mutations in these genes were isolated using reverse genetic methods, plants mutated for one or two of these genes show no obvious floral organ defects (Pelaz et al., 2000). When all three genes are defective, the triple mutant flowers produce only sepals, similar to the *ap3 ag* double mutant, indicating that these genes together are required for both B and C functions. Based on the triple mutant phenotype, the *AGL2, AGL4,* and *AGL9* genes have been renamed as *SEPALLATA1 (SEP1), SEP2,* and *SEP3.* In addition to different combinations of ABC genes, when a *SEP* gene is expressed ectopically, leaves are converted into petals or stamens (Honma and Goto, 2001; Pelaz et al., 2001b), supporting the idea that floral organs are homologous to and may have evolved from leaves, as originally proposed by Goethe (1790). *In vitro* studies suggest that SEP proteins provide transcriptional activation to complexes containing the ABC proteins.

The analyses of ABC and *SEP* genes illustrate elegantly the power of both forward and reverse genetic approaches, which uncovered a set of transcription factors that in various combinations control a series of organ identities. The challenge for the near future is to identify the targets of these transcriptional regulators and their roles in carrying out different developmental programs in each floral organ type.

Floral Meristem Identity Genes

In many plants, a reproductive shoot or inflorescence. consists of a stem with multiple flowers arranged in a group or series. Both *Arabidopsis* and *Antirrhinum* have the same type of inflorescence—a raceme—which has a long stem with flowers spiraling from bottom to top. At the tip of the inflorescence stem, the inflorescence meristem maintains itself with a pool of undifferentiated cells, giving rise to an indefinite number of floral meristems on its flank. This property of indefinite meristem activity is called *indeterminacy.* In contrast, the floral meristem is determinate and produces four whorls of floral organ primordia, terminating with a pistil.

A few *Arabidopsis* and *Antirrhinum* genes are known to genetically control the identity of reproductive meristems (Ma, 1998; Zhao et al., 2001a). *Arabidopsis LEAFY (LFY)* is required for specifying floral meristem identity (Weigel et al., 1992; Zhao et al., 2001a). In *lfy* mutants, early appearing floral meristems at lower positions of the inflorescence are completely converted to inflorescence meristems, resulting in the replacement of flowers with inflorescences. Subsequently, *lfy* mutant flowers exhibit characteristics of inflorescences such as partial indeterminacy and spiral arrangement of organs. In *Antirrhinum,* mutations in the *LFY* homologue *FLORICAULA (FLO)* cause a replacement of flowers by branch shoots with additional branches, indicating a complete conversion of floral to inflorescence meristem (Coen et al., 1990). The LFY protein activates the expression of *AP1* and *AG* directly and also regulates *AP3* and *PI* expression (Parcy et al., 1998; Busch et al., 1999; Wagner et al., 1999; Zhao et al., 2001b).

In addition to *LFY* and *FLO,* the *Arabidopsis* A function gene *AP1* and its *Antirrhinum* homologue *SQUAMOSA (SQUA)* also play critical roles in determining floral meristem identity (Ma, 1998; Zhao et al., 2001a). *Arabidopsis ap1* mutants produce inflorescences and branched flowers instead of normal flowers, and *Antirrhinum squa* mutants also produce shoots in place of some flowers. Normal *AP1* expression depends on *LFY* function and *AP1* in turn positively regulates *LFY* expression. In addition, ectopic expression of either *LFY* or *AP1* can promote floral meristem identity in inflorescence meristem, resulting in the conversion of inflorescence to flower. Moreover, genetic studies revealed that *CAULIFLOWER (CAL),* a MADS-box gene closely related to *AP1,* plays a redundant role with *AP1* in promoting floral meristem identity (Ma, 1998; Zhao et al., 2001a).

In contrast to *LFY* and *AP1,* the *Arabidopsis TERMINAL FLOWER1 (TFL1)* gene inhibits the floral meristem identity (Ma, 1998; Zhao et al., 2001a). Mutations in *TFL1* cause the production of flowers at the apex of shoots or in place of inflorescences, indicating that the inflorescence meristem has been converted to floral meristem. This arrangement is similar to that of the *Antirrhinum centroradialis (cen)* mutant, which produces a terminal flower at the inflorescence apex. *LFY* and *AP1* are ectopically expressed in the *tfl1* mutant apex, suggesting that *TFL1* exerts its effect by preventing *LFY* and *AP1* expression at the center of the inflorescence meristem. This negative interaction is mutual because *LFY* and *AP1* overexpression also represses *TFL1* expression (Liljegren et al., 1999; Ratcliffe et al., 1999; Zhao et al., 2001a). *CEN,* the *Antirrhinum* homologues of *TFL1,* also antagonize *FLO,* but *CEN* expression is later than and depends on *FLO* expression (Zhao et al., 2001a).

The C function genes, *AG* and *PLENA,* play crucial roles in controlling floral meristem determinacy (Ma, 1998). In both *ag* and *plena* mutants, flowers are indeterminate and produce many more floral organs than the normal flower (Mizukami and Ma, 1997; Zhao et al., 2001a). In addition, *ag* mutant flowers can revert to inflorescence-like structures under short-day and other conditions that do not favor flowering in *Arabidopsis.* Furthermore, ectopic expression of *AG* at the inflorescence apex can also lead to the formation of flowers and termination of the inflorescence. *LFY* and *AP1* are required for activation of *AG,* suggesting that at least part of *LFY* and *AP1* function in controlling floral meristem identity is mediated through *AG.* This conclusion is further supported by the observation that ectopic *AG* expression promotes flower formation, partially rescuing the floral meristem defects of *lfy* and *ap1* mutants.

Recently, new insights into the mechanism controlling floral meristem determinancy have been obtained from analyses of the *LFY, AG,* and *WUSCHEL (WUS)* genes (Lenhard et al., 2001; Lohmann et al., 2001). *WUS* encodes a homeodomain protein and is critical for maintaining the stem cell pool in both inflorescence and floral meristems (Mayer et al., 1998; Brand et al., 2000; Schoof et al., 2000). It turns out that WUS is required for activation of *AG* expression by LFY. This makes sense because *AG* expression begins when the floral meristem is still active. AG then negatively regulates *WUS* expression to terminate the floral meristem activity, thereby causing determinacy of the floral meristem. In the inflorescence meristem, the absence of LFY and AP1 means that *AG* is not expressed there and *WUS* is not turned down, allowing indeterminacy. The *SEP* genes are also redundantly required for floral meristem determinacy (Pelaz et al., 2000); therefore, it is possible that one or more SEP proteins also interact with AG to down-regulate *WUS* during later stages of floral meristem.

Genetic and molecular studies in the model plants *Arabidopsis* and *Antirrhinum* have identified many floral homeotic genes and provided much insight into how they control development. Molecular cloning has revealed that, unlike homeotic genes in animals, most floral homeotic genes are members of the MADS-box gene family, with additional ones being plant-specific transcription factors. Other molecular studies have also provided considerable evidence that these genes are generally conserved in derived eudicots, the closest relatives of the model plants (Ma and dePamphilis, 2000). At the same time, functional conservation in basal eudicots and basal angiosperms is still uncertain (Kramer and Irish, 1999; Ambrose et al., 2000; Ma and dePamphilis, 2000). Additional genes will probably be identified with the completion of *Arabidopsis* genomic sequencing and the availability of numerous reverse genetic tools. A major challenge is to determine whether floral genes identified in model species are conserved in all angiosperms, including the basal angiosperms.

DAVID B. WAKE

Homology and Homoplasy

The word *homology* is used in diverse ways, always with an implication of similarity. The concept of homology is ancient, though Owen (1843) is credited with developing a precise definition of a homologue: "the same organ in different animals under every variety of form and function" (p. 379). The word "same" meant something very different to Owen than to evolutionary biologists today, which is why homology has remained a topic of discussion for over 150 years. Owen, an outstanding anatomist and powerfully influential member of the British scientific establishment who became Director of the British Museum, Natural History), resisted Darwinism and evolution in general to his death (Rupke, 1994). Evolutionists seized Owen's term and made it their own, going so far as to give their homology a central place in evolutionary thinking. Their adaption of the form led to problems, some of which persist today.

Owen envisioned a Platonic archetype, at first in the form of a Platonic idea inherent in what he termed the *organizing principle*. In his later work, the archetype itself became the idea (Rupke, 1994; Panchen, 1994). Owen's explanation of homology is key to understanding how we have reached the present state of argument on the subject. Two structures in different organisms, such as the seventh cervical vertebra of a mouse and of a monkey, are *special homologues* as two versions of the same structure. They are also each and collectively *general homologues* of a vertebra in the archetype that includes vertebrae but not necessarily the seventh cervical vertebra. The seventh cervical vertebra was to Owen the *serial homologue* of the first thoracic vertebra, the eighth caudal, and so on in the same organism.

Owen also developed criteria for recognizing homologues. The critical criteria were relative position (e.g., the last vertebra pierced by a spinal nerve) and connections (e.g., lying between the sixth cervical and first thoracic vertebrae). He defined an *analogue* as a part or organ in one animal that has the same function as another part or organ in a different animal (Owen, 1843). Some organs could be both homologues and analogues. The flippers of whales and lateral fins of sharks, for example, are homologues because they are both paired appendages; these structures are also analogues because they are used for swimming though they are not used in the same way and they do not contain the same parts. The flipper of a whale more closely resembles a limb of a land mammal than the fin of a shark.

While Owen was explicitly nonevolutionary in his approach, other early nineteenth century biologists were more ambiguous in their views on evolution. The word "affinity" was widely used with respect to homologues. Toward the end of the nineteenth century, in the wake of developing Darwinism, biologists accepted the reality of homology but were uneasy with Owen's interpretation and sought an alternative explanation—evolution. Lankester (1870) attempted to sort out problems associated with homology. He used "homogeny" as a replacement name for homology, arguing that an evolutionary rather than a Platonic philosophical framework required a more technical term. Homogenous organs in two species are present in their most recent common ancestor; all other resemblances are "homoplastic." Thus, serial homologues within an organism are homoplastic; and while the whale flipper and the shark lateral fin are homologues (i.e., homogenous) as paired appendages, they are homoplastic as swimming organs. Analogous organs that are not homologues—such as the whale flipper and a crustacean swimmeret—are also homoplastic. Lankester's concept of homoplasy survives, but his definition of homogeny was rarely used and has effectively disappeared. Perhaps had it survived with its original meaning, we could have avoided years of disputatious argument (reviewed by authors in Hall, 1994).

Lankester was not alone in recognizing the evolutionary foundation of what earlier scientists had tried to capture in their use of the term homology. In the same year the German anatomist Carl Gegenbaur embraced evolution as the explanation for homology (Gegenbaur, 1870). Of course, Owen would have none of it! Confusion and debate ensued.

Van Valen (1982) defined homology as correspondence resulting from continuity of information. In the spirit of this view, Wake (1999) argued forcefully that the homology debate is the result of biologists attempting to save an ancient, vague concept. Homology is not evidence for evolution, as has often been claimed. Rather, once evolution is understood to have occurred, homology is the "anticipated and expected consequence" (Wake, 1999, p. 27) of common ancestry. There is no reason to seek a naturalistic explanation for instances of homology; biologists should instead turn their attention to questions that can be resolved. Wake emphasized the opportunities inherent in homoplasy for understanding similarity.

Homology

Study of homology is in essence the study of evolution, and it can be pursued at many levels of biological organization and with respect to virtually all organismal traits. The central questions include: Is an organ, part, or trait in different organisms the same organ, part, or trait? How have these apparently homologous organs come to differ? Related questions focus on homoplasy: How have independently derived organs come to resemble each other? Do the organs have a common developmental basis? Or, is the similarity related to the ultimate function, such as swimming, flying, or lekking, rather than to detailed form? To make any progress in answering these questions, scientists need a phylogenetic hypothesis, the more robust the better. Although much progress has been made in defining terms and in developing general explanations for the phenomena, there remain many issues related to the nature of the questions being asked and the criteria for assessing sameness.

There are two general categories of approaches to studying homology. The first is a *taxic approach* (Rieppel, 1994), which involves generating phylogenetic hypotheses from character data. This approach asks: Are two features in two different taxa the same thing? All statements that two features represent the same character are hypotheses that are rejected or tentatively accepted depending on some criteria of tree topology and reliability. There is a large, technical literature dealing with taxic homology, with most researchers operating within the cladistic framework of Hennig (1966), who devised a technical terminology including apomorphy, synapomorphy, symplesiomorphy, and so on. Taxic homology uses criteria to establish hypotheses of homology, but only a phylogenetic analysis that rejects the hypothesis of common ancestry serves as a test.

With the advent of molecular systematics and ever-growing molecular databases, a commonly encountered question is: What is to be compared? On the one hand, there are long sequences of base pairs. On the other is the issue of alignment, which determines what the unit characters will be. A growing technical literature deals with this topic (e.g., Giribet, 2001). Alignment of multiple sequences often involves introducing gaps to make unequal lengths of sequence equal, whereas optimization alignment creates a unique set of homology hypotheses for each tree topology in an analysis (e.g., Wheeler, 2001).

The second general category of approaches to homology is *transformational homology,* which deals with the evolution of homologues, tracing the changes and perhaps the causes of the changes that have occurred. Transformational homology asks questions such as these: Are whale flippers homologues of ungulate forelimbs and how have they been derived? How have limbs been derived from pectoral fins of ancestral forms that have persisted relatively unchanged in sharks? Often transformational homology tries to infer from diverse criteria whether two features are homologues. Such inferences usually are treated not as testable hypotheses but as conclusions.

Regardless of the approach used, criteria are required to identify homologues. Owen had relatively few criteria; but by the time of Remane (1952), the list had grown. The goal (an ancient one, predating even Owen's work) is to establish "correspondence" among the parts or phenomena being compared, and all agree that relative position (in the case of structure) is important. Remane added a second criterion, special quality; because of its imprecision, this criterion has led to much confusion. For example, common development is a special quality, but it could be extended to any special nature that two features in different organisms share. Remane also used intermediate forms as a criterion. For example, the three inner ear ossicles of mammals are said to have homologues in outgroups because of the evidence of intermediacy in fossil taxa.

Patterson (1988) proposed three "criteria" (from our perspective, one criterion and two tests): similarity, conjunction, and congruence. *Similarity* is the starting reason for even thinking two things might be homologues. *Conjunction* refers to the presence of two features thought to be homologues in the same organism, and is thus a test of homology. Conjunction rejects a homology hypothesis (the characters were incorrectly delineated). *Congruence* is the failure of a phylogenetic hypothesis to reject the homology hypothesis. Patterson's focus was taxic homology and so he was trying to determine which homology hypotheses should be viewed as sufficiently robust to be considered characters for phylogenetic analyses.

A persistent problem in questions relating to homology and homoplasy involves hierarchy (Lauder, 1994). Whale flippers and shark lateral fins are homologues at the level of paired vertebrate appendages, but whale flippers and porpoise flippers are homologues at the level of mammalian anterior limbs in which the phalanges are enclosed in a pad-like structure used for aquatic locomotion.

Homology relates to something biological that is inherited and shared by two or more taxa, whether with absolute fidelity or not. Keeping the level of analysis clearly in mind is important. Thus guanine in a particular position in a particular polypeptide in two organisms is homologous if it occurred in their most recent common ancestor even if a mutation in one molecule has led to a silent codon. Lekking behavior might be homologous in all grouse, if it can be inferred that their common ancestor lekked, assuming that we have defined lekking at an appropriate level—grouse and their immediate relatives—and with appropriate delimitations (e.g., males gathering in a particular area and displaying solely to attract mates). So delimited, lekking might also be studied in drosophilid flies, in which the character is homoplastic relative to grouse.

Although the term *functional homology* is misused when it is intended to denote mere similarity of function for traits, particular functions may be homologous as characters if shared among taxa due to common ancestry. In Figure 1, gene function "role 2" is homologous for taxa 1 and 2, but not taxa 1 and 5 (Mindell and Meyer, 2001). Potential confusion regarding the level at which homology is implied by authors can be avoided by stating explicitly whether the homology is genic, structural, functional, or behavioral. Further hierarchical subdivision might be needed within any one of these major levels.

Although these considerations would seem to make the study of homology clear, problems arise from every direction! Many different

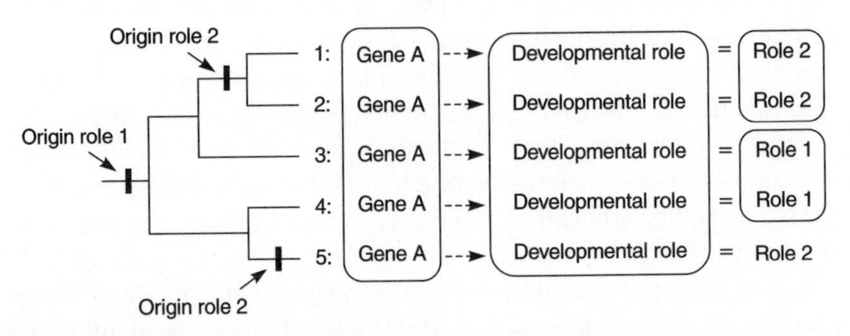

FIGURE 1. Application of common ancestry as the criterion for homology in a hypothetical case of dissociation between a regulatory gene and its role in development. Homologous characters and states are enclosed in boxes. Regulatory gene A first plays a developmental role (role 1) in the common ancestor for taxa 1 to 5. Dissociation events occur such that role 2 is substituted for role 1 in the most recent common ancestor for taxa 1 and 2 and in taxon 5. The following homology relationships can be described. Both gene A and developmental role are homologous in all five taxa. However, the state of the character developmental role changes such that character state 1 (role 1) is homologous in taxa 3 and 4 and character state 2 (role 2) is homologous in taxa 1 and 2. Character state 2 (role 2) in taxon 5 is not homologous to that in taxa 1 and 2, as character state 2 is not shared owing to inheritance of character state 2 (role 2) from the most recent common ancestor of taxa 1 and 2. After Mindell and Meyer, 2001, with permission of the publisher).

reasons are adduced for wanting to "know" homologies and to differentiate them from homoplasies. For example, in studies of molecular evolution the apparent similarity of macromolecules has led to many explanations, including homology.

Several terms are in common use to explain molecular similarity (Fitch, 2000; Mindell and Meyer, 2001). Two similar macromolecules (e.g., hemoglobins differing only in a few amino acids) might arise simply as a result of sister taxa evolving independently; this is *orthology* (lineage splitting, perhaps just species formation, or even cladogenesis if other than immediate sister taxa are involved), or the independent evolution of both taxa from an inferred ancestor. *Paralogy* is the result of gene duplication, which produces two versions of the original macromolecule. These versions can exist in the same taxon, thus violating the conjunction test and disclosing a problem—in this instance, a levels problem. Just as cervical and thoracic vertebrae are homologues at the more general level "vertebra," so also α and β hemoglobin in whales are homologous at the level "hemoglobin," but not at the lower levels of α and β (which are inferred on phylogenetic grounds to have evolved by gene duplication in a remote ancestor). Iterated parts in an organism, or serial homologues, are essentially paralogues, but at the level of organs.

Homology as a general issue in modern biology can be viewed from three perspectives (Butler and Saidel, 2000; Wagner, 1989): historical, biological, and generative.

1. *Historical homology* is the domain of phylogenetic systematics. Following the revolutionary work of Hennig (1966), a consensus developed to base phylogenetic hypotheses on shared derived states of characters, termed *synapomorphies*. When the characters are directly observable in living and fossil taxa, procedures exist for segregating synapomorphies from *symplesiomorphies,* which are shared ancestral states of characters. Criteria for polarization include use of outgroups and sequence of ontogenetic appearance (a highly controversial issue). The criteria are all problematic, and with the advent of large volumes of DNA sequence data, the problems become acute. Thus, *a priori* assessments have largely been abandoned. Historical homology involves the procedure known as *optimization;* once a phylogenetic hypothesis is proposed, states of characters are mapped back onto a tree using explicit criteria (such as maximal parsimony) to give the homology hypothesis for each character.

2. *Biological homology* focuses on the character itself and how it has evolved. Preservation of what might be called "design" has led those in-

terested in biological homology to seek the essence, or cause, for such preservation—the "biological" basis of homology. An example comes from the structure of the wrists and ankles of salamanders (Shubin and Wake, 1996). Variation displayed in a large sample of a newt population discloses a dominant pattern of arrangement of the carpal and tarsal elements, one that is characteristic of the taxon and its close relatives. However, symmetrical variant patterns are found that are recognizable to those who have studied the entire salamander radiation and fossils of outgroups (Shubin et al., 1995). Some of these are atavistic, restoring conditions found in phylogenetically more basal salamander taxa or in remote fossil outgroups. Others are identical to derived patterns that are dominant in species of the same clade, or of related clades, thus revealing the potential to evolve in specific directions. Rather than focusing on the question of biological homology, a potentially productive research program might seek the biological reasons for the conservation of form over many millions of years. Such reasons are likely to lie in the mechanics of morphogenesis.

3. *Generative homology* focuses on the genetic and developmental basis of characters; the approach recalls the archetype. The idea of an animal *zootype,* a genetic groundplan, has been postulated to unite all animal body plans that were founded on a shared pattern of regulatory gene expression (Slack et al., 1993, Schierwater and DeSalle, 2001). The zootype model is essentialistic; it implies that underlying morphology is an even more universal basal set of genetic mechanisms that is highly conserved. The idea that body plan evolution has a fundamental genetic and developmental basis harkens back to earlier ideas of Haeckel and even Owen. In the framework of this general view, the Hox gene cluster is envisioned as the central organizing principle in the evolution of form in complex metazoans (for criticism, see Schierwater and DeSalle, 2001).

The idea that organisms during their development pass through a *phylotypic stage* at which they most closely resemble one another, although they diverge before and very much after this stage, is similar to the zootype concept but at a different organizational level. Reasons for phylotypic stability may derive from the many interactions that take place throughout the embryo at this stage, as well as from the general modularity of development (Raff, 1996). Homology of developmental process (e.g., Gilbert and Bolker, 2001) is another approach that lies within this perspective. Once one accepts that process can be homologized, the possibility of *partial homology,* evident at the level of macromolecules (Hillis, 1994), must be considered at higher levels of organization as well (Minelli, 1998).

Implicit in studies of biological and generative homology is a quest for a naturalistic or mechanistic explanation for the causes for homology (Laubichler, 2000). However, the ultimate cause for the appearance of sameness is evolution itself. Rather than seeking an elusive and ultimately circular explanation, a more productive approach would explicitly state which explanation is sought. Why has a character or set of characters remained static? Why might a phylotypic stage be a biological reality rather than an idea? How has a given process evolved (such as the Hox gene system in relation to appendages or the brain stem, or the Pax gene system in relation to vision)? To call all of this "homology assessment" is to detract from efforts to understand central biological problems (Wake, 1999).

Homoplasy

Homoplasy travels intellectually with homology and the associations can be confusing. From a phylogenetic perspective, homoplasy is false homology (derived similarity that is not the result of immediate common ancestry), and productive research programs entail study of the reasons for the apparent similarity. Study of homoplasy may elucidate the questions so many biologists have sought to understand in the debate over homologies. Chief among these is the controversy in transformational homology over biological homology (e.g., Wagner, 1989, 1999). Homoplasy has been the subject of much attention recently (e.g., Sanderson and Hufford, 1996; Wake, 1999; Meyer, 1999; Hall, 2002), but its study has been an integral part of evolutionary biology from the beginning.

The least controversial kind of homoplasy is *convergence,* when organisms attain "sameness" in different ways or by following different phylogenetic routes. Wings of flies and birds are convergent, as are eyes of squids and sharks. Lekking is convergent in flies and grouse. Fossoriality and associated elongation and attenuation in tropical salamanders has evolved independently in two different segments of a single clade, one by evolving more trunk vertebrae and the other by making each of the vertebrae (the same number as in inferred ancestors) longer (Parra-Olea and Wake, 2001). This kind of evolution can be envisioned as being founded on elaboration of different developmental pathways or more generally by having different generative systems.

More problematic are the other kinds of homoplasy: *reversals* and *parallelisms*. Reversal is a term from taxic homology; the comparable term for transformational homology is atavism. Reversal is the phylogenetic reappearance of the "same" organ. Frogs typically have no maxillary teeth, but most frogs have mandibular teeth. Only one frog, *Gastrotheca guentheri* from Ecuador and Colombia, has maxillary teeth (some others have convergently evolved tooth-like bony fangs on the maxillary bone). *Gastrotheca guentheri* is deeply nested not only within *Gastrotheca* but within the Hylidae, itself a deeply nested clade of anurans. These maxillary teeth are biologically identical to those of the mandible except in position, and probably develop identically as well. It is this kind of phenomenon, sometimes termed *latent homology,* that has led scientists to propose a kind of continuum from homology to this kind of homoplasy. Butler and Saidel (2000) coined the term *syngeny* for this kind of generative homology, and they contrast it with generative homoplasy, which they term *allogeny,* in which different generative pathways lead to the apparently same feature. Hall (2002) does not care for these terms; he goes further to suggest that, while different developmental pathways generate convergent characters, parallelism is founded on similar or even identical developmental mechanisms and should be considered a form of homology. Examples include the evolution of elongate fossorial salamanders by the same method of increasing the numbers of trunk vertebrae in separate lineages of tropical and temperate salamanders (Wake, 1991), and the repeated evolution of a sword-like tail in a clade of xiphophorine fishes (Meyer, 1999). Atavistic traits might be either.

A theme in comparative developmental work is the possibility of what might be termed *deep homology.* For example, there are many similarities in the development of appendages of arthropods and vertebrates (e.g., Shubin et al., 1997), which may be an example of *generative homoplasy.* As outgrowths controlled by the same genes, the structures are the result of syngeny; from a historical perspective, the organs are clearly convergent and result from homoplasy. The gene *Pax6* is involved in the production of eyes in very diverse taxa, some derived from taxa that may have been eyeless. What is troubling about the designation of organs produced by the same genes as sharing generative homology is the reductionist perspective. Perhaps we should view visual organs as the level of focus. This perspective would interpret *Pax6* genes as the most immediately useful tools that organisms deploy in order to evolve eyes.

There is no reason to believe that homologues share some kind of essence just because they use some of the same genetic tools in their development.

Many questions about homology can be resolved by making certain that one has selected the correct hierarchical level for analyzing the characters in question, and by making certain that one is proceeding within the correct phylogenetic framework. Technical debates over definitions of homology doubtless will continue; but if we understand that homology is a necessary component of any theory of evolution, rather than something to be understood on its own, evolutionary biology as a discipline will be well served.

JAN SAPP

Inheritance: Extragenomic

Extragenomic mechanisms of inheritance, from those based on self-perpetuating states of gene regulation to those based on pre-existing cell structure, are sometimes lumped together and discussed under the rubric of *epigenetic inheritance*. This term distinguishes them from the more well known inheritance based on nucleic acids, which is a genetic system (Jablonka and Lamb, 1994). However, because the term epigenetic is an antonym of preformation and the phenomenon of structural inheritance is an argument precisely for preformation, this entry refers to structural inheritance as such and distinguishes it from both epigenetic inheritance and nucleic acid inheritance. The recognition of a plurality of mechanisms of hereditary reproduction and transformation calls for a broader, more inclusive definition of heredity that includes horizontal gene transfer—the inheritance of acquired genes and genomes. Each of these different mechanisms of hereditary change counters classical neo-Darwinian tenets.

Epigenetic Inheritance

In the course of development, once a cell is determined or differentiated, this state is transmitted to daughter cells (cell heredity): epithelial cells give rise to epithelial cells, fibroblasts to fibroblasts. Yet the genome of those somatic cells is identical. Historically, this paradox of cellular differentiation in the face of genomic equivalence was resolved with the recognition that genes can be turned on or off, as illustrated by B-galactosidase synthesis in *E. coli.* In human somatic cells, for example, about 3,000 genes may be on all the time, while the others are regulated through either gene replication or gene expression. In terms of the former mechanism, one hypothesis claims that after the timing of DNA replication in a chromatin region is established, it becomes self-perpetuating by a type of steady state system (Riggs, 1990; Riggs and Pfeifer, 1992). During the period of DNA synthesis (the S-phase), proteins that are necessary to assemble the newly replicated DNA into an active chromatin confirmation are preferentially produced. Thus early DNA replication is self-perpetuating. DNA that is late replicating would not be active because there would be no proteins left to assemble it into an active chromatin confirmation. Because of its inactive conformation, it would be late replicating in the next cell cycle. Thus late DNA replication is also self-perpetuating.

One of the best-understood mechanisms of cell heredity involves the regulation of gene expression resulting from changes in chromatin structure due to methylation. In many eukaryotes, some of the cytosines in DNA can be modified by the enzymatic addition of a methyl group. The methyl group does not change the coding properties of the base, but it does influence gene expression; highly methylated DNA is usually transcriptionally inactive. Methylation may affect different levels of chromatin organization (Levine et al., 1991, 1992), and DNA methylation patterns are inherited from one cell generation to the next. Methylation is an important mechanism of somatic cell heredity, and there is suggestive evidence that the directed changes are transmitted through sexual reproduction (Jablonka and Lamb, 1994). Thus, methylation may be an instance of the inheritance of acquired characteristics.

Genomic imprinting provides another example of epigenetic changes that can be transmitted from parent to offspring. Normally, it does not matter whether a gene or chromosome is transmitted from mother or father. However, an increasing number of studies in the 1980s found that the transmission and expression of a gene, a part of a chromosome, a whole chromosome, or a whole set of chromosomes depend on the sex of the parent from which it was inherited. The process that establishes the differences between paternally and maternally inherited chromosomal genes is known as imprinting. In some insect species, during the formation of sperm, all chromosomes derived from the paternal line are eliminated. Offspring inherit only maternally derived chromosomes (Brown and Chandra, 1977). Imprinting is also widespread in mammals, and may be important for understanding a number of human diseases, including some cancers (Monk, 1988; Hall, 1990; Reik, 1992). Homologous chromosomes or genes may be differentially methylated depending on the parent from which they were derived: in some tissues only the gene inherited from the mother is active; in other tissues, only the gene inherited from the father. In fact, a substantial part of the mammalian genome carries imprints of its parental origin (Surani et al., 1984; Jablonka and Lamb, 1994).

Structural Inheritance

Cells, Rudolph Virchow proclaimed in the nineteenth century, never arise spontaneously but invariably descend from a pre-existing cell. Virchow was right. Cells never arise by aggregation, but by the growth of preexisting cells; they model themselves on themselves. Although molecular biologists and popular science writers have often spoken of a program in which the organism is specified by nucleic acids in the genome (e.g., Jacob, 1974; Dawkins, 1989; Ridley, 1999), this is misleading. The organization of the cell or organisms neither occurs as a bottom-up self-assembly of parts orchestrated by the genome, nor by the intrinsic molecular structure of DNA. As a cell grows and makes itself, the macromolecules specified by the genes are released into a context that is already spatially structured. Certainly some intricate biological structures arise by self-assembly of preformed protein or nucleic acid subunits, with little or no input of additional energy or information: viruses, ribosomes, bacterial flagella are fairly directly specified by genes that encoded their components (Harold, 1990). However, many other organelles such as chloroplasts, mitochondria, and centrioles cannot be formed anew solely from the information in DNA.

Mitochondria and chloroplasts possess their own genomes of about 50 to 100 DNA-based genes, respectively. However, proteins coded by these genes and others in the nucleus cannot make a new mitochondrion or chloroplast: these organelles are formed only from preexisting organelles. Still other organelles such as centrioles lack DNA, but still need to be inherited. How centrioles reproduce is not fully understood (Sapp, 1998). In the cells of some species, centrioles seem to arise *de novo;* in the cells of others, centriolar reproduction requires cytologically visible preformed centrioles. New cartwheel structures develop from amorphous material, in a stepwise orderly sequence tubule by tubule, usually at right angles to preexisting centrioles. In other words, the preexisting structures may provide crucial spatial information: they are the sites at which proteins made elsewhere are assembled. The endoplasmic reticulum, the major sites of the synthesis of membrane lipids and proteins, also seems to be derived from preexisting structures; there is no apparent mechanism to assemble these structures *de novo* (Shima and Warren, 1998).

The same is also true for the overall structural organization of the cell itself. Embryologists and cytologists have long used the term *cell polarity* to refer to the visible directionality of cellular processes (see Sapp, 1987). Cell polarity is also especially noticeable in tip-growing organisms such as fungal hyphae, pollen tubes, yeast, and germinating algal zygotes as well as growing nerve cells. These cells orient their activities toward a unique site, called the *apex.* Building blocks of the cell, made elsewhere, are translocated to the apex where they are assembled into new plasma membrane, new cell wall, and new cell cortex. Cellular organelles also become oriented toward the apex and migrate to it along tracks supplied by the microtubules and microfilaments. The cytoskeleton, a network of microtubules and microfilaments in the cytoplasm of cells, gives the cell its characteristic shape. In no organisms has the morphogenetic role of preexisting cell structure in guiding the elaboration of new cell structures been more studied than in ciliates.

The complex patterns that make up the surface (the cortex) of ciliates are their most impressive and readily observable structural features. The cortex is composed of linear arrays of a large number of fundamentally similar "ciliary units," complex structures that include kinetosome (centriole), cilium, a variety of subcortical fibers, and specialized membranes. These complex ciliary units are arranged in a precise repeating pattern. On cytological observation alone, the cortical pattern appears to reproduce faithfully through a regular sequence of events during growth and fissions. Thus, a cytoarchitecture that embryologists repeatedly postulated as necessary to explain the orderly development of eggs is easily observed in ciliates and in fact plays a cardinal role in their morphogenesis (Lwoff, 1950; Tartar, 1961).

If the continuity of spatial organization is so crucial in the propagation of form, one might expect that a structural modification could be passed from one generation to the next without any alterations in genes. Tracy Sonneborn and his colleagues demonstrated that gross differences in cortical organization did indeed breed true to type through both sexual and asexual reproduction, independently of DNA. In the first classical grafting experiments, Beisson and Sonneborn (1965) inverted a small patch of ciliary units and showed that the progeny inherited the inverted row or rows for hundreds of generations in strains isogenic with their normal fellows. Subsequent studies with other ciliates have produced many examples of such structural inheritance: existing cell structural arrangements constrain and mold the structural arrangement of cell progeny (Nanney, 1980, Frankel, 1989). Sonneborn (1963, p. 202) coined the term *cytotaxis* to designate "the ordering and arrangement of new cell structure under the influence of pre-existing cell structure."

In addition to local preexisting cell structure, i.e., the ciliary unit, ciliatologists have concluded that there is another unknown spatial property pervading the cell as a whole. This spatial property is also passed on from cell to cell and is responsible for the perpetuation of morphogenetic patterns. The most spectacular illustration of this is the propagation of experimentally produced double monsters, which are ciliates that have not completely divided. They are fused back to back by aborting cell division or by surgical intervention. Doublet cells can propagate indefinitely as doublets, even though no gene has mutated. What maintains the doublet organization seems to be something other than the visible ciliature itself. Supporting evidence comes from experiments with the ciliate *Oxytricha fallax*. When *Oxytricha* is starved, it forms cysts with no visible preformed cortical structures (Grimes, 1982). Neither cilia nor kinetosomes—nor other microtubular structures—can be seen by electron microscopy. Nevertheless, doublets emerge from their cysts as doublets and singlets emerge as singlets, which means the emerging cells "remember" their original architectural pattern. Some other aspect of structural inheritance or cortical organization persists through the cyst stage (Frankel, 1989).

Thus ciliatologists argue that there are at least two kinds of hereditary mechanisms operating to maintain cell structure. One is a local constraint involving microscopically visible cell structures acting as a scaffold (e.g., ciliary units), and the other is a global mechanism that pertains to cell structure as a whole (Frankel, 1992). Embryologists have long spoken of a coordinating field of forces that pervades the organism. the morphogenetic field concept applies to the cell as well; moreover, fields are inherited. The inheritance of a particular structural arrangement in the cell cytoskeleton can have far-reaching consequences for morphogenesis, such as the shaping of an organism's adult body plan (Horder, 1983; Frankel, 1989). However, what this field is exactly remains obscure; it stands where the problem of inheritance stood before the rediscovery of Mendel.

The proposition that cellular organization involves mechanisms of heredity distinct from that encoded in genes disturbs the conceptual foundations of the Modern Synthesis. For example, several biologists have interpreted cortical inheritance as experimental proof of the inheritance of acquired characteristics (Lwoff, 1990, Landman, 1991). Neo-Darwinian evolutionists have occasionally admitted their concern about structural inheritance: John Maynard Smith (1983, p. 39) commented, "There are a few well-established exceptions, of which the phenomenon of 'cortical inheritance' in ciliates is perhaps the most important. Neo-Darwinists should not be allowed to forget these cases because they constitute the only significant threat to our views." Far from an exception, cytotaxis is a general property of all organisms, because all use preexisting cell structures for the reproduction of their cytoskeleton and cell surface (Hjelm, 1986; Bray, 2001). Any cell structure passed on from one cell to its daughters without being totally deconstructed to its component molecular parts could act as a seed for the propagation of similar structures. Structural inheritance is pervasive, from prions, to centrioles, to cell cortex, and to cell structure as a whole.

Symbiomics

The important role of symbiosis in the origin of species, of new organs, and of other adaptive changes has been proposed since the late nineteenth century (Sapp, 1994), but its full importance is only beginning to be appreciated today. Comparative molecular analysis of nucleic acid sequences has led to a revolution in the study of eukaryotic cell origins. Comparing nucleic acids of the chloroplast, mitochondrion, and nucleus with each other and with different kinds of bacteria closed the main controversy over the venerable idea that mitochondria and chloroplasts originated as bacteria (Margulis, 1993; Dyer and Obar, 1994). Mitochondria and chloroplasts are of eubacterial origin (alpha-proteobacteria and cyanobacteria, respectively) and became incorporated in a primitive host cell some 1,800 million years ago (Gray, 1992; Gray et al., 1999; Kurland and Andersson, 2000).

Several models have been proposed for the origin of the nucleus, and the origin of centrioles remains far from certain. Although in the early 1990s it was thought that centrioles possessed a circular linkage group (chromosome), like mitochondria and chloroplasts, these claims were refuted a few years later (Hall and Luck, 1995). The absence of DNA in centrioles does not, however, completely refute the idea that they evolved from symbiosis (Chapman et al., 2000). It has also been suggested that peroxisomes arose from aerobic bacteria that were adopted as endosymbionts some 2 billion years ago (de Duve, 1996).

Evidence is accumulating that bacterial symbiosis is of greater generality than previously supposed. Bacteria are the most biochemically diverse organisms on earth, and they carry out many chemical reactions impossible for their hosts; they can photosynthesize, fix nitrogen, metabolize sulphur, digest cellulose, synthesize amino acids, provide vitamins and growth factors, and ward off pathogens. The inheritance of acquired microbial genomes is easy to imagine in the case of protists in which symbionts are easily transmitted from one cell generation to the next. There are abundant examples of the inheritance of acquired microbes in protists, from the diverse photosynthetic microbes of foraminfera to the algae and bacteria of ciliates (Margulis and Fester, 1991).

Neo-Darwinian evolutionists have asserted that "the transmission of symbionts in the host egg is unusual" (Maynard Smith and Szathmary, 1999, p. 107). However, molecular techniques for screening nucleic acids, including the use of the polymerase chain reaction, have dramatically increased the facility of detecting symbionts (Higgins and Azad, 1995). One of the most striking results of this approach is research on *Wolbachia*. Bacteria of this genus are maternally inherited through the egg cytoplasm of their host insects and disseminated throughout the hosts' body cells. Surveys based on molecular phylogenetic techniques for screening have, so far, found *Wolbachia* in more than 16% of all known insect species, including each of the major insect orders (Werren, 1997); their complete distribution in arthropods and other phyla is yet to be determined. *Wolbachia* are alpha-proteobacteria, like mitochondria, and they appear to have evolved as specialists in manipulating reproduction and development of their hosts. They cause a number of profound reproductive alterations in insects, including cytoplasmic incompatibility between strains and related species, parthenogenesis induction, and femininization: they can convert genetic males into reproductive females (and produce intersexes). *Wolbachia* are thought to be the most common hereditary infection on earth, and evolutionists are beginning to recognize the implications of their prevalence for important evolutionary processes, especially in terms of a mechanism for rapid speciation.

Wolbachia are certainly not the only microbes to modify their host animals. All eukaryotes are superorganisms, each with a complex intracellular symbiome composed of chromosomal genes, organellar genes, viruses, and often other symbionts perhaps on their way to becoming organelles. Moreover, the symbiome, the limits of the multicellular organism, extend beyond the activities of its *own* cells. All plants and animals are made up of complex ecological communities of microbes, some of which function as commensals, some as mutualists, and others as parasites, depending on their nature and context. The relationship between some bacteria and our own cells may be intimate indeed. Our cells may be fashioned physiologically and morphologically by bacteria; our bacterial community regulates many of our own genes and is crucial for warding off pathogens (Hooper et al., 1998, 2001).

There is also increasing evidence that horizontal gene transfer has been rampant in the evolution of bacteria, making phylogenetic analysis hazardous. Though usually excluded from discussions of symbiosis and morphogenesis, viruses have a crucial role as vehicles for evolutionary change. Endogenous retroviruses (RNA-based viruses), relics of ancient germ cell infections, are harboured in 1% of the human genome (Löwer et al., 1999). Viruses may have been involved in key evolutionary events leading to the rise of placental mammals from egg-laying ancestors (Harris, 1998).

Hereditary symbiosis has played a most important role in macroevolution (Schwemmler, 1989; Margulis and Fester, 1991; Saffo, 1991; Sapp, 1994); because microbial symbiosis draws genomes from the entire biosphere, the resulting changes are often far greater than those that arise through recombination, gene mutation, and hybridization (Jeon, 1983). Thus horizontal gene (and genome) transfer from bacteria to eukaryotes is a cause of saltationist evolution, a mechanism for the discontinuous appearance of species seen from the fossil record.

In summary, recognition of horizontal gene transfer, and the inheritance of acquired genomes (whole or in part), contradicts a number of tenets of the classical neo-Darwinian synthesis (Sapp, in press): (1) transfer of genes across species and higher taxonomic groups occurs regularly; (2) early evolution is not tree-like, it is reticulated; (3) speciation does not require geographic isolation; (4) evolution may occur very rapidly by horizontal gene transfer; (5) the changes that result may be quite dramatic compared with those due to single gene changes; and (6) by implication the mechanisms for macroevolution are different from the mechanisms for microevolution.

NORMAN MACLEAN

Inheritance: Genomic

When living systems arose, the storage of genetic information was entrusted to DNA at an early stage, although probably RNA originally fulfilled this role. DNA offered the advantages of good thermal stability and a built-in capacity for faithful replication by semiconservative means. However, the ready transmissibility of DNA after replication must have guaranteed its crucial role as the custodian of genetic information, and DNA has continued to fulfil this pivotal role in most viruses and all prokaryotes and eukaryotes to this day.

Although the focus of this chapter is on eukaryotic animals, some observations on the situation in prokaryotes are relevant. The early observations of Fred Griffiths (1928) made it clear that DNA transmission in bacteria can be very simple indeed and does not even require that the origin of the DNA be a living cell. However, other more complex transmission is possible, either via phage transduction or, in the case of sex factor, via specialized extragenomic plasmids. Through all of these molecular exchanges and replication cycles, DNA efficiently retains its genetic information in the carefully ordered base sequences. It is surely no coincidence that these sequences are on the inside of the double helix backbone. Indeed, one of the astonishing features of living systems is how the relatively inaccessible information is so readily replicated and transcribed on demand by relevant polymerase enzymes. It is also now clear that although DNA retains the genetic code in its ordered structure, there is a powerful backup system of DNA repair through protein repair enzymes, and that access to and expression of the genome through transcription and translation is facilitated at every stage by quite elaborate protein interactions.

The evolution of multicellularity and cell specialization leads to really serious difficulties for the transmission of the genome. First, the genome itself has become much larger, in part because cell specialization requires many more genes. Also, perhaps to facilitate transmission of the DNA and to ensure its complete fidelity, the genome comes to be subdivided into separate chromosomal packages, each endowed with some autonomy in terms of the separate provision of centromeric and telomeric sequences. A number of very particular complications and specializations of genomic inheritance are found in eukaryotes, some of which are considered here.

All the Genes Are in All the Cells—Almost

In general every gene (and, in most cases, two allelic copies of every gene) is present in every cell. This requires a sophisticated system of gene regulation in which each gene is independently regulated in every cell. This is indeed the basic organization in almost all eukaryotes, but there are noteworthy exceptions.

When Theodor Boveri (1862–1915) was studying the chromosomes of *Parascaris aequorum* throughout development, he stumbled on an amazing phenomenon. Although this nematode worm has only two large chromosomes per haploid gamete, when the first two embryonic blastomeres divide, the two chromosomes in one of the blastomeres fragment into many smaller chromosomes, most of which are subsequently lost (thus the name for the phenomenon—*chromosome diminution*). At further rounds of cell division, diminution recurs, so that at the sixteen-cell stage, only two cells have undiminished chromosomes. One of these cells proceeds to form the germ line cells, and the other undergoes chromosome diminution and joins the other cells in the somatic lineage (see Gilbert, 2000a).

It would be interesting to know details of the portion of the genome persisting in the somatic cells and what genes are present only in the germ line cells. There is as yet no evidence that the particular portions of lost chromosome differ in different lines of somatic cells. In almost no other case does germ line/somatic line differentiation involve actual breakup and unequal segregation of the genome. The nematode exception is a situation, therefore, in which sexual inheritance is complete but the inheritance of the genome by somatic cells is incomplete. The only equivalent situation so far discovered occurs in some insect species such as the cecidomyid fly *Wachtiella,* in which thirty-two out of the forty chromosomes are lost in the nongerm cell tissues. Here, chromosome diminution is of number rather than size; the lost chromatin contains important genes expressed during germ cell differentiation (Kunz et al., 1970).

Germ Plasm and Germ Cells

The conservation of a distinct line of germ cells is a key feature of animal but not plant evolution. The retention of a germ line in animals is presumably a result of the frequent loss of totipotency in differentiated tissues. In animals such as *Drosophila* and *Xenopus,* the determination of the germ cell lineage depends on cytoplasmic determinants in the fertilized eggs, the so-called germ plasm. In most other animals, including mammals, there is no obvious equivalent to germ plasm, and mammalian germ cells are not morphologically distinct from other cells during early development.

Despite the fact that all the components of germ plasm have not been identified, an interesting hypothesis is that all are inhibitory of transcription and translation (Wylie, 1999) and so cells become germ cells precisely because they are prevented from differentiating in any other direction.

Why do animals go to such lengths to isolate germ cells from somatic cells? Although nuclear totipotency persists in the somatic tissues of many or most animals, cellular totipotency does not, and so germ cells ensure that gametes can be produced. A gene called *Oct4,* which codes for a transcriptional regulator, is a strong candidate for the role of controlling totipotency in mammals (Nichols et al., 1998). So what is so special about gamete production? It requires a reduction from diploidy to haploidy and, in so doing, involves a meiotic division. Meiosis is extraordinary in genetic terms primarily because it provides an opportunity for quite extensive crossing-over between homologous chromosomes. And, although it has been postulated that chiasma formation evolved to ensure that the homologues are kept together, it also provides an opportunity for the quite extensive mixing of alleles, which is often of selective advantage, between the homologous chromosomes. Of course, meiosis and chromosome crossing-over between homologues also occurs in plants, in which there is no permanent segregation of germ line cells. Both vegetative and reproductive structures of the shoot derive from the shoot meristem, and the work of McDaniel and Poethig (1988) emphasizes that no cells are set aside in the meristem that are earmarked for reproductive function.

Despite the fact that most animals segregate germ cells and the hereditary material within them from an early stage, it is hard to identify an important biological purpose. Many biologists have hinted at advantages for germ line separation in terms of reducing exposure to mutation, conserving telomere length, regulating DNA methylation, and preserving totipotency, but none of these has emerged as a causative or convincing driver of the evolution of germ line separation. Rather, such separation seems to be simply a convenient mechanism in animal evolution to compartmentalize the extreme differentiation needed to produce gametes. In male mammals, it allows close temperature regulation of the gonads, but this is a very limited situation.

Sex, Asexuals, and Genetic Recombination

The reasons for the emergence of sexual differentiation in reproduction are by no means clear (Hurst and Peck, 1996; West et al., 1999), although there seem to be remarkably few examples of highly successful asexual animals. Because recombination generates genetic diversity that is inherent in the transmitted genome, the topic cannot be ignored in this entry. Sexual reproduction involves alternation of haploid and diploid states, while the combination of diploidy and meiotic recombination allows the subtlety of multiple alleles to arise in a population and for an individual to be heterozygous at specific loci. The incidence of recombination is not only variable between species but it is especially variable between chromosomal regions, and highly dependent on chromosomal distance. This is why gene linkage is so significant.

In the context of genomic inheritance, allelic assortment occurs at meiosis—but also, more rarely, at mitosis. Therefore, the reduction divisions necessary for the production of gametes are important generators of genetic diversity. The success of at least some ancient asexual organisms such as rotifers thus tantalizes students of evolutionary theory (see review by Judson and Normark, 1996). The onset of a sexual cycle in otherwise asexual organisms such as *Volvox* or some aphid species is frequently associated with the arrival of deteriorating environmental conditions such as low temperature or food scarcity, which serves to underline that the diversity afforded by sexual reproduction offers a selective advantage.

In the present context, it is simply important to stress that sexual reproduction helps to ensure that the transmitted genetic message is open to the allelic scrambling that results from meiotic recombination. This, coupled with mutation and DNA methylation, offers the nonuniformity that characterizes the transmitted genome.

DNA Methylation and Genomic Imprinting

In all eukaryotic organisms, the genome is closely associated with protective proteins, the histones, to form chromatin, which, in most cases, is divided into discrete chromosomal packages. Chromatin structure allows not only transcription and replication of the genome, but also modification of the genome through DNA methylation. Methylation is the only permitted DNA modification apart from mutation and is very, though not absolutely, stable. Methylation involves the addition of a methyl group to a C residue within a CpG base sequence, thus producing the new base 5 methylcytosine. Although *Drosophila,* nematodes, and some other invertebrates have little or no methylated DNA, all vertebrates have extensive methylation; about 5% of the cytosines in mammalian DNA are methylated (Bird, 1992). Methylation is indirectly related to gene activity, in that the methylation of certain C residues in promoters can prevent or impede transcription by RNA polymerase.

Recent evidence suggests that a protein, MeCP2, binds specifically to methylated DNA and also binds histone deacetylase, therefore tending to prevent the nucleosome destablization that is presumed to be a prerequisite for transcription (Ng and Bird, 1999). However, it is also true that much DNA methylation has no obvious consequences for gene activity, especially when it occurs in coding sequences. There is some selection pressure against the production of 5 methyl C, because it is open to deamination to thymine, a significant mutation. This probably explains why CpG sequences are rather scarce in eukaryotic genomes apart from the frequently transcribed coding sequences of housekeeping genes. The constant activity of such genes seems to effectively prevent the methylation of the CpG sequences within them.

Most DNA methylation, so-called maintenance methylation, takes place on hemimethylated sequences after DNA replication, ensuring that the methylation pattern on the original DNA is copied onto the daughter strands. However, extensive *de novo* DNA methylation occurs during gametogenesis. The primordial germ cells are essentially wiped free of methylated DNA (Driscoll and Migeon, 1990), proving that the methylated status is not absolutely resistant to reversal. When gametes are differentiated from germ line cells, methylation occurs; at least in eutherian mammals, the pattern differs between sperm and egg in the case of at least forty genes. This is styled *genomic imprinting.*

About half of the imprinted sequences are methylated only in the sperm, and the other half only in the eggs. This curious arrangement results in androgenotes (with two paternal genomes), gynogenotes, or parthenogenotes (with two maternal genomes) that develop abnormally (Tilghman, 1999). The majority of imprinted genes act in the fetal growth pathway of insulin-like growth factor II (IGFII) or in the cell cycle. The precise way in which the methylation pattern affects transcriptional activity varies from gene to gene and, in some imprinted genes such as *IGFII,* the methylation pattern that inhibits expression of the maternal allele is not in the promoter but many kilobases away (Wutz et al., 1997).

The precise evolutionary pressures that have led to genomic imprinting remain hard to define (Tilghman, 1999); but from the viewpoint of genomic inheritance, the essential message is that in some organisms there is not complete genomic equivalence between male and female gametes. A "battle of the sexes" is ongoing at a genomic level (Reik and Walter, 2001).

DNA methylation also plays a crucial role in the silencing of one copy of the X chromosome—though the silencing is incomplete; at least some genes on the inactivated X remain active (Brown et al., 1999). This topic is not covered here because the methylation modification occurs after genomic transmission.

Mitochondrial Inheritance

In most animals mitochondrial inheritance is uniparental (maternal) and multiple, so that the egg carries a large number of mitochondria that go on to form all of the mitochondria in the new organism; sperm mitochondria are preferentially degraded. The breakdown of the male-derived mitochondria is not absolutely effective. In mice about 1 out of every 10,000 mitochondria is sperm-derived (Gyllensten et al., 1991)—maybe 1 in every 50 cells, because most cells carry about 200 mitochondria. However, there are noteworthy exceptions to exclusively maternal origin of mitochondria in the marine mussel *Mytilus galloprovincialis* and the freshwater mussels *Unionidae,* which have two distinct types of mtDNA, an F type that is maternally derived and an M type transmitted through sperm (Quesada et al., 1998; Zouros, 2000). Some individuals have mitochondria of purely maternal origin while others are heteroplasmic, having some mitochondria of female origin and some of male origin. This situation has been used to explore possible recombination between the two types of mtDNA, and this does indeed occur at quite a high rate (Ladoukahis and Zouros, 2001).

The inheritance of multiple maternal mitochondria from the egg by the embryo must provide some buffering against the rather high mutation rate of the mitochondrial genome (Brown et al., 1997). Perhaps the movement of genes from mitochondria to nucleus in the course of eukaryotic evolution is partly a result of the mitochrondria being a mutational hot spot, possibly as a result both of the free oxygen radicals that often derive from respiratory chain metabolism and of the lack of proofreading DNA repair in these organelles.

Although most eukaryotic cells have about 200 mitochondria per cell, the number is highly variable, reaching about 1000 in liver cells and even 10^7 in the *Xenopus* egg, in which a huge mitochondrial cloud overlies the nucleus and the mtDNA accounts for over 90% of the total cellular DNA. Most mitochondria have between five and fifty copies of the genome per organelle.

The high rate of mitochondrial mutation, already alluded to (Brown et al., 1997), is clearly compensated for by the many copies of the mt genome per mitochondrion and the many mitochondria per cell. However, some mtDNA mutations do affect the phenotype and are recognized as muscle-specific human diseases. These include Kearns-Sayre syndrome and Lebers hereditary optic neuropathy (see Vanbrunt, 1991, for a useful review), the former being a somatic mutation and the latter an inherited mutation, although all affected patients are male (see discussion in Strachan and Read, 1999).

The mitochondrial genome, although tiny, with fewer than fifty genes in the human and 17 kb of DNA, is clearly a law unto itself in terms of replication, mutation, and transmission. However, it remains highly dependent on the nuclear genome for most of its proteins and even for the subunits of enzymes such as NADH dehydrogenase, some subunits of which are coded on the mt genome and some on the nuclear genome. So in many respects, the inheritance of the mt genome is more complex than that of the nuclear genome, although the two are strikingly interdependent.

Mutation and DNA Repair

The transmitted genome is subject to two kinds of change in addition to gene reassortment: methylation of individual bases and mutational changes. Here the latter of these two changes is considered. Mutational changes to DNA can be gross, such as deletions, inversions, or translocations, or fine scale, such as point mutations. The latter are much the most frequent and result from replication errors or, to a lesser extent, various kinds of DNA damage by radiations or chemical mutagens. Because all eukaryotic cells have elaborate systems of proofreading and DNA repair enzymes, mutations result from unrepaired or misrepaired DNA damage. Gross mutations may result from recombinational activity, transposition, or other factors.

Regarding mutation and repair, one or two recent pieces of evidence are noteworthy. The first results from a remarkable study by Ayala and his collaborators (Ayala, 2000) on the rates of evolutionary change in two proteins, a glycerol-3-phosphate dehydrogenase (GPDH) and superoxide dismutase (SOD). The sequences of these two proteins were compared between species, genera, families, phyla, and kingdoms. Astonishingly, two consistent trends were observed. GPDH evolves very slowly in *Drosophila* species, but several times faster in mammals, plants, and fungi, while SOD shows the exactly opposite pattern. Ayala used these data to argue cogently that molecular clocks based on neutral theory (see, for example, Kimura, 1983) cannot be relied on and must be used with extreme caution.

The conclusions that stem from this observation are either that significant and variable selection bears down on protein evolution, or that highly variable and gene-specific rates of DNA repair are operating. Of course these conclusions are not mutually exclusive. It is already well known that mutation rates vary greatly between species, being much higher (2×10^{-5} to 2×10^{-6} per gamete per generation) in maize *(Zea mays), Drosophila melanogaster,* and mammals than it is in viruses or bacteria (1×10^{-7} to 1×10^{-9} per replication or cell division) (Klug and Cummings, 2000). However, it is striking to find that different rates apply to different proteins in the same organism. Rates of DNA repair can be highly variable between organisms, being very efficient in *Dinococcus radiodurans* but very inefficient in human diseases such as xeroderma pigmentosa, in which nucleotide excision repair is ineffective due to a mutation. If rates are also variable between different genes in the same organism, the genome would be even more dynamic than previously supposed.

A further aspect of mutation worth emphasizing is in the context of trinucleotide repeat diseases such as fragile X, myotonic dystrophy, Huntingdon's disease, and others (see Strachan and Read, 1999). These diseases are characterized by expansions in trinucleotide repeats, often CAG or CCG or CTG, in the promoters or coding sequences of the relevant genes. The mutational aspect that is remarkable in all these diseases is the phenomenon of anticipation. The severity and age of onset worsens with successive generations due to expansion of the repeat. Such expansion is presumably the result of unequal crossing-over, and highly expanded repeats eventually cause loss of gene function (Akarsu et al., 1996).

Recent observations of mutation rates and trinucleotide repeat expansions therefore underline the emergence of a new appreciation of genomes—not only as highly dynamic entities, but with the capacity to be dynamic in directional and nonrandom ways.

Horizontal Transmission via Viruses

Horizontal transfer of genes is possible between eukaryotic individuals through the activity of retroviruses. The retrovirus life cycle is characterized by the integration of proviral DNA into the host genome, as well as by proliferation and infective transfer of further virions. The recent human HIV epidemic underlines the remarkable effectiveness of this viral proliferation cycle. However, the oncogenic retroviruses demonstrate a further remarkable property, namely the occasional ability of an integrated retrovirus to incorporate a human gene into its replicative genome and thus come to possess the ability to spread genes horizontally from person to person (or possibly species to species). Of course most of this transfer is somatic, but it can become germ line also, as shown by the large number of retroviral genomes that have accumulated in the human genome over our evolutionary history (Löwer et al., 1996). Thus horizontal as well as vertical transmission of parts of the genome does occur, although rarely.

Genomic inheritance is more complex and dynamic than had been imagined in the early days of molecular genetics. For the most part, DNA is stable and resistant to change, but we should not conclude that genomic inheritance is therefore intolerant of transmitting change. The truth is very much the reverse.

GERD B. MÜLLER AND GÜNTER P. WAGNER

Innovation

The evolution of phenotypic characters proceeds by a two-stage process: the spontaneous generation of heritable phenotypic differences and the fixation of these changes in a population. This applies to any form of phenotypic evolution, including adaptation, innovation, and speciation. To make a distinction between innovation and adaptation implies that evolutionary innovation represents a specific class of phenotypic change that is different from adaptive modification. This is the case with the origin of new body parts, such as feathers, eyes, or limbs—often called "evolutionary novelties"—and with major organizational transitions, such as the origin of multicellular organisms. Ever since the formulation of a theory of evolution by natural selection, the missing explanation for the origin of such new traits has represented a point of criticism (Nitecki, 1990).

Arguments in favor of the distinctness of innovations from other forms of phenotypic evolution have been made on two grounds: (1) the origin of novelty may include different mechanisms than the mutations underlying variation and adaptation; and (2) certain phenotypic changes may have more important and long-lasting consequences for the dynamics of evolution. Innovations that introduce new entities, units, or elements into phenotypic organization are called *novelties.* Innovations that have a major impact on adaptive radiation are called *key innovations.* We now briefly summarize the main theoretical problems associated with the explanation of innovations.

Evolutionary Problems Posed by Innovations

The innovation concept makes sense only if it refers to a set of evolutionary processes that raise research questions different from those pertaining to other forms of evolutionary processes. Here we identify four problems specifically related to novelty and other innovations.

Levels of selection problem. One class of major innovations consists of the integration of smaller replicative units into a higher-level replicator. Examples are the origin of the eukaryote cell involving symbiosis with bacteria, the association of cells resulting in multicellular organisms, as well as the origin of colonies and eusocial communities in which only a few individuals reproduce, as in bee colonies or naked mole rats (Buss, 1987; Maynard Smith and Szathmary, 1995; Michod, 1999). All these transitions involve cooperation among replicative units (cells or individuals) to form higher-level units that sacrifice the reproductive success of the lower-level units. In essence, the innovation consists of the creation of a higher level of selection. This poses a special problem for the innovation's explanation—adaptation by natural selection is usually based on competition among a given set of replicators, be they genes or individuals. Hence, special conditions must be fulfilled to overcome the competition among the lower-level units and allow the origin of the higher-level ones.

Origination problem. Origination refers to the causality of the generative mechanisms that underlie the origin of new morphological characters (Müller and Newman, 2002). The origin of new body parts or of entire new body plans is often difficult to explain because of the sheer phenotypic distance between the ancestral and the derived phenotype. Some of these problems may be solved by the discovery of transitional forms—as, for example, in the transition from the primary jaw joint of reptiles to the secondary jaw joint of mammals. Other examples have mechanistic implications because the transitions often require developmental modifications that are not within the mutational reach of the ancestral character state. An example is the origin of feathers, a radically different morphology that cannot be explained by a smooth shape change of the reptilian scale (Prum, 1999). In such cases, natural selection alone is of limited explanatory value. The real problem lies in understanding the mechanisms for the origin of the new phenotypic state. Many innovations may involve a breaking up of developmental or functional constraints that prevailed in the ancestral lineage. Whether this is in fact the case is an unresolved empirical question (e.g., Eberhard, 2001; Wagner and Müller, 2002).

Functionality problem. New body parts and characters are often associated with radically new functions not present in the ancestral lineage. It is thus tempting to assume that the new function is to some extent responsible for the origin of the novelty. The problem is that the derived function often requires a highly developed form of the new organ. The insect wing is an informative example. Small or rudimentary wings do not allow any form of flight and thus cannot be perfected for flight by natural selection. The solution to this problem has been proposed to lie in a transfer of function, meaning that the new structure initially serves a different function and becomes secondarily co-opted into the derived function (Mayr, 1960). For example, one theory for the origin of insect wings assumes that wings were originally external gills of aquatic arthropods (Averof and Cohen, 1997) that may have been used for other purposes than flight, such as serving as "sails" for surface-skimming (Marden and Kramer, 1994).

Integration problem. The functional interdependency of morphological characters leads to various kinds of connections and couplings among the characters that will influence their evolutionary fate. These couplings often have negative effects on the evolvability of characters. Here, then, the problem is how evolutionary innovations can be accommodated into the preexisting, tightly coupled character system of a body plan (Mayr, 1960; Roth and Wake, 1989).

Phenotypic Novelty

Several definitions of phenotypic novelty have been proposed; in order to be sensible, a definition must be firmly rooted in a character concept that is adequate for the respective level of organization. We have argued that for the morphological level, the most useful definition is based on the homology concept. Accordingly, a morphological novelty is *a new constructional element in a body plan that neither has a homologous counterpart in the ancestral species nor in the same organism (serial homologue)* (Müller and Wagner, 1991). This definition excludes characters that lie within the normal range of variation, or deviate only quantitatively from the ancestral morphological condition. But it includes those cases in which a new homologue has arisen through individualization of a preexisting, serial element. Novelty always represents a qualitative departure from the ancestral condition, not merely a quantitative one (Müller, 1990).

Examples of novelties that satisfy this definition exist in all tissues and organ systems of plants and animals. Frequently cited cases in vertebrates include the nervous system (eyes, the corpus callosum), feeding system (jaws, teeth), locomotor system (limbs, the autopod, additional digits), skin appendages (feathers, hair, glands), and others. Because of its capacity to fossilize, the skeletal system provides a particularly rich field of examples. New cartilaginous and osseous elements were added at all stages of vertebrate evolution and were accompanied by equally important innovations of joints, muscles, and the nervous system. Next we discuss the case of feathers, an example of novelty in which advanced developmental insight permits the formulation of a scenario for its evolutionary origin.

Feathers. Feathers are branched epidermal appendages that consist of a specialized form of beta-keratin, the so-called phi-keratin (Brush, 1996). Beta-keratin exists only in reptiles and may have no homology at all with the alpha-keratins found in the integument of all vertebrates. Feather precursors as well as fully formed feathers arose prior to the most recent common ancestor of birds and *Archaeopteryx* (for a review see Prum and Brush, 2002). An argument is often made that feathers are derived from archosaurian epidermal scales, but the developmental commonalties of feathers and scales are limited (Brush, 1996). Even though it is plausible that scales and feathers have a common root, feathers clearly qualify as novelties. The range of character states accessible to feathers is almost completely nonoverlapping with that of epidermal scales. Some overlap may exist in spike-like states that both scales and feathers can attain.

Feathers thus represent a radical departure from the ancestral character (Arnold et al., 1989) and also represent a new biological homologue, in the sense of possessing distinctly different developmental constraints in its natural variation (Wagner, 1989). In addition, they allow for new functions (Mayr, 1960). The morphological, variational, and functional individuality of feathers is tied to derived developmental features that are unique to feathers (Prum, 1999). Feathers thus exceed the limits of the natural variation of scales, i.e., the origin of feathers is caused by evolutionary events that overcome ancestral developmental constraints (Müller and Wagner, 1991; Wagner and Müller, 2002). Therefore, an explanation of the origin of feathers must identify the developmental innovations that overcame the variational limitations of archosaurian scales.

Scales and feathers are both epidermal differentiations that develop as a result of epithelial-mesenchymal interactions (Chuong et al., 1996). They derive their anterior-posterior polarity from the Shh/BMP2 signaling system (Harris et al., 2002). In feather development, the epidermis then forms a finger-like papilla. This character state is similar to elongated scales; beyond that stage, feather development proceeds along completely different and novel pathways.

The first event that deviates from that of scales is a rapid proliferation at the base of the papilla, creating a cylindrical invagination of the epidermis (Maderson, 1971), which becomes the follicle from which all further feather growth takes place. At the base of the follicle cell, divisions create a tube of keratinizing cells, which is pushed outward from the socket of the follicle to form the inherently tubular feathers. Differential keratinization leads to the formation of barb ridges, which eventually become the barbs of the mature feather. The characteristic branching pattern of the barbs and rachis is created by the helical growth of barb ridges around the tubular feather germ and fusion on the anterior margin of the follicle. Interestingly, many of the pattern formation events in feather development are realized through the repeated redeployment of the Shh/BMP2 pathway (Harris et al., 2002).

The important evolutionary point in this scenario is that the innovation of the follicle allowed the prolonged growth of the epidermal papilla, which in turn provided the opportunity for the formation of barb ridges and further patterning processes that led to the morphology of the mature feather. Later developmental events are conditional on the earlier events, e.g., barb ridges only differentiate if there is a follicle, which implies a specific order in which these developmental innovations must have occurred in evolution. Furthermore, this model predicts a series of morphological character states for ancestral feathers (Prum, 1999). The earliest stages of feather evolution in this model are: (1) a simple hollow tube resembling the sheath or calamus of a modern feather; (2) barb ridges forming in the tube leading to a tuft of unbranched barbs; and (3a) a helical displacement of barb ridges resulting in the pinnate feather with unbranched barbs. Simple tubular stage 1 feathers appear to be present in the phylogenetically basal *Sinosauropteryx* (Xu et al., 2001). Epidermal appendages that match the predicted morphologies 2 and 3a have recently been found in new fossils of the "feathered" dinosaur *Sinornithosaurus* (Xu et al., 2001), a nonavian theropod closely related to birds (Gauthier, 1986).

This model for the origin of feathers not only postulates a limited set of developmental innovations but also a specific sequence in which these innovations must have occurred. Both the specific stages and their hierarchical sequence are supported by the data on *Shh/Bmp2* expression patterns (Harris et al., 2002). Furthermore, this sequence of developmental innovations predicts a specific sequence of morphologies that can be tested by comparative analysis of fossil forms.

Evolutionary Mechanisms Underlying Novelties As Innovations

The examples already discussed show that, mechanistically, morphological novelties can be based on developmental innovations at the molecular or cellular level, or on new developmental modes and interactions. But they are as likely to arise from the redeployment of any of these components at a new location in the embryo, through subdivision or combination of earlier structures, or through developmental individualization of serial structures, such as teeth (canines, molars, tusks), vertebrae (cervical, thoracic, lumbar, sacral), or hair (whiskers, spines).

These changes of developmental mechanisms should not be confounded with the ultimate evolutionary factors that elicit novelty by acting on developmental systems. Novelty, as a deviation from quantitative variation, is not easily explained by the conventional mechanisms of evolution. Neo-Darwinism is primarily concerned with the variation, fixation, and inheritance of existing characters and less with the mechanisms responsible for generating new ones. It is difficult to implicate selection as the direct cause of novelty because selection can act only on what exists already. Three kinds of evolutionary mechanisms through which phenotypic novelties arise can be postulated (Müller, 2002): (1a) rare mutations that cause the immediate realization of the derived novelty; or that (1b) generate structural effects by stepwise accumulation of mutations; (2) symbiotic unification of previously separate genetic and developmental systems or components thereof; and (3) epigenetic byproducts caused by selection acting on other properties of the organism.

The first scenario (1a) is not realistic because of the many deleterious pleiotropic consequences of large-effect mutations. An exception might be structural novelties in plants, which are much more flexible in accommodating large phenotypic changes. Some radical structural modifications in plants may have been caused by a single mutation (Gottlieb, 1984). The second scenario (1b) could underlie such cases as the origin of feathers, which seem to be based on a limited set of specific, new developmental events.

Another well-documented example of cumulative genetic effects is the origin of the eyespot pigmentation pattern on nymphalid butterfly wings (Nijhout, 1991). The ground plan of eyespots consists of three so-called symmetry systems, which are closed-loop or elongated pigmentation patterns that are symmetrical either around a central point or along an axis. This pattern on butterfly wings is most likely derived from irregular pigment patterns that consist of developmentally equivalent spots and blotches found in moths and trichopteran flies (Nijhout, 2001). Eyespots develop from a central organizer called the *focus*. The phylogenetic origin of eyespots is thus inferred to require the evolution of the organizing focus and of concentric pigmentation patterns around it. The initiating developmental change here seems to consist of a number of genetic changes in the hedgehog signaling pathway on which pigmentation rings are based (see Keys et al., 1999; Brunetti et al., 2001).

Symbiosis, the third scenario, represents a radically different mechanism for generating novelty. It assumes the successful combination of genetic material from distantly related organisms (Margulis and Fester, 1991). The occurrence of symbiosis has been documented extensively and is recognized as a source of innovation in the evolution not only of bacteria, fungi, algae, and eukaryotic cells, but also in certain evolutionary instances of more complex organisms, although its importance seems to have receded in advanced forms of plants and animals. Symbiosis and other horizontal transfers might not only lead to the combination of pieces of DNA from different sources but could include the transmission of entire genetic or developmental modules between species.

The fourth scenario proposes that novelties are largely epigenetic in origin, both in early pre-Mendelian and in advanced Mendelian stages of evolution (Newman and Müller, 2000). Selection acting on other features than the novel character itself—such as organismal proportion, function, or behavior—is thought to elicit epigenetic by-products that arise from generic or ancestral properties of the modified developmental systems. For instance, a continuous directional selection affecting parameters such as cell division rates, cell and tissue interaction, position of inductive tissues, and timing of processes, can bring the affected developmental system to a threshold point at which incipient new structures arise automatically, as a consequence of the specific properties of the involved cells, tissues, and their interactions. Examples include novel skeletal elements that form at new locations, based on changes of cell number, blastema size, and mechanical load that accompany proportional modifications of a body part (Streicher and Müller, 1992). The concept advocating this mode of novelty generation has been called the side-effect hypothesis (Müller, 1990). According to this view, genetic modification and fixation are rarely the cause of novelty but rather a secondary consequence of natural selection favoring the maintenance of the novel traits.

Key Innovation

The concept of key innovation postulates a connection between the origin of specific structural novelties and the phylogenetic consequences for the lineage (Liem, 1974, 1990; VanValen, 1971; Rosenzweig and McCord, 1991). Specific innovations, whether based on novelty characters or physiological or functional innovations, are thought to open up possibilities for the exploitation of new ecological niches and trigger the adaptive radiation of the innovation-bearing group.

An example of the role of key innovation is the radiation of cichlid fish in East African lakes. The cichlids exploit a wide range of feeding niches based on the innovative versatility in their pharyngeal jaw apparatus. The species diversity of cichlids is much higher than that of sister groups that do not possess this construction. This example suggests that the major constructional event that triggered the adaptive radiation of cichlids resided in the innovation of their jaw apparatus (Galis and Drucker, 1996).

A similar example exists in snakes. The key innovation in the ophidian feeding apparatus involves the anterior separation of the left and right lower jaw bones and the establishment of various kinds of joints. This innovation permitted a tremendous new versatility of movements, enabling the capture and swallowing of very large prey, which in turn triggered a cascade of other specialized prey capture mechanisms in the mouth area, such as mobile upper jaws, fangs, and venom glands. The ophidian jaw innovation is therefore thought to represent the key event that facilitated the use of new food resources and was at the origin of the extensive diversification of snakes (Greene, 1983).

Endothermal homeothermy, the organism-regulated maintenance of constant body temperature in mammals and birds, is an example of a physiological key innovation. Homeothermy facilitates the independence from external temperature and provides a more efficient metabolism, which permits the exploitation of colder climatic regions, nocturnal activity, sustained rapid movement, avian flight, and so on. Endothermy thus was a key event in the explosive radiation of mammals and birds, itself possibly triggered by environmental changes (Liem, 1990). This example underlines an important feature of key innovation—namely, that the innovation itself is a consequence of events that occur in the ecological domain, such as changing climatic conditions. Key innovation is therefore a concept that postulates a reciprocal link between external environmental (energy related) factors, internal constructional (genetic and developmental) factors, and population (reproduction and fitness) factors of evolution.

Innovation in Developmental and Evolutionary Theory

The notions of evolutionary innovation and particularly of evolutionary novelty make sense only if they support a distinct research program. Evolutionary biology already includes three well-established clusters of research programs. One, taxonomy and phylogenetics, seeks to understand the genealogical relationships among organisms and associated phenomena, such as the geographical distribution of taxa. The second, adaptationist research, seeks to understand the evolution of adaptations of existing characters to the environment. And finally the third, largely independent research program is concerned with the origin of species. While research in all these areas has a bearing on the themes discussed in this entry, the question arises: In what respect does research on the origin of novelties and innovations go beyond these other agendas?

The research program described here belongs to the agenda of evolutionary developmental biology. Sterelny (2000) proposed a useful way to distinguish this program from the adaptationist program. While in both programs the spontaneous generation of heritable phenotypic differences and natural selection are likely to be involved, the relative role of these two processes differs. In adaptationist research, the focus of explanation is on natural selection, while the existence of heritable phenotypic differences is considered a boundary condition that does not enter the explanatory story. In the case of novelties, the assumption of ample natural variation of the kind necessary for evolving the derived character state is no longer possible. Instead the focus is on understanding which specific modifications in the ancestral developmental process were necessary to reach the derived character state. Here the existence and efficacy of natural selection is considered a boundary condition and does not explicitly enter the research program, while the mechanisms for the origin of the new generative possibilities is the real focus. This difference in perspective is congruent with the notion that innovations and novelties represent deviations from functional or developmental constraints acting in the ancestral lineage.

The concept of key innovation invokes causal connections between the acquisition of certain derived character states and the evolutionary dynamics of a whole clade. These consequences can neither be considered the direct result of natural selection nor the result of speciation per se. By implication, research on key innovations requires methodological tools that go beyond those used in the adaptationist and the origin of species programs. Hence research on the origin of evolutionary novelty and innovation represents a genuine expansion of the scope of evolutionary biology, primarily in terms of the inclusion of development.

MICHAEL HART

Larvae and Larval Evolution

The emergence of evolutionary developmental biology as a distinct discipline can be traced in part to the gradual acknowledgment that all stages of organismal development, including embryos and larvae, are subject to natural selection and adaptation. Examples of larval adaptation and evolution have been the focus of substantial attention since the nineteenth century (Lillie, 1898). However, this realization was captured most succinctly by Bonner's (1965) more recent insight that "the life cycle . . . is the unit of evolution" (p. 13). As a result, the larval forms of animals became objects of evolutionary study on par with adults. The functional and ecological study of larval evolution is complementary to, and sometimes at odds with, developmental biology. Evolutionary biologists have emphasized diversity of larval forms and hypotheses to explain the origin and maintenance of this diversity, whereas many developmental biologists have largely emphasized unity of mechanisms and processes among distantly related animal embryos.

For some developmental biologists, a larva is a distant abstraction that follows the more interesting events of fertilization, cleavage divisions, segregation of morphogenetic determinants, cell movement, inductive interactions, and other early events in the formation of cell lines, tissues, and organs. However, most major animal taxa include members that develop initially into distinctive larval forms in which the definitive adult structures are restricted to small, nonfunctional rudiments consisting of set-aside cells (Davidson et al., 1995). As a result, many of the early and important features of animal development appear to reflect natural selection acting on variation among larval forms, not adults. In this light, the analysis of larval diversity and evolution among higher taxa and among closely related species is important for a fuller understanding of evolutionary developmental biology.

The analysis of larval evolution is a taxonomically idiosyncratic affair: researchers often specialize on single phyla or classes (though usually not on single model species). Much of the literature on larval evolution is focused on three distinctive taxonomic groups: amphibians, insects, and that paraphyletic assemblage of lineages called "marine invertebrates." This entry largely reflects my own knowledge of and preference for marine invertebrates and particularly echinoderms at the expense of many well-known examples from other taxa.

There are at least four distinct but overlapping approaches to the study of larval diversity and evolution. The approaches use radically different technologies and often employ different exemplar species, but the problems analyzed within each approach have many elements in common. Each of the approaches is explicitly comparative.

The first approach is *developmental.* Practitioners in this area attempt to understand larval evolution in terms of familiar developmental processes (e.g., Raff and Sly, 2000). However, the specific goal is not to understand the general cellular mechanisms controlling these processes. Instead, the goal is to understand how these processes produce divergent larval forms whose morphological or functional features vary among groups. Much of this research is devoted to terminal larval structures or to parts that are not derived from set-aside cells in taxa that have maximally indirect development.

The second approach is *morphological.* Larval bodies bristle with structures adapted for life in a distinctive pre-adult habitat, including many transient structures lost at metamorphosis. Larval morphologists focus on the functional properties of larvae for swimming, feeding, defense against predators, or other kinds of mechanical and physiological interactions with the environment (e.g., Hickman, 1999). These interests overlap broadly with those of larval developmental biologists in cases in which both the developmental origins and the functional properties of particular structures can be studied.

The third approach is *ecological.* Larval ecologists focus on interactions between sperm and egg, between larvae and other organisms, and/or between larvae and physical features of the larval habitat (McEdward, 1995). In many study systems, this habitat is the marine plankton. These interactions influence fertilization rates, physical dispersal, growth rates and size at metamorphosis, mortality in the larval stage, selection of a site for settlement and metamorphosis, and responses to physical or chemical signals that influence settlement and metamorphosis. Dispersal includes dispersion of individuals away from their parents and diffusion of siblings away from each other. The goal of this research is an understanding of the effects of such interactions on population dynamics, community structure, biogeographical distribution, and other ecological phenomena. The outcome of these interactions depends to a large extent on morphological (or other phenotypic) variation among gametes or larvae, and so the interests of morphologists and ecologists also overlap broadly.

The final approach is *evolutionary.* Direct evidence for larval evolution comes from the fossil record. Remarkably, some small and delicate larval forms (or at least their hard parts) are occasionally found in well-preserved deposits. Other direct evidence comes from the fossilized skeletons of some adult molluscs and echinoderms in which a few characteristics of the larval stage can be inferred, such as the presence or absence of a larva that fed and grew in the plankton (e.g., Emlet, 1989). Indirect inferences of larval evolution come from explicit phylogenetic (as opposed to merely comparative) studies in which evolutionary changes in larval characteristics are optimized on a well-established phylogenetic hypothesis for groups with diverse larval forms (Hart, 2000). The phylogenetic approach leads to tests of hypotheses concerning the order, frequency, and rate of changes in larval forms. These phylogenetic analyses use information from all of the other approaches.

Perspectives on Larvae

The choice of approaches to larval evolution depends in part on the perspective of individual researchers on the role of larvae in a life cycle. There are at least three quite different perspectives on larvae in life cycles.

Larvae as paths to adulthood. In this view, larval forms are transient morphological way stations en route to the adult form. Their morphological differences from adults could merely reflect the constraints on building an adult phenotype from a single zygotic cell. Some developmental biologists consider the larvae of most arthropods and chordates from this point of view (Davidson et al., 1995) because these lineages turn the larval phenotype directly into the adult. The evolution of larvae in such a life cycle reflects processes, including selection, acting on mechanisms to produce the adult form.

Larvae as digressions in a formerly direct path to adulthood. Larval phenotypes from this point of view would be considered specializations for the larval habitat or lifestyle. Such specialization is sometimes considered to arise via selection favoring ecological separation between larval and adult stages of the same life cycle by evolving different lifestyles and niches before and after metamorphosis. Alternatively, such transient forms could arise by selection to take advantage of an empty niche that can be exploited by small larvae but not by large adults. Evolutionary changes among larvae viewed from this perspective suggest natural selection acting on the phenotype of the larva and not the adult. Selection favoring larval specializations is thus viewed as the source of both the distinctive and temporary morphological, behavioral, or physiological features of larvae, and of the dramatic metamorphosis seen in many such groups.

Larvae as ancestral adult forms without gonads. A few larval forms resemble the adults of related lineages but lack reproductive organs and behaviors. In these groups the main question is the order in which the two forms first appeared in their common ancestors. If the larval form appeared first, then such larvae represent ancestral adults in which gonad formation and reproduction have been delayed into a postlarval stage with a distinctive phenotype (the adult). Evolution of larvae from this perspective represents both changes in the path to adulthood and heterochronic changes in the timing of gonad maturation (see Gee, 1996).

The particular view of larvae depends in part on the approach taken by particular researchers and in part on the degree of specialization and metamorphosis found in the larvae they study. Developmental biologists tend to view larvae as paths to adulthood or as ancestral adult forms. Many developmental biologists study nematodes, chordates, and arthropods in which most of the larva is turned into the adult and metamorphosis is relatively mild. Morphologists and ecologists are more likely to view larval phenotypes as digressions in development in which larval specializations are also adaptations to a novel larval habitat or way of life. This point of view seems to be strongly informed by the catastrophic metamorphosis of many marine invertebrates in which much of the larval body is cast off (e.g., brittle stars), eaten by juveniles (e.g., nemerteans), or histolyzed (e.g., bryozoans) at metamorphosis.

Diversity

Larval diversity occurs on two taxonomically and morphologically different scales. The first is much less well understood than the second.

Diversity among phylotypic stages. Most phyla have a morphological stage in which important developmental events produce definitive features of the phylum (Raff, 1996). Well-described examples of such phylotypic stages include the pharyngula of vertebrates, the dipleurula of echinoderms, the nauplius of crustaceans, and the veliger of gastropods (Levin and Bridges, 1995). More controversial examples include the various "trochophore" larvae of annelids and their relatives. These phylotypic stages are icons of the taxa they characterize. However, the origins of this diversity among major taxonomic groups are not known. Why do classes and phyla have dramatically different characteristic larval forms?

Diversity within phylotypic stages: variations on a theme. Among members of the same phylum or class, larval forms also vary, but much of this variation on the phylotypic theme concerns variation in life history traits and reproductive tactics related to ecological modes of larval development (Strathmann, 1985).

Larval ecologists typically recognize four or five distinctive modes of development:

Planktonic development with feeding larvae
Planktonic development without feeding larvae
Benthic development in egg masses
Benthic development with external brooding
Internal brooding with live-born offspring

Many groups have evolved combinations of these modes (such as barnacles, in which early embryonic stages are brooded but later develop a feeding planktonic nauplius stage).

These modes of development differ in some important morphological or ecological features. *Egg size* of species with feeding development is typically small ($<500\ \mu m$) and eggs are yolk-poor but can be produced in clutches of 10^5–10^7. Eggs that become nonfeeding larvae are usually yolk-rich and large but often produced in smaller clutches of dozens to thousands. Exceptions include the relatively small eggs of some live bearers that are nourished indirectly by the parent or by ingestion of siblings.

Morphological complexity is typically greater for larvae that feed in the plankton or for larvae that swim and disperse (whether or not they feed). Brooded or live-born embryos and larvae are simplified (in comparison to their close relatives that live in the plankton) by the loss of many structures used for swimming, feeding, or defense against predators (Wray, 1995). Vestigial remnants of these structures in nonfeeding or nondispersing larvae imply that the direction of this change is usually from complex to simple and that these losses are probably not often reversed.

Rates of growth and development of feeding larvae are usually slower than for closely related larvae that are endowed with all of the materials and energy needed to turn a larger egg into a juvenile. Thus the risk of mortality before metamorphosis is probably greater for larvae that are required to grow in the plankton, or at least for larvae that have some obligate period of planktonic development before metamorphosis is possible.

Planktonic larvae receive little *protection from parents* other than the choice of time and place for their release into the plankton. In addition to avoiding the risks of the plankton, brooded or live-born offspring may be protected by capsules or jelly, by diurnal or migratory movement of the parent away from physical or biological threats, and by choices of microhabitats in which to deposit or brood clutches of offspring.

Consequences of Larval Evolution

The functional or evolutionary consequences of diversity *among phylotypic stages* are not known because there are relatively few explicit comparisons. One study has shown that the design of a ciliary suspension feeding apparatus is constrained by the body form to produce relatively low feeding rates in bryozoan cyphonautes larvae, but can be expanded to produce much higher feeding rates in echinoderm dipleurula larvae (McEdward and Strathmann, 1987). Many other functional comparisons among phylotypic stages would help shed light on the potential causes or consequences of this diversity. Some larval forms characteristic of higher taxa might be much better solutions than are others to the problems of life in the marine plankton, and the differences most likely represent true phylogenetic constraints on the potential for optimizing the performance of a phylotypic stage.

The consequences of variation *among modes of development* are better known because the morphological differences are more easily linked to functional and ecological processes. One important difference is the nature of trade-offs among life history traits. Species with small eggs, large clutches, and feeding larvae trade high fecundity against high mortality rate in the plankton. Variation in mortality rates among cohorts may produce frequent reproductive failures but occasional massive recruitment events in such species. Alternatively, species with large eggs, small clutches, and nonfeeding or nonplanktonic development trade low fecundity for higher survival rate of individual offspring. Recruitment rates of such larvae are often lower but more predictable.

The evolutionary loss of a planktonic larval stage—via encapsulation or brooding—should lead to reduced dispersal ability of larvae and less frequent movement of larvae and their genes between discrete populations of sedentary adults. On average, such evolutionary decreases in dispersal lead to more highly structured populations with larger allele frequency differences between populations (Bohonak, 1999). Localized recruitment of offspring should in turn allow enhanced adaptation to local conditions.

One major prediction following from the population genetic consequences of changes in developmental mode is higher rates of speciation and also extinction among lineages that have evolved low dispersal larval forms (Jablonski, 1986). These species-level differences in the fitness of lineages may be an example of true species selection, and could influence long-term trends in the frequency of species with different modes of development. However, a corollary of this prediction is the discovery of large clades of species with low dispersal larval forms in phylogenetic analyses. There are only a few examples of such phylogenies that could be used to test the species selection hypothesis (Hart, 2000), but they do not usually show rare loss of planktonic dispersal followed by large numbers of speciation events (Duda and Palumbi, 1999). Instead, most of these phylogenies show relatively large numbers of parallel losses of larval feeding or of planktonic dispersal. Phylogenetic analyses of larval evolution, combined with comparisons of genomic (e.g., Cameron et al., 2000) and genetic (e.g., Jeffery et al., 1999) variation among larval forms, represent one of the fastest-growing fields of larval evolution.

MARVALEE H. WAKE

Life History Evolution

Analyses of life history strategies focus on the demographic attributes of populations, often correlated with habitat-based ecological phenomena. Recently, such analyses have become comparative, within and across lineages. Because life history analysis is inherently comparative, this advance is both logical and progressive. Most features of life histories are presented in textbooks as representing elements of dichotomies (e.g., life span as long vs. short, numbers of reproductive events as semelparous vs. iteroparous, population growth patterns as "r" vs. "K" strategies), though in reality these are the extremes of continua. More attention is now being paid to the evolution of life history strategies; scientists currently place such analyses in a phylogenetic context so that directions of change (evolution) can be assessed.

During the last decade, analyses of life history evolution have begun to integrate genetic, developmental, and ecological components; and the studies have changed from the largely descriptive (though these are still useful and necessary) to the experimental, testing the effects of ecological parameters on life history attributes, the genetic bases of life history features, and the nature of plasticity of such attributes. Such research was stimulated by two landmark volumes: Dingle and Hegmann's *Evolution and Genetics of Life Histories* (1982) and Stearns's *The Evolution of Life Histories* (1992). Further, a body of theory has been generated that allows new insight into the evolution of life histories, including those features that are intrinsically developmental (Stearns, 1992; Charnov, 1993). It has become apparent that most life history strategies, and the features of life history evolution that are subject to study, have developmental bases.

Analyses of life history strategies typically focus on some or all of the following features in a population-based framework: size at birth; growth pattern; age and size at maturity; size, number, and sex ratio of offspring; size- and age-specific mortality; and lengths of generations. These features are traditionally analyzed within the context of characteristic population growth patterns—the so-called r and K strategies—with most species having some of the characteristics of both elements that define the spectrum. The r strategists have a high intrinsic rate of population growth characterized by exponential growth followed by crashes, and tend to live in unpredictable or rapidly changing environments. They reproduce early and have large numbers of small young that mature rapidly. Generation times are short, and there is little or no parental care. K strategists, in contrast, are fairly slow-breeding and live in stable, predictable environments. Their growth curve is sigmoid, and their numbers maintain at or just below the carrying capacity (K) of their environments. K strategists characteristically reproduce late, have few, large offspring that mature slowly, have long generation times, and effect parental care.

The simplest kind of life history analysis, as done for many bird and mammal species, explores and describes the features of birth, reproduction, and death; but many more complex life histories exist. Such patterns are generally grouped and considered as life history or reproductive modes. Alternation of generations among sexual and asexual organisms, complex genetic or chromosomal configurations in gametes, developmental stages (instars, larvae) that have different ecologies from the adults, oviparity vs. viviparity—all characterize different lineages of animals, and most have evolved several times in different lineages (convergent and parallel evolution). Similarly, patterns of development such as parthenogenesis, direct development, and viviparity vary among and within lineages and have evolved many times. Analysis of such reproductive and developmental patterns in a life history context emphasizes the role of correlation with the environment of the populations being studied and the populations' responses to ecological factors. Life history analysis is intrinsically developmental or ontogenetic, but this aspect is often not emphasized directly. For example, egg size and number, fecundity, peak egg production, time to metamorphosis or maturity, age at first reproduction are considered, but usually construed as demographic features rather than developmental variables.

The general goal of life history theory and analysis is to understand what combination of demographic traits is maximized by natural selection (see Miaud and Guyetant, 1998). The key current questions in the evolution of life histories, reproductive modes, and developmental patterns have to do with: (1) genetic basis of such features as egg size, longevity, and developmental patterns; (2) ecological parameters that influence the evolution of life history traits; (3) phenotypic plasticity and physiological substrates that underlie the expression of traits; and (4) interactions and trade-offs that occur in the evolution of life histories. Considerable experimental research involves controlling for single genes or single ecological features (e.g., temperature) that affect particular features of life histories; now more synthetic approaches that examine multiple components of life history evolution (e.g., natural selection and the genetic basis of temperature-dependent phenomena) are emerging. The stage is set and the tools are available for integrative studies of life history evolution for a broad array of species. Hypotheses based on predictions of life history theory can be tested in field and laboratory situations for constructed or natural populations, and their analysis, when comparative, can be examined in a phylogenetic context to assess the pattern of change within and across lineages (= evolution).

A number of case studies illustrate the developmental and evolutionary elements of life history analysis. Those selected for discussion here include such areas as (1) the interaction of physiology, developmental plasticity, ecology, and evolution; (2) the evolution of life history features in predation or parasitism systems; (3) the ecological and developmental life history components of the evolution of social systems, including those of humans; and 4) the evolution of alternative modes of reproduction, especially direct development and viviparity.

Physiology, Plasticity, Ecology, and Evolution

Sinervo and his colleagues are exploring several facets of life history evolution (Sinervo and Doughty, 1996; Sinervo and Lively, 1996; Sinervo and Svensson, 1998; Svensson and Sinervo, 1998). In a decade-long field study, they manipulated hatching density and egg size in a species of lizard to assess the theory that in density-dependent natural selection, intraspecific competition favors juveniles that have high competitive ability (Svenssen and Sinervo, 2000). The strength of selection on egg size changes significantly in the absence of older competitors. The effect of selection on egg size among later-clutch hatchlings released in areas without older intraspecific competitors was nearly twice that of hatchlings released where there were such competitors, as measured by size and survivorship.

Sinervo (1999) manipulated clutch size, egg size, and total clutch mass so that offspring size and number varied. He found that fecundity selection favored females that laid large clutches of small offspring, but was balanced by survival selection for large offspring; offspring number and quality trade-off had a stabilizing effect on mean egg size over several generations. The field- and lab-based experimental manipulations of life history parameters by this group are providing significant insights into the genetic basis of life history features, the nature of plasticity, and the effects of natural selection.

Qualls and Shine (1996) demonstrated the effects of phenotypic responses to incubation temperature in a species of lizard that is apparently an intermediate between oviparous and viviparous reproduction. They found that the range of reaction norms (phenotypically plastic responses) suggest that the shift from oviparity to viviparity must have induced changes in hatching times as well as hatchling morphology and behavior in the transitional form. This work examines the basis for a major life history shift and establishes a means of reconstructing ancestral reaction norms.

Gasser et al. (2000) examined life history correlates of evolution under high and low adult mortality regimes in *Drosophila melanogaster*, finding a number of physiological correlates of life history trade-offs. Not only did development rate and duration change and early fecundity evolve in the high-density group, but high adult mortality affected three life history traits that are expressed early in development: body size, growth rate, and ovariole number. Conversely, there was little or no effect on body composition, viability, metabolic rate, activity, or resistance to starvation or desiccation. They concluded that there were trade-offs between early and late fecundity, and starvation resistance, mediated by differential allocation of lipids. Such studies illustrate the need for a broadly integrative approach in order to understand life history evolution.

Predation and Parasitism Systems

Johnson and Belk (2001) examined life history phenotypes in a Costa Rican live-bearing fish, *Brachyraphis rhabdophora*. Populations that co-occurred with fish predators reached maturity at a smaller size and produced more and smaller offspring relative to populations from predator-free environments. Clear differences in the covariance structure of life history traits (which were highly correlated) allowed populations to be characterized by predation category.

Richner (1998) examined host-parasite interactions and the evolution of life history parameters. He noted that selection pressures are reciprocal and life history traits are often negatively intercorrelated so that a change in one trait occurs at the expense of another. One effect is that parasites can shift life history trade-offs in the host from an optimum without parasites to a new optimum with parasitism. The trade-offs occur at morphological, physiological, and behavioral levels. Richner tested his hypotheses by examining a bird-ectoparasite system, and showed that the parasite's host specificity and virulence evolved, effecting changes in the host's parasite avoidance and reproductive effort, and that there were parasite-induced maternal effects. The change in reproductive effort included a trade-off with resistance to *Plasmodium*, a blood parasite that causes malaria in birds. Richner provides a model for the consequences of host-parasite interactions for life history evolution, and especially the trade-off between current and future reproduction in iteroparous organisms.

Trouve et al. (1998) assessed the evolution of life history traits in parasitic and free-living platyhelminths. They used the independent contrasts method to look at patterns of interspecific covariation in adult size, daily fecundity, number and size of progeny, total reproductive capacity, age at first reproduction, and longevity among several species of worms. Increased longevity favors delayed reproduction, growth pattern determines adult body size and age at maturity, and total reproductive capacity is determined by body size. They found no effect of parasitism on the basic patterns of life history evolution in platyhelminths.

The Evolution of Social Systems

Svenssen and Sheldon (1998) assessed the way the social context in which a life history occurs can influence the evolution of that life history. When organisms interact, conflicts of interest occur. These are prominent in reproduction. Conflicts between the sexes and between parents and offspring influence many life history traits. Frequency- and density-dependent selection can arise from social interactions, which can influence the evolution of life histories. Svenssen and Sheldon state that studies of sexual selection have rarely included a life history perspective, and combining them might give new insight. They posit that including the roles of social interactions might resolve such questions in life history research as the maintenance of high levels of additive genetic variance for traits related to fitness.

Some studies that involve behavior and life history parameters do exist. For example, a major life history feature in eusocial insects is the longevity of the queens. Carey (2001) examined longevity patterns among bees, wasps, ants, and termites, in order to assess the nature of the evolutionary association of longevity and sociality in insects. Carey found that broad interactions of the physiology of the queen and the support provided by the social system, such as by provisioning the queen, had evolved. He provides a foundation for further research on the physiological, ecological, and evolutionary trade-offs involved, and the general implications for the evolution of longevity in all groups.

The evolution of life history parameters in primates, especially humans, has received considerable attention. For example, Alvarez (2000) examined life history traits in sixteen primate species in the context of life history theory. The adaptive significance of such traits as midlife menopause in human females has long been debated, but rarely in the context of primate life histories. In humans, late ages of maturity and higher-than-expected birth rates are associated with extended postmenopausal longevity. Alvarez concluded that links among "adjustments" in the primate pattern can explain how selection could slow senescence without extending the period of fecundity. She illustrates the utility of life history theory for examining questions of adaptive evolution of primate, especially human, life history traits.

Cooperative breeding has evolved in a number of lineages, both in vertebrates (e.g., birds and mammals) and in invertebrates (e.g., social insects). Pen and Weissing (2000) explored the interaction of ecology and life history traits in the evolution and maintenance of cooperative breeding. The mechanism of density regulation determines the factors that promote cooperative breeding. Low adult mortality favors cooperation, because it enhances the direct benefits of helpers. If fecundity is high, lower probability of obtaining a territory and lower survival of dispersers facilitates cooperation. Pen and Weissing found differences between birds and social insects in the covariance between cooperative breeding and life history traits, and they concluded that this is a consequence of different mechanisms of density regulation in different groups of animals. Similarly, Hatchwell and Komdeur (2000) examined the association of ecological constraints and the prevalence of cooperation in birds, using data from intraspecific studies. Low adult mortality and low dispersal appeared to predispose certain lineages to cooperative behavior, in specific ecological conditions. However, they emphasized that life history traits and ecological factors likely act in concert, rather than the former predisposing cooperation and the latter facilitating cooperation.

Alternative Modes of Reproduction

Nearly all features of life history study relate directly or indirectly to developmental patterns, whether considered for a single population or species, or in an evolutionary, lineage-based, comparative approach. One area of life history evolution that merits more attention is the evolution of derived modes of reproduction. The state usually considered ancestral for most animals is that of producing numerous small ova that are fertilized externally, typically in water, often with rapid development of small larvae or instars that have high mortality. Survivors metamorphose, and adults typically have a distinctly different morphology and ecology.

Conversely, derived reproductive modes often produce fewer young but have features, such as large eggs and/or parental care, that assure increased survivorship. Such derived reproductive modes as direct development and viviparity have evolved in many lineages of animals. Direct development—in which embryos complete metamorphosis before hatching from the egg membrane and there is no free-living larval stage—occurs notably in echinoderms and amphibians. Characteristically, fecundity is reduced but survivorship increased, because of the increased parental care often provided and the obviation of a free-living but nonfeeding period of metamorphosis. In direct-developing animals such as amphibians that lay their eggs on land, a fully metamorphosed juvenile hatches ready to feed in the adult mode and the whole world of aquatic predators is avoided. In many direct-developing species, features of larval morphology appropriate to the biphasic life of their congeners either never develop or are modified in terms of developmental pattern and regulation. Again, the association of life history parameters with development and evolution is profound.

Viviparity, the condition of live-bearing, usually by the female parent and often with maternal nutrition after the yolk provision of the egg is exhausted, has evolved in several animal lineages. Among invertebrates, viviparity occurs in at least thirteen orders of insects, in mites and scorpions, and in several groups of crustaceans among arthropods, onychophorans, annelids, molluscs, echinoderms, and even coelenterates. Some molluscs, insects, and onychophorans even have a placental analogue that facilitates nourishment. The range in live-bearing modes includes simple egg retention in or on the body of a parent to complex mechanisms for maternal nourishment of the embryos that involve elaborate modification of both maternal and embryonic tissues.

Characteristically, the evolution of viviparity involves a number of trade-offs in life history attributes. Fecundity is usually reduced, but developing young are protected. In forms with maternal nutrition, yolk supply is diminished and egg size reduced, but young are nourished after the yolk is exhausted. A well-documented example of such trade-offs occurs in thrips (Crespi, 1989). Within a single species, *Elaphrothrips tuberculatus,* females are facultatively viviparous. The proportion of viviparous females in a population varies in spring and summer, with a nearly 50:50 ratio in summer, but 25:75 viviparous in spring. In both cases, fecundity in viviparous females is half that of oviparous females, but relative survivorship to pupation is much higher for viviparous females. Further, viviparous females produce male offspring, and oviparous females produce female offspring; the mechanism is not known.

Viviparity has evolved in all vertebrate lineages except agnathans and birds. Teleost fishes and amphibians in particular exhibit many "natural experiments" in modes of egg retention—including carrying the eggs on their backs, in their mouths, and in their gill pouches—internal fertilization, and maternal and even paternal nutrition of the embryos they are carrying. Fish embryos exhibit different modes of pseudoplacentae to effect nutrient transfer; most amphibians do not have pseudoplacentae, but embryos/fetuses actively ingest nutrients orally. Among amniotes, the extraembryonic membranes may be modified to form a placenta, but both nonplacental and placental modes of live-bearing have evolved—more than 100 times in lizards and snakes (see summary in Wake, 1989). Marsupial and eutherian mammals have simple to complex placental arrangements. Given this complexity of morphological and physiological evolution, it is notable that the same kinds of life history trade-offs characteristically occur: decreased fecundity, increased survivorship, iteroparity, longer life spans, and in some lineages the evolution of social systems.

These examples of the interaction of life history features, the predictivity and testability of life history theory, and the interaction of development and evolution to modify life history parameters provide abundant evidence of the need for more extensive and integrative research on ecology, physiology, development, morphology, and evolution in order to understand the pattern and process of life history evolution.

DAVID R. LINDBERG AND ROBERT P. GURALNICK

Lineages: Cell and Phyletic

Differentiation into cell lineages is one of the earliest expressions of organismal morphology. While the undivided, fertilized egg is a relict structure from its parents, cell divisions followed by subsequent differentiation of cell lineages give rise to an unique individual. As early as 450 B.C., embryonic differentiation was seen as reflecting the "cosmogonic" processes (Gould, 1977), and this connection between ontogeny and phylogeny has been with us ever since.

The formation of cell lineages can restrict the developmental potential of descendant cells following differentiation (also referred to as specification). In different taxa, the timing and strength of differentiation vary widely; in general, it is divisible into two basic developmental patterns: *mosaic* and *regulative*. Mosaic developing embryos are characterized by early differentiation of cell lineages, while embryos under regulative development do not differentiate cell lineages until later in their development. These categories show a very high degree of correlation with both cleavage patterns (spiral versus radial cleavage) as well as the protostome (mouth first)–deuterostome (anus first) dichotomy. However, there are taxa that combine these characters to different degrees, making these distinctions neither absolute nor evolutionarily fixed.

In the late 1800s cell lineage studies were pioneered by a group of researchers at the Marine Biological Laboratory at Woods Hole (Stern and Fraser, 2001). Using direct observations of embryos with distinctive early cell morphologies (pigmentation patterns, cell size, etc.), C. O. Whitman, E. B. Wilson, and E. G. Conklin followed cell lineages in several marine invertebrate taxa, including Mollusca, Annelida, and Tunicata. Increasing use of photography allowed scientists to broaden studies to organisms without distinct cellular morphologies, but many taxa remained problematic because of their opaqueness or the movement of descendant cells deep into the developing embryo (Stern and Fraser, 2001). Marking individual cells with vital dyes and following their descendants subsequently reduced many of these problems. Today molecular techniques including the injection of retroviral vectors and monoclonal antibodies into early cells have joined vital dyes and transplantation as methodologies for reconstructing cell lineages. Regardless of the techniques used to follow and document cell lineages, the scheme for encoding cell lineages still follows that proposed by Conklin (1897) (Figure 1a).

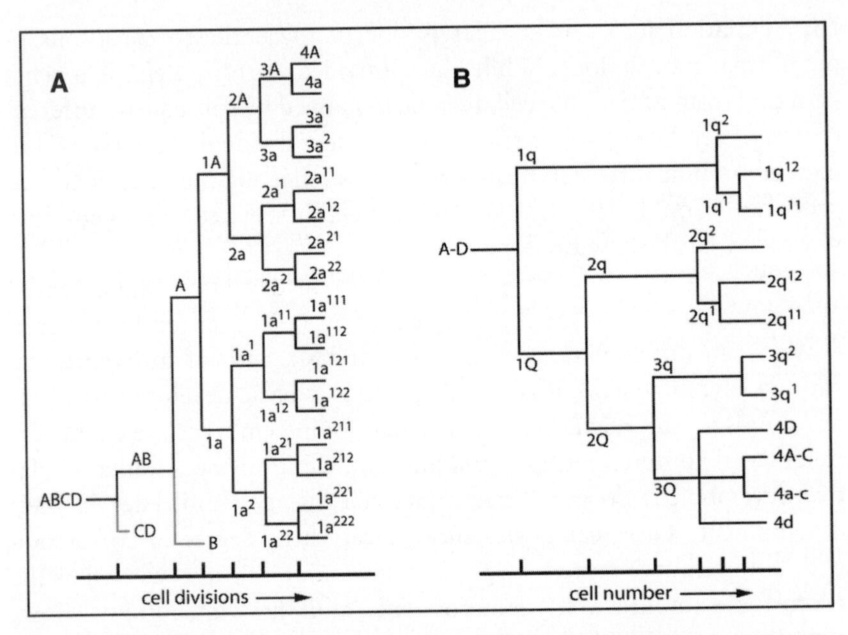

FIGURE 1. Cell lineage trees. a. Conklin's (1897) notation for the designation of cell lineages. For brevity, only six cell division cycles are shown and for only the A macromere; identical representations exist for the B, C, and D macromeres. While this node-based view of a strictly bifurcating tree accurately reflects the number of cell divisions at which a given lineage is formed, cell lineage data in this format is not generally useful for comparative studies between taxa because of its invariance. Alternatively, cell lineage trees can be scaled by the number of cells present in the embryo at the time of appearance of a given cell lineage, as in b. In this notation, Q = macromere quartets A–D, q = micromere quartets a–d, and the timing of the appearance of each lineage is defined relative to cell number. For example, although both the 2A and $1a^1$ cell lineages (represented by 2Q and $1q^1$ in Fig. 1b) arise at their respective fourth cell divisions (Fig. 1a), 2A forms when eight cells are present, while the $1a^1$ cell lineage does not appear until the stage of twenty-four cells. When substantial variation exists within a quartet (such as the acceleration of the 4D and 4d origination shown at the lower right), cell number can be calculated for individual or subsets of cells.

The invariant aspect of many spiralian cell lineages, especially within the Annelida and Mollusca, as well as their correlation with the protostome-deuterostome dichotomy led many early scientists (1892–1907) to consider these characters as important and conservative phyletic markers. However, these views were primarily based on Haeckel's biogenetic law and prospective significance—what the cell became, not what it was. Even on issues like adaptation, most lineagists argued teleologically from the adult backward. By 1900, most of those working with cell lineages were to varying degrees mechanists, experimentalists, and recapitulationists simultaneously. The exception was Conklin, whose views were more akin to those of a Darwinian evolutionist. Lineage work eventually declined and, by 1907, published accounts of new lineages had almost stopped. Both established and younger researchers stopped taking on cell lineage projects—possibly because the general patterns were the same for all the spiralians while the specifics showed too much variation. It was theoretically hard to encompass or analyze the minutiae of variation in a recapitulationist or mechanist framework.

With the advent of the Modern Synthesis, neither cell lineage studies nor the relationship between ontogeny and phylogeny were regarded as modern topics in evolutionary biology. However, that all changed with the discovery of regulative genes (e.g., Gerhart and Kirschner, 1997). The role of maternal cytoplasmic determinants as well as zygotic transcription factors and signaling systems, which can be passed from cell to cell during cell divisions, is critical to our understanding of evolutionary developmental biology today. Phyletic comparisons of cell lineage trees and their contributions to different components of the phenotypes promise to revolutionize comparative morphology.

Much of this renaissance of cell lineage data in evolutionary biology has occurred during the last ten years and is directly related to the revolution in our understanding of regulatory genes and their networks (Davidson, 1990, 1991). In another approach, the timing of origination of cell lineages has also been argued to be informative in phylogenetic reconstruction and studies of adaptation. For example, Freeman and Lundelius (1992) used general embryological information, including the timing of origination of the D-quadrant, to examine relationships and patterns in the evolution of the life history characters in eutrochozoan taxa.

A similar approach was taken by Wray (1994) in an analysis of cell lineage evolution in the echinoderms. Wray reconstructed and mapped character transformations onto an echinoderm tree to produce an evolutionary history of echinoderm cell lineages. Many of Wray's characters were specification events in cell lineages relative to cleavage number. From this analysis, Wray concluded that different cell lineage characters showed contrasting evolutionary patterns. While some predated the Cambrian origination of the Echinodermata and were broadly conserved across echinoderm subclades, other character transformations were more variable. Extensive reorganization was possible in relatively short periods of time, often associated with changes in life history features (Wray and Raff, 1989).

Van den Biggelaar and Haszprunar (1996) also made both phylogenetic and adaptive arguments using cell lineage data from twelve gastropod mollusc exemplars. They tested the hypothesis that cell lineage patterns in the Gastropoda were informative as to taxonomic and evolutionary ranking. Their focus, like that of Freeman and Lundelius, was on the formation of the 3D macromere and the 4d mesentoblast, which play a critical role in the determination of the developmental fates of a large number of blastomeres and, to a lesser extent, on the retardation of the appearance of the 1a-d cell lineages (Figure 1b). However, unlike Freeman and Lundelius, who mapped cell lineage data onto a putative phylogeny, Van den Biggelaar and Haszprunar constructed four alternative trees from cell lineage timing data, all of which were reasonably congruent with contemporary gastropod phylogeny (e.g., Haszprunar, 1988). More recently, Boyer and Henry (1998) re-examined the evolution of spiralian development, focusing on cleavage patterns, cell lineages, and the formation of mesoderm.

Guralnick and Lindberg (2001) used an approach similar to that of Van den Biggelaar and Haszprunar: they used the timing of cell lineage formation to produce phylogenetic trees rather than map the characters onto a preexisting tree. Unlike Van den Biggelaar and Haszprunar, they used multiple cell lineages in their analysis and expanded the study group across the Spiralia to include sixty-seven taxa. Their results showed that cell lineage data reconstructed a phylogenetic hypothesis that was similar in several respects to patterns found in other more traditional analyses. They also looked at the range of variation present at equivalent nodes within quartet formation, the variation in timing of cell divisions in cell lineages across taxa, and how variation was sequestered between clades.

In addition to the presence of phyletic information present in the cell lineage data, Guralnick and Lindberg found that the mesentoblast (4d) was unique compared to other cell lineages in that it sped up and slowed down relative to other cell lineages in taxa with both unequal and equal cell sizes. They also showed that some cells that form in the same quartet, at the same point in the cell lineage hierarchy, have less variation than other analogous cells. This result argues for architectural constraint or stabilizing selection acting on those cells. In addition, the timing of formation of progeny cells in the first quartet had lower variation than the parent cells, suggesting that some "regulation-like" behavior might also be present. Overall, the distribution of variation had a phyletic component as well as a cell lineage component; although variation in cell timing generally increased during development, it also varied between taxa (Figure 2).

The finding that cell lineage data have phylogenetic value should not be surprising in light of the earlier work of Wray (1994) and Van den Biggelaar and Haszprunar (1996). As expected of any single character suite analysis, different levels of resolution are found in different clades. For example, taxa such as the Clitellata, Echiura, and Gastropoda were monophyletic, well resolved, and often congruent with trees based on both molecular and morphological data. Other groups such as Bivalvia and Polychaeta were not well behaved relative to traditional groupings. As Wray argued for the Echinodermata, some cell lineage characters are remarkably conserved while others are highly variable, and such an assortment of character transformations contributes both apomorphy and homoplasy to phylogenetic analyses. This is especially true when the character set comes from single systems or features and remains unbuffered by characters from other systems with different evolutionary transformation patterns.

Overall, the rate of cell division, relative rates of division among different cell lineages, and total number of cell divisions for specific cell lineages can vary widely within related taxa (Freeman, 1979; Jeffery and Swalla, 1992; Martindale and Henry, 1992; Wray, 1994), providing an important perspective for evolutionary developmental biology. One of the key questions is: What is the contribution to phenotypic diversification of these relatively small offsets and onsets in early cell lineage formation? In some cases, these changes appear directly related to life history evolution (Freeman and Lundelius, 1992; Van den Biggelaar and Haszprunar, 1996; Wray and Raff, 1989). In other taxa, changes appear to be associated with diversification of body plans irrespective of life his-

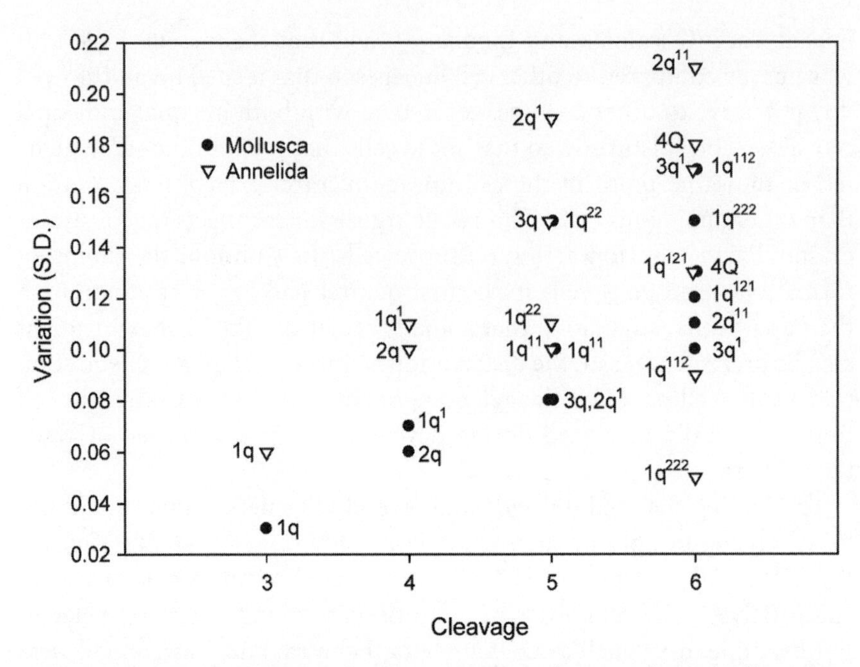

FIGURE 2. Variation in cell lineage formation relative to log-transformed cell number plotted against cell division cycle for the Annelida and Mollusca. Variation (represented by standard deviation) was calculated for 14 cell lineages in 22 annelid and 29 molluscan taxa (data from Guralnick and Lindberg, 2001). Variation increases with cell division cycle in both taxa, but in substantially different ways. While third and fourth cleavages show roughly symmetrical increases in variation in the Annelida as compared to the Mollusca, beginning with the fifth cleavage cycle, this symmetry is absent. For example, although variation in the $1q^{11}$ cell lineage is identical in both the Mollusca and Annelida, other cell lineages show opposite patterns (e.g., 3q and $2q^1$); and by the sixth cleavage cycle, all correspondence in variation is lost between the two groups. Overall, the Annelida have higher levels of variation in cell lineage formation than the Mollusca.

tory (Jeffery and Swalla, 1992; Wray, 1994; Ponder and Lindberg, 1997; Guralnick and Lindberg, 2001). Regardless of the ultimate causality, changes in the timing of formation of cell lineages can alter the spatial and temporal patterns of cells within the developing embryo. These new alignments will undoubtedly affect subsequent specific inductive interactions between cells while shifting other specifications from local to global patterning signals, or vice versa (Freeman, 1979; Davidson, 1990; Wray, 1994). Moreover, temporal changes in cell lineage formation might disrupt regulatory gene activity that previously constrained variation in certain quartets by delimiting specific periods during which gene expression or signaling occurred.

Cell lineage patterns represent important data in the study of metazoan evolutionary development. They contain important phyletic information that can aid in the construction of phylogenetic hypotheses, especially among spiralian taxa. In other cases, mapping of these characters onto robust phylogenies may prove more insightful. In addition, comparative assessment of cell lineage data in a phyletic framework can provide insights into major transformations in the morphology of embryos, larvae, and adults across clades. Other aspects of cell lineage data, such as fate maps and spatial differences in formation, may prove as crucial and informative as timing data when examined in a modern, comparative context.

WALLACE ARTHUR

Micro-, Macro-, and Megaevolution

Evolution is a continuous process. Any extant organism could in principle (but not in practice) be traced back through a chain of ancestors to a unicellular prokaryote living some four billion years ago. Throughout the history of all multicellular lineages, the genetic material has been altered by mutation, the ontogenetic trajectory changed by developmental reprogramming (Arthur, 2000), and the resultant organisms subjected to the scrutiny of natural selection. Given this continuity of descent and uniformity of mechanism—Leroi's (2000) "scale independence"—it would seem that there is just evolution, rather than the three fundamentally different levels of that process implied by the prefixes "micro," "macro," and "mega." However, this unitary view of evolution has been challenged from two perspectives: developmental-genetic and paleontological. Here, I focus on the former. To pursue the extensive paleontological debate, see Stanley (1979), Erwin (2000), Jablonski (2000), and references therein.

Origins and Meanings

The terms *micro-* and *macroevolution* were first used in English by Dobzhansky (1937), following on from Filipchenko's Russian equivalents (Adams, 1980). Dobzhansky characterized microevolutionary changes as alterations in the composition of populations "observable within a human lifetime," in contrast to macroevolutionary changes "which require time on a geological scale" (p. 12). *Megaevolution* was added by Simpson (1944), who restricted "macro" to evolution at the species and genus levels, using "mega" to apply to "large-scale evolution," or "the differentiation between families, orders, classes and phyla" (p. 98).

A problem exists here. Two very different things are being confounded—level (= rank) of taxon and length of time. To ask questions such as "Do macro- or megaevolution require additional mechanisms beyond those involved in microevolution?" we must eliminate this confusion and use the terms unambiguously. This is especially necessary given the frequent noncorrespondence between length of time and amount of evolutionary change, which is manifested in two ways: long periods of time with negligible change in many lineages; and rapid, explosive, or bush-like radiation of many high-level groups in comparatively short periods of time. An example of the former is the remarkable similarity between the fossil scolopendromorph centipedes of the Mazon Creek Formation, which date from some 300 million years ago (Mundel, 1979), and their modern-day equivalents (Lewis, 1981). Possible examples of the latter include the rapid radiation of many orders of mammals around 65 Ma and the Cambrian explosion (Valentine, 1991, 1994), although molecular studies have questioned the timing and explosiveness of both events (Hedges et al., 1996; Wray et al., 1996; Bromham et al., 1998; Kumar and Hedges, 1998).

In this entry, I use the terms micro-, macro-, and megaevolution to refer to levels of taxon rather than lengths of time for the following two reasons. First, Goldschmidt (1940) brought this whole issue to prominence, some might say to notoriety. Although he started from Dobzhansky's time-based definitions, he clearly departed from them and ended up focusing on levels of taxa. For example, when Goldschmidt starts dealing with macroevolution, he talks of "the lower level of macroevolution" as being "the evolution of species, genera, and even families" (p. 184). His higher level, orders to phyla, equates with Simpson's megaevolution.

Second, one of the most interesting issues in the whole of evolutionary biology, at least in my view, is the origin and divergence of radically different types of organism—that is, different body plans or *Baupläne* (Arthur, 1997). This is perhaps especially intriguing in the case of sister phyla of radically different design, e.g., the deuterostome phyla Echinodermata, Hemichordata, and Chordata; see Bromham and Degnan (1999) on the relationships between these phyla. In contrast, the question of whether evolutionary changes taking hundreds of millions of years involve different mechanisms from those taking thousands of years is less meaningful. The answer is likely to be highly lineage-dependent. In slow-evolving lineages, like the scolopendromorph centipedes, the answer may simply be "no."

Toward a Probabilistic Approach

Goldschmidt (1940) took what might be called an absolutist approach. He believed that there was a qualitative distinction between micro- and macroevolution. He concluded his book in characteristic style: "There is no such category as incipient species. Species and the higher categories originate in single macroevolutionary steps as completely new genetic systems" (p. 396). Simpson (1944) also suggested a qualitative distinction, but of a different kind and at a different level: "The paleontologist has more reason to believe in a qualitative distinction between macroevolution and megaevolution than in one between micro-evolution and macroevolution" (p. 98).

If either of these proposed qualitative distinctions were valid, the whole of population genetics would be irrelevant to an understanding of the major events in evolution. However, the accumulated work of the last half-century provides little support for qualitative distinctions, and I do not believe in their existence, except for the obvious distinction that macro- and megaevolution by definition include reproductive isolation, whereas microevolution does not. In other words, I do not believe that micro-, macro-, and megaevolution are fundamentally different processes powered by distinct mechanisms and with clear boundaries between them. The approach I take, therefore, differs from that of Goldschmidt (1940), who looked for mechanisms that operated in macro- (and mega-) evolution but not within the confines of the species; see his now-discredited ideas of chromosomal repatterning and systemic mutation. Instead, I ask a single, deceptively simple probabilistic question, involving the distribution of evolutionary changes in ontogeny along the dimension of developmental time—from zygote to adult.

To visualize this question, a thought experiment is useful. Suppose we wish to compare 100 pairs of congeneric sister species with 100 pairs of sister classes or other higher taxa. I use "class" very loosely; it has no precise definition anyhow. For each pair of species, starting from the first cleavage divisions and moving sequentially through development toward the adult, we ask when we first encounter a different pattern of behavior in the proliferating cell population between the two species being compared, for example, different in terms of division, differentiation, or movement. Providing that each species has a more-or-less fixed ontogeny, it should be possible to identify a particular time in development when a difference is first observable. Given 100 such comparisons, there will be a frequency distribution, perhaps something like that shown in Figure 1(a), though in reality the distribution will be much more complex.

Comparisons between sister classes pose a much greater problem, because of the considerable variation in the pattern of ontogeny *within* any one class. For the moment, we adopt the notion of looking at an extant representative member of each class—though, as will become clear, what is really of interest (but not observable) is the ontogeny of the ancestral species of each class. So, for example, we may choose the mouse as a representative mammal and the chick as a representative bird, simply because these have been used as model systems and there is a wealth of developmental data available for both. Note, however, that extant mammals and birds are not sister groups; the living sister group of birds is the order Crocodilia, while for mammals it is, arguably, the rest of the Amniota.

As with the comparison of a pair of sister species, we can ask, for each comparison of the representatives of a pair of sister classes, when in developmental time there are first observable differences between them in cell behavior, again in terms of division, differentiation, or movement. Given 100 such interclass comparisons, we will again get a frequency distribution, perhaps as shown in Figure 1(b), but again with many complications.

Ignoring complications for the moment, and treating the two distributions in highly idealized form as normal or Gaussian curves, as in Figure 1, there are three possibilities regarding the relationship between the intercongener and interclass distributions:

1. The distributions are indistinguishable; i.e., the two curves are precisely superimposed on each other.

2. The distributions overlap, but nevertheless one is significantly shifted relative to the other. (This is the situation in Figure 1, with the interclass distribution shifted towards earlier ontogenetic divergence.)

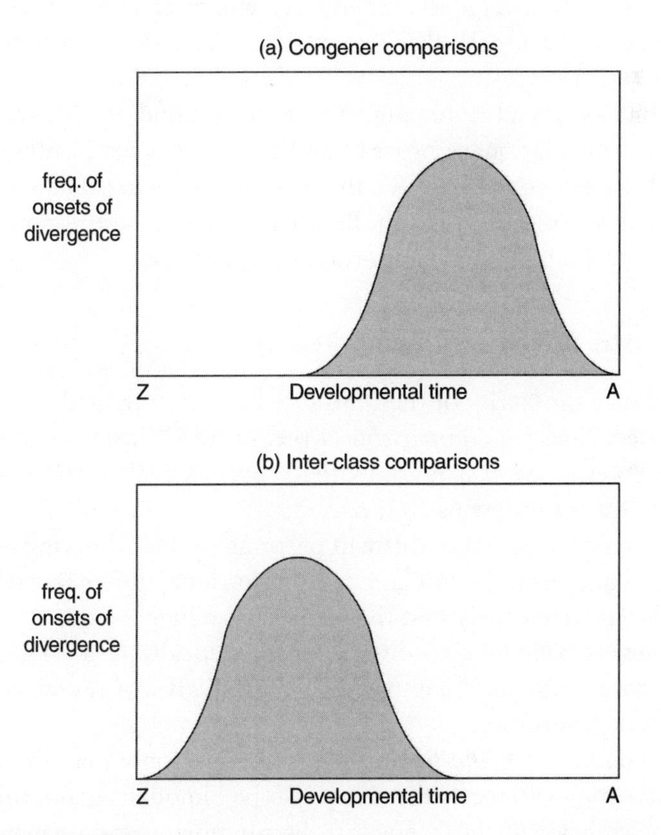

FIGURE 1. Possible difference between intercongener and interclass frequency distributions of initial ontogenetic divergence times (Z = zygote, A = adult), shown in simplified form before taking into account a series of complications. Compare with Figure 2.

3. The distributions are disjunct. There is no overlap. Representatives of different classes *always* diverge earlier in ontogeny than do congeneric species (always later is also theoretically possible, but implausible).

This is just a thought experiment with no known answer. There have not, in the entire history of comparative embryology, been 100 detailed comparisons of either kind. But *some* comparisons of each kind are possible on the basis of existing data, and they enable at least the elimination of the third possibility, above. The distributions cannot be disjunct because congeneric species can be found whose embryogeneses differ markedly at the very earliest stages—for example, at the first cleavage division for sinistral and dextral gastropod species (Verdonk and van den Biggelaar, 1983; Robertson, 1993). Many interclass comparisons also show very early divergence. Therefore, whatever else is true of their shape and position, both distributions must include the very beginning of development.

Possibilities 1 and 2 remain. These correspond to different world views of the evolutionary process, and there are as yet insufficient data to enable a clear choice between them—though my personal view is that the second is closer to the truth. Before rehearsing these arguments, it is necessary to examine some important complications.

From Abstract Pictures toward Reality

To maximize the clarity of the central issue, I have treated this problem as if it were simply a comparison of the means of two normal distributions. However, the distributions that apply in reality are much more complex, for the following four reasons.

1. *Mean versus variance.* Instead of, or as well as, differing in means, the intercongener and interclass distributions may differ in variance. In particular, it seems likely that the former has a higher variance, with sister congeners beginning to diverge ontogenetically at a wide range of points, while interclass comparisons almost always reveal very early ontogenetic divergence.

2. *Bimodality and the phylotypic stage.* Not only may the distributions have high variances, they may also be bimodal. Again, this is particularly likely for the intercongener distribution. There are many examples of congeners diverging very early in development—for example, the gastropod case noted above and the sea urchins *Heliocidaris erythrogramma* and *H. tuberculata* (Wray and Raff, 1989). Equally,

there are many cases in which congeners appear not to exhibit differences until much later in development, well after the onset of organogenesis, though it is harder to be certain of this because we are relying on negative evidence. If there is a dearth of instances in which intercongener ontogenetic divergence begins in between these extremes, as may well be the case, the distribution will be bimodal (Figure 2a).

This bimodality connects with the idea that there is, at least within some phyla, a stage in development that is highly conserved. The phylotypic stage, sandwiched between more variable early (egg, zygote, cleavage, gastrulation) and late (organogenesis, allometric growth) stages, is a stage that gives an overall egg timer or hourglass shape to a group of ontogenies (Sander, 1983; Slack et al., 1993; Duboule, 1994). Although it is clear that the idea of a precise phylotypic stage such as the tail bud stage of vertebrates is too simple (Richardson et al., 1997), a less well defined phylotypic period undoubtedly exists.

Whether the existence of this relatively invariant period renders the interclass distribution bimodal is a moot point. If all interclass comparisons reveal very early divergence (i.e., before the phylotypic period), the interclass distribution is unimodal with low variance (Figure 2b). However, at times sister classes may not be distinguishable until after the phylotypic period. Whether this is so brings us to another problem: Precisely what is being observed? For example, the eggs of birds and placental mammals could hardly be more different, but the early embryos that develop from or within those eggs are similar.

3. *Evolutionary changes in developmental timing.* So far, I have treated all comparisons as if the total span of developmental time is the same in the species or classes being compared. Clearly this is not so, but this problem can be minimized by dealing with recognizable stages, e.g., gastrulation, rather than with absolute time, e.g., days since fertilization. Thus divergence in genital arch structure between a pair of *Drosophila* species that is not manifested until metamorphosis is much later in development than a change in the embryonic proliferation of segments in a geophilomorph centipede, which occurs about halfway through the embryonic period, despite the fact that the latter occurs about a week later than the former in postfertilization real time (Coyne, 1983; Kettle et al., 2002).

A more difficult problem arises when an evolutionary divergence causes the development of some particular feature to be shifted relative to others in one of the descendant lineages *(heterochrony).* Suppose the ancestral state was for character X to appear late in development, but the derived state in one descendant lineage is for the same character to appear much earlier. Does this represent an early or late divergence in the pair of ontogenies concerned? It seems best to retain the approach of treating the onset of divergence as the time at which the ontogenies being compared can first be distinguished, which is thus the *early* point in developmental time at which character X can be observed in one lineage but not in the other.

4. *Complex life histories.* The evolution of complex life histories (Nielsen, 1998) is beyond the scope of the present entry. However, their existence potentially affects the kind of comparison under discussion,

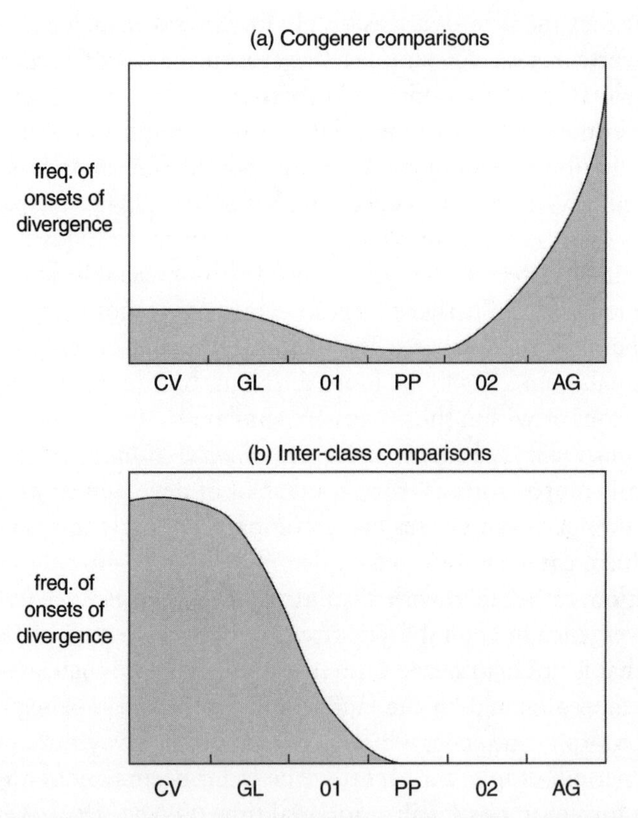

FIGURE 2. Possible difference between intercongener and interclass distributions, as in Figure 1, but now allowing for possible bimodality—manifested in (a) but not (b)—associated with the existence of a phylotypic period. (CV = cleavage; GL = gastrulation; O1 = beginnings of organogenesis; PP = phylotypic period; O2 = continuing organogenesis; AG = allometric growth.)

i.e., the time of onset of ontogenetic divergence. If the comparison is between two rather similar complex life histories, such as in the case of two *Drosophila* species, or between two closely related marine gastropods with trochophore larvae, then there is no problem. But, especially when comparisons are made between representatives of different higher taxa, we may end up comparing a complex life history with a simple one, for example, insect with centipede or frog with mouse. In such cases, the alignment of stages becomes difficult, at least over some parts of the developmental time axis. Luckily, in many such cases, the onset of observable ontogenetic divergence occurs sufficiently early that the two ontogenies are still broadly comparable; the comparison is thus unaffected by the fact that the correspondence of stages may break down later on. This is true, for example, for the insect/centipede comparison, in which the embryos can be readily distinguished at the embryonic segmentation stage or even earlier, and thus long before the developing insect undergoes metamorphosis.

Alternative World Views of the Evolutionary Process

There is no evidence for a qualitative distinction of mechanism between microevolution on the one hand and macro- or megaevolution on the other. If such a distinction is rejected, there remain two very different views of the ways in which new higher taxa originate in evolution.

1. The origins of higher taxa require no special explanation at any level. The subset of speciation events involved in the establishment of what later can be seen as a higher taxon is effectively a random sample of such events. Higher taxa become distinguishable only with the passage of time, perhaps aided by extinctions in flanking lineages. The idea of a body plan is fanciful. Embryonic and postembryonic development undergo small changes during most or all cladogeneses (and during anagenesis too). The developmental changes involved in the speciation events associated with the appearance of a lineage leading to a new higher taxon are routine.

2. Development is an essentially hierarchical process. Some developmental genes act early and make key decisions (e.g., specifying the anteroposterior axis); others act later and have more minor effects, either in the sense of operating within a restricted spatial domain of the embryo, or in the sense of making subtle quantitative (e.g., allometric) changes to the whole, but without radically altering the relative spatial

arrangement of tissues and organs. The developmental hierarchy (or morphogenetic tree: Arthur, 1988) is, however, a messy one with no clearly defined levels. Also, some genes are expressed for longer periods than others, and some are switched on and off twice or more as development proceeds, sometimes having different functions corresponding to their different expression periods (Salser and Kenyon, 1996). Nevertheless, despite such complications, the genetic control of development is best seen as hierarchy + noise rather than as an amorphous tangle of effects.

There is a connection, albeit only a statistical one, between the structure of the developmental hierarchy (itself evolutionarily labile) and the appearance of higher taxa at different levels. The rare changes that go deep into ontogeny and involve the genes controlling early key decisions are more strongly associated with the appearance of lineages leading to new higher taxa than are the more frequent "shallower" changes involving later-acting genes. However, no single speciation event creates, all at once, a new body plan in the way that Goldschmidt (1940) imagined. Also, mutations of genes that make key developmental decisions are not necessarily large-effect mutations (Stern, 1998). But overall, there is a statistical connection between the developmental and phylogenetic hierarchies. There are not three distinct processes—micro-, macro- and megaevolution—but there is a continuum from micro to mega.

A few points of clarification may be useful. First, these two alternative views are hypotheses or, perhaps more accurately, sets of interrelated hypotheses. It is not yet clear which is closer to the truth, and even with the accumulation of more data they will not be easy to distinguish. Second, I have deliberately avoided naming the two views. It is tempting to label the first the neo-Darwinian view, because a reading of Dawkins (1986) or other general neo-Darwinian accounts of the evolutionary process engenders a mental picture corresponding to that view. However, I suspect that some neo-Darwinians would support one view and some the other. It is also tempting to label the second view the evo-devo view because several scientists in the field of evolutionary developmental biology have proposed something very like the scenario depicted in the second view (e.g., Thomson, 1988). But again, I suspect that the evo-devo community is not characterized by unanimous support for either view. Third, these two views do not connect cleanly with the paleontological debate about micro-, macro-, and megaevolution. That debate has a very different focus—for example, the role of species selection. Finally, the way in which these views connect with the kinds of distribution depicted in Figures 1 and 2 requires careful consideration, especially in relation to the distinction between representative and ancestral species.

Representative versus Ancestral Species

There are pragmatic reasons for selecting particular species, such as the mouse or chick, to represent higher taxa (mammals or birds). Because so much is known about the development of such model organisms, it is easy to make detailed embryological comparisons and to be fairly certain exactly when phenotypic differences between the embryos are first apparent. However, there is a problem with this approach. Two representative species of this kind are separated by a very large number of speciation events (perhaps thousands) since their last common ancestor. Each of these events will have involved embryological changes. Some of these will be early, and others late. The variable "earliness of onset of ontogenetic divergence" acts in the same way as a max/min thermometer. That is, it reflects the most extreme occurrence, in this case the *earliest* developmental changes. But it is not known in which speciation event these occurred. As a result, a contrast such as that shown in Figure 2, between intercongener and interclass distributions, could be compatible with *either* of the two world views described earlier.

The problem derives from the fact that we would like to, but cannot, compare the embryogeneses of the two extinct ancestral species leading to the two higher taxa concerned—that is, the two daughter species of the last common ancestor of the two clades. Although no single speciation event creates a new higher taxon in itself, because multiple accumulating changes follow later, there may be something special about these speciations at higher taxon branch points. The argument I am advancing is that these speciations are not in any sense unique, but nevertheless are a very nonrandom sample of speciation events, both in relation to the *timing* of developmental changes and the *type* of changes (though this entry deals only with the former).

Given the impossibility of looking at this problem directly, rather than through representative species, supporters of this view will have difficulty providing conclusive evidence in its favor, as will dissenters attempting to disprove it. In a sense, this illustrates the problems that have beset the debate about micro-, macro-, and megaevolution since its beginnings in the 1930s and 1940s. In order to make progress, the best strategy for the future is to examine in detail the embryology and developmental genetics of a much wider range of extant taxa than currently available.

The idea that there are qualitative distinctions in mechanism between micro- and macroevolution, or between macro- and megaevolution, is in my view incorrect. Nevertheless, approaching the issue through comparisons of extant species acting as representatives of higher taxa suggests that macro- and megaevolutionary divergences may represent highly nonrandom subsets of speciation events generally. However, there is a difficulty in this approach. It is indirect and may be misleading in relation to the comparison we would ideally like to make but cannot: between the ancestral species of each clade. Therefore, neither of the two conflicting world views described can be conclusively confirmed or disproved. However, a greater taxonomic spread of species for which there is detailed developmental information will help in future attempts to distinguish between them.

This debate about developmental-genetic aspects of micro-, macro-, and megaevolution is quite distinct from the corresponding paleontological debate, which focuses on other issues such as species selection.

GILLIAN L. GASS AND JESSICA A. BOLKER

Modularity

Modularity, a biological approach that views organisms as the integration of partially independent, interacting units at several hierarchical levels, has been described as "a conceptual framework for evo-devo" (von Dassow and Munro, 1999, p. 307), and "a meeting place for evolutionary and developmental biologists" (Bolker, 2000, p. 770). Such declarations recognize that the relationship between the organism and its component parts is a central concern both within each field, and at their intersection.

For developmental biologists, the modularity question centers on the process by which an organism can be reliably constructed through the differentiation and integration of cells, tissues, organs, and organ systems. Evolutionary biologists seek to understand how species can respond to selective forces by a change in only a specific character under selection, with minimal effect on other physiological processes or on the rest of the body plan. Evolutionary developmental biologists ask how organisms (embryos, larvae, and adults) can combine their developmental resources to generate variation and thus, potentially, novel adaptations. A view of modularity that envisions modules as resources has much to offer this field, which seeks to address both how modularity in development enables modularity in evolution and how modular ontogenies have themselves evolved.

Evolutionary biologists and evolutionary developmental biologists may appear to be asking nearly the same question about parts and wholes. The critical difference is that a traditionally evolutionary outlook addresses only the adult organism. That is, a strictly evolutionary definition of modularity need not reference underlying generative processes. The key feature of an evolutionary module is its ability to change independently of the rest of the body. The evolutionary definition of modularity thus focuses on the dissociability of different elements of the phenotype, which permits processes such as mosaic evolution (in both morphology and genes) and whole gene duplication and divergence (Raff, 1996; Wagner, 1996; Gerhart and Kirschner, 1997; Brandon, 1999).

This same dissociability allows us to isolate and compare discrete characters, such as the limbs of mammals as disparate as bats and whales, and assess their homology. The hierarchical nature of homology (Roth, 1991; Striedter and Northcutt, 1991; Hall, 1994; Dickinson, 1995; Bolker and Raff, 1996; Abouheif, 1997) reflects that of the modules themselves: for instance, within the limb the skeleton and musculature may vary independently, as may individual skeletal elements. The divisibility of the overall phenotype (at both the structural and the genetic levels) into relatively independent, hierarchically organized subelements allows natural selection to act locally to shape specific adaptations, thus enhancing evolvability of the phenotype as a whole (Jacob, 1977; Lewontin, 1978; Wagner, 1996; Wagner and Altenberg, 1996; Baatz and Wagner, 1997; Gerhart and Kirschner, 1997; Brandon, 1999; Schank and Wimsatt, 2001).

Echoing the evolutionary focus on the independence of modules, developmentalists' modularity concept emerges in part from the observation that developmental processes are not completely interdependent. Experimental removal of salamander limb buds, for example, does not interfere substantially with development of the rest of the embryo, and a transplanted bud can form a limb in another area of the body (Harrison, 1918). Similarly, mutation of embryonically expressed genes does not always completely derail development: some structures may be profoundly affected while others develop normally. But development also has an important interactive component, exemplified by the pleiotropic effects observed in *Drosophila* carrying mutations at the *decapentapelegic (dpp)* locus: a range of developmental processes require the *dpp* gene product, and subsequent events require the correct formation of structures made by *dpp*-dependent processes (Padgett et al., 1993, and references therein).

Developmental and evolutionary perspectives differ in that developmental biologists focus on generative processes, rather than on their final results. Much work has gone into identifying and describing developmental structures such as the limb bud or the dorsal lip of the blastopore, structures that differ significantly from those that concern evolutionary biologists. In particular, embryonic structures are often transient and may comprise a changing population of cells—for example, as cells involute around the blastopore lip during amphibian gastrulation (Gilbert and Bolker, 2001). This dynamism is best expressed by T. S. Eliot (who was *not* a developmental biologist): "What is actual is actual only for one time/and only for one place" (1930, p. 53).

An evolutionary module is usually a structure, a gene sequence, or some other phenotypic character; in contrast, *a developmental module is less an object than it is a resource.* This resource may be a gene (see Moss, 2001, for a discussion of genes as resources), a set of gene products integrated into a "gene net" (Bonner, 1988) or a signaling pathway, or a transient developmental structure such as a cell condensation or other determined cell population (Atchley and Hall, 1991; Gilbert et al., 1996), which can be drawn upon in the construction of some element of the embryonic, larval, or adult phenotype. Modules may be arranged hierarchically according to the extent to which their effects are pleiotropic; one way to measure this property is to determine experimentally how profoundly their absence disrupts development as a whole.

Although the focus of evolutionary and developmental definitions differ, they nonetheless share a number of criteria (von Dassow and Munro,

1999; Bolker, 2000; see also Winther, 2001). Dissociability of modules is a critical feature of any modularity concept: modules are discretely specified and function as a unit in their interactions with other modules (Raff, 1996). An entity need not exist in a particular location nor even be a physical structure to qualify as a module (Gilbert and Bolker, 2001; but see Raff and Sly, 2000). It must, however, have properties and capacities for interaction that are special to that unit and not characteristic of other units or of its component parts. This is why the similarity/iteration criterion brought into some, particularly ecological, definitions of modularity (e.g., Acosta et al., 1997; Bayer and Todd, 1997; Honkanen and Haukioja, 1998; Maillette et al., 2000; see Winther, 2001 for some discussion) is unnecessary and even misleading in an evolutionary developmental context.

We define the relationship between evolutionary and developmental modules as follows: *An evolutionary module is the phenotype resulting from both the composition and the connectivity of the particular suite of developmental modules involved in its construction.* The suite of modules provides all that is needed to construct the phenotype with a minimum of outside interaction, which gives the resulting character its evolutionary modularity. This is not to say that the relationship between a given evolutionary module and the modules that participate in its development is fixed or immutable. Repeated use and co-option of such modules as genes and signaling pathways is a recognized evolutionary pattern and constitutes an important explanation for the looseness of the relationship between phenotype and genotype, as well as for the development of similar structures in different species.

Phenotypic changes can result from changes within individual modules in the group, or from shifts in the interactions between modules (Raff, 1996; Bolker, 2000). Changes in phenotypes can also be due to environmental factors acting during development (Gilbert, 2001), but a discussion of this aspect of phenotypic evolution is beyond the scope of this entry. Over evolutionary time, the recruitment of developmental modules originally associated with a particular phenotype into new groups, or shifts in their time or place of participation, may involve them in the formation of other characters as well. This combinatorial potential makes the evo-devo module an *evolutionary* resource as well as a developmental one: modules are the source of variability (*sensu* Wagner and Altenberg, 1996), made visible to selective pressures indirectly through the phenotypes in whose formation they participate.

How might we identify the modules of evo-devo? Both morphogenetic fields (Gilbert et al., 1996) and cell condensations (Atchley and Hall, 1991) have been proposed as such units, and both certainly qualify as identifiable units, with individualized properties not characteristic of their component parts. Both interact with other fields and condensations, and variation in the properties of either a field or a condensation can affect the form of the structure in whose development they are participating (e.g., Atchley and Hall, 1991). Both also seem to include an implicit criterion that a module exist in one place at one time; however, our definition allows modules to be any individualized developmental and evolutionary resource units. Cell condensations and morphogenetic fields qualify as modules because of their discrete specification, individualized properties, and interactivity.

Because development is hierarchical, a variety of entities can act as evo-devo modules. We offer three examples that illustrate both the diversity of evo-devo modules and the common attributes that can render a given unit a potential participant in development, and thus in evolutionary changes related to development.

The leaf primordium can be regarded as both a developmental and evolutionary resource: a true evo-devo module. Leaf development begins with a primordium, a set of cells derived from the apical meristem and specified to become a leaf. As the developing leaf grows and extends from the shoot, it develops species-specific morphology through differential growth and localized cell death (Singer, 2000). The primordium is an evo-devo module because, as the source of cells determined to become a leaf, it is a highly internally integrated developmental unit (comparable to a morphogenetic field in animal development). The leaf primordium also satisfies the interactivity or combinatorial criterion: the type of leaf that forms from the primordium cells is influenced by the participation of other developmental resources, especially signals from other parts of the plant.

For instance, the short, wide adult leaves of *Pseudopanax crassifolius* (a heterophyllous tree native to New Zealand) differ remarkably from the long, narrow leaves observed in juvenile specimens (Clearwater and Gould, 1994). The early leaf primordia in juveniles and adults are morphologically indistinguishable, and the dramatic phenotypic differences appearing later in development probably stem from the effects of interaction between the primordium cells and other developmental inputs (Clearwater and Gould, 1994): different module combinations participate in juvenile and adult leaf development. In this case, the interactions are age-specific; similar combinatorial variability could be invoked to explain species-specific leaf shapes. Both age- and species-specific leaf shape are evolutionary outcomes, with age-specific leaf shape as a developmental feature representative of the life cycle evolution that is of special interest to evolutionary developmental biologists.

Signal transduction pathways, which play the vital developmental role of translating extracellular information into changes in gene expression, represent another form of evo-devo module. Each transduction pathway is composed of a ligand-specific receptor, a number of intermediate proteins, and a transcription factor. The signaling module consists of the full set of elements present in a single cell (Hartwell et al., 1999): the binding of ligands to the receptor triggers changes in the receptor and intermediate proteins, activating the transcription factor, which then alters the pattern of gene expression. Multiple signaling pathways may be present in a given cell, and each requires only its own elements to function. However, interaction between pathways is also an element of cell signaling: multiple signals may be required to achieve a change in cell state, as in the mammalian immune system (Kuby, 1997; Gilbert, 2000).

Further, pathways may interact through a shared protein element: for instance, the wnt and hedgehog pathways share the intermediate protein slimb, which inhibits both the hedgehog pathway transcription factor ci, and the wnt pathway transcription factor ß-catenin (Jiang and Struhl, 1998). Binding of the hedgehog ligand results in inhibition of slimb function, allowing ci to enter the nucleus. When the wnt and hedgehog pathways are present in the same cell, they may cross-talk through slimb, and ß-catenin inhibition may be attenuated without the wnt ligand binding the receptor (Figure 1). Signal transduction pathways are thus developmental resources that can be used alone or together to enable higher-level modules (such as cell populations) to interact. The nature and interactions of the higher-level modules can be affected by evolutionary changes in the composition or arrangement of the underlying pathway.

At the gene level, modularity is frequently taken for granted. The coding region of a gene may be the source of a variety of transcripts; by virtue of the gene's regulatory region, the transcripts may be used in different tissues at different points in development. Not all of the coding region sequence is present in each transcript, and the variety of different combinations of sequences that are used during development, as well as the potential for novel sequence combinations, make the gene a versa-

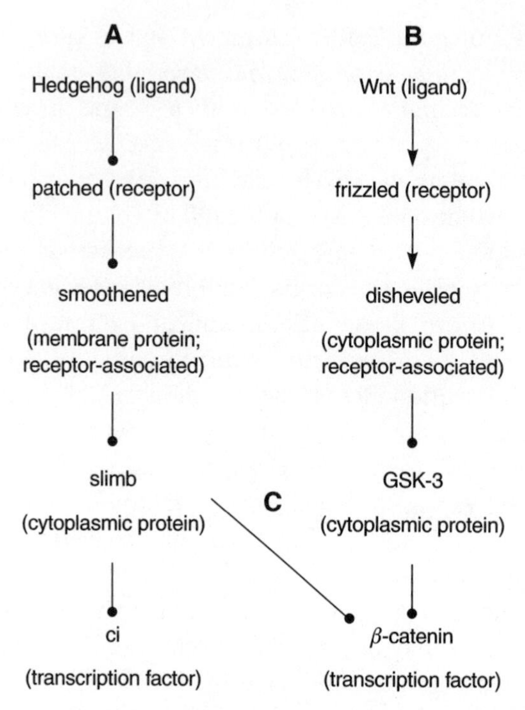

FIGURE 1. Two signal transduction pathway modules, and their interaction when both are present in the same cell. Regulatory relationships are illustrated between adjacent protein elements in their activated form, *not* as the sequence of events following receptor binding. Round-headed arrows indicate an inhibitory relationship between two protein elements; pointed arrows an activation relationship. A. The Hedgehog pathway. When the hedgehog ligand is present, the patched receptor is inhibited from inhibiting smoothened; smoothened is thus able to inhibit slimb, which is then unable to inhibit ci, and ci can thus bind DNA and affect the rate of transcription. B. The Wnt pathway. When the wnt ligand is present, the frizzled receptor is activated and is therefore able to activate disheveled. Activated disheveled is able to inhibit GSK-3, and inhibited GSK-3 cannot inhibit beta-catenin, so beta-catenin is active as a transcription factor. Hedgehog and Wnt comprise two sets of independent processes capable of affecting transcription rate following ligand binding. The two modules interact via C. The slimb protein is inhibited by smoothened and inhibits ci in the Hedgehog pathway, but also inhibits beta-catenin of the Wnt pathway. One consequence of this interaction is that wnt ligand binding is not always sufficient to release beta-catenin from all inhibition: for maximum beta-catenin activity, the hedgehog ligand may need to be present as well.

tile evolutionary and developmental resource. For instance, the *methyl-CpG-binding protein-2* gene (recently implicated in Rett syndrome, a neurodevelopmental disorder) is the source of several transcripts, ranging in size from 1.8 kb to 12.1 kb, that show tissue- and stage-specific distributions (Coy et al., 1999). The approximately 10 kb size difference between the smallest and largest transcript covers a vast region of unused potential transcripts, fertile ground for evolutionary novelty.

A complete gene includes not only the coding region, but also the variety of *cis*-regulatory sequences that bind transcription factors, controlling the level of transcription (Arnone and Davidson, 1997). These *cis*-regulatory regions make a gene available as a developmental resource by driving its expression in specific tissues with appropriate combinations of transcription factors. Expression of the mammalian *Pax3* gene in the neural crest and somites is controlled by separate *cis*-regulatory elements (Li et al., 1999). The cascade of transcription factors in which *Pax3* participates during development is a good example of the combinatorial potential of modules in development. In addition, by reducing the pleiotropic effects of mutation, thus decreasing the likelihood that a mutation will be deleterious, *cis*-regulatory regions help render genes usable sources of evolutionary variability (Stern, 2000).

Evolutionary developmental biology, like any emergent entity, should be more than the sum of its parts. While addressing problems of interest to both evolutionary and developmental biologists, evo-devo must recognize the historical limitations of each field. In particular, we must be prepared to re-examine and in some cases redefine central concepts and terms, incorporating the strengths of previous usages while striving to address what has formerly been under emphasized. In our discussion of evolutionary developmental modularity, we have sought to integrate evolutionists' emphasis on independence and recombination with developmentalists' focus on the importance of ontogenetic processes. Both aspects of modules are central to their role as resources for construction and change, on both ontogenetic and evolutionary time scales. Ideally, the modularity concept can encompass part of the relationship between evolution and development in a single theoretical construct, illustrating both the utility and the necessity of studying the two in concert.

MARVALEE H. WAKE AND ADAM P. SUMMERS

Morphology

Morphology is the shape and arrangement of structure, the phenotypic representation of development (proximately) and evolution (ultimately). Morphology reflects the formation of organisms and lineages. Its variation is part of the stuff on which selection operates. Its features provide "characters" by which we identify members of taxa, generate phylogenetic hypotheses, and infer function and role. Robust analyses of patterns of morphological variation and transitions use an hypothesis of phylogenetic relationships. The integration of disparate, rapidly evolving fields—phylogenetic reconstruction, molecular genetics and development, functional morphology and biomechanics—makes the current study of evolutionary and developmental morphology an arena for great advancement in information, experimental tests of hypotheses, and theory.

Morphologists and morphology are at center stage in the synthesis of evolution, development, and analysis of structure-function relationships during ontogeny and evolution. For example, D. Wake (1982) proclaimed the "renaissance" of morphology in modern biology; M. Wake (1992) examined the field of "evolutionary morphology" and recognized five then-current subareas of analysis (functional morphology, biomechanics, ecomorphology, developmental morphology, systematics, and phylogenetics in morphology) and predicted a sixth (integrative evolutionary morphology). Minelli and Schram (1994) and Koehl (1996) appraised morphology in various contexts; Sattler and Rutishauser (1997) and Endress et al. (2000) proclaimed that morphology and morphogenesis have fundamental relevance to plant research. These are exciting times as new tools from related areas of study, such as molecular biology, physiology, genetics, systematics, physics, and engineering, reveal the interactions and causal relationships among genes, development, morphology, function, and evolution in animal and plant biology.

The subareas of evolutionary morphology discussed by M. Wake (1992) have been transmogrified by synthesis, input from other fields of biology, and technical revolutions. Evolutionary developmental morphology is a recognized and identifiable area of study (Holland, 2000) that integrates the genetics of development with analysis of pattern and process of evolution. We are approaching an understanding of gene regulatory processes in development and evolution (Davidson and Ruvkun, 1999). Comparative developmental biology is used to infer evolutionary mechanisms (Raff, 1996), developmental quantitative genetic models are used to predict patterns of morphological evolution (Atchley et al., 1994), and the effects of developmental factors on morphology and performance are being tested (e.g., Van Der Muelen and Carter, 1995). Classic questions in morphology are receiving new attention because novel techniques and approaches promise to provide long-awaited answers.

To illustrate the interplay of morphology with development and evolution, we recognize four synthetic areas of study through a case study approach.

Enhancements by Evolutionary Developmental Genetics

Quantitative genetic models have been applied to assess patterns of evolutionary change of morphology and its underlying development (Atchley et al., 1994). The appropriate model must reflect the relevant genetics and development of the organism, but still be computationally available so that deductions can be made and hypotheses generated. While models vary in complexity and scope, in general they predict whether evolutionary developmental/morphological change can occur, its range and direction, and whether changes in other features will be correlated with those of the feature under study. An example is the examination of mammalian cranial evolution by Cheverud (1994), who explored the morphological integration of the cranium, including developmental, genetic, functional, and evolutionary aspects. Patterns of genetic variation produced by pleiotropic mutations and stabilizing selection for traits that interact developmentally and functionally result in their co-inheritance with other traits, through coordinated response to selection. Cheverud concluded that functional and developmental integration lead to genetic integration, which then fosters evolutionary integration.

Another thrust of the genetic analysis of morphogenesis is exemplified by research on body segmentation genes. Burke et al. (1995) examined expression boundaries of suites of Hox genes to assess the relationship between homeobox expression patterns and developmental morphology in animals (chick and mouse) in which the regulatory genes are homologous, but in which vertebral morphologies differ. Hox anterior expression boundaries were transposed in synchrony with morphological boundaries, suggesting a mechanistic basis for the presumed homology of the anteroposterior order and variation in regional vertebral morphology. This raises the possibility of examining gene interactions in terms of sequence (temporal order) and timing of effects that produce different morphologies within individuals and across lineages, despite the conservation of the genes responsible for the patterns.

New Approaches to Old Questions

Classical issues in morphology—such as the basis for homology (morphological, developmental, or phylogenetic), the evolution of body form (is there a phylotypic stage? How does variation arise?), and whether structural change drives behavioral modification, or vice versa—are receiving new attention. Richardson et al. (1997) looked carefully at the tailbud stage of vertebrate embryos, which had long been considered the phylotypic stage—a conserved stage characterized by great similarity among all vertebrates. They found considerably more variation than had been assumed since Haeckel's time and concluded that at least some developmental mechanisms are not stage-constrained. Minelli and Schram (1994) expanded the concept of the zootype as a spatial pattern of gene expression and the phylotype as a stage of development that expresses the zootype, and they proposed several phylotypes for invertebrate animals. A comparison of approaches to vertebrate and invertebrate "phylotypes" would be instructive.

Collazo's (2000) expression of the importance of neural crest and placodes to analysis of vertebrate morphological evolution points to the prescient insight of Carl Gans and Glenn Northcutt (Gans and Northcutt, 1983, 1985; Northcutt and Gans, 1983) that the origin of neural crest is the key to the evolution of the vertebrate head. This idea has given rise to a cottage industry of two decades of exciting research on the contribution of neural crest and placodes to many vertebrate cranial and postcranial structures, and on the way that cell numbers and genetic and chronological timing influence developmental and adult morphological variation. A synthesis of development and evolution in the context of morphology is occurring incrementally (see Hall, 1999a).

The issue of *morphological homology* has long been debated; evolutionary developmental biology, through the integration of cellular and molecular approaches, allows new insights. Many analyses do not consider that there may be multiple developmental mechanisms that result in the same, or similar, morphological forms (Collazo and Fraser, 1996). Variation in the formation of a structure as a consequence of evolutionary changes in ontogeny requires further study. Examining seemingly different developmental and genetic bases for homologous structures will lead to a clearer understanding of homology. Indeed, modern cell, molecular, and genetic techniques will provide a broader criterion for assessing homology.

Two cases in which different developmental trajectories result in the same morphological structure follow. Page (2000) examined the development of protoconch shell form in nudibranch and cephalopod gastropods. She found diverse patterns of shell growth, shell dissolution, and loss versus retention of the protoconch, all influencing the uncoiled body form. Bellairs and Gans (1983) examined the development of the orbitosphenoid bone in the skulls of groups of reptiles and found that, in most, the bone is derived from the orbital cartilage, but in amphisbaenians, it is a membrane bone. This phylogenetically homologous element has distinctly different modes of development among reptiles.

The age-old problem of the generation of morphological variation is ripe for attack using genetic and molecular tools. Stern (2000) laid a framework as he discussed the kinds of mutations that generate phenotypic variation that is phenotypically relevant, and the kinds of molecular changes involved. The long-held view that mutations of important developmental genes always have large pleiotropic effects is seen as a misapprehension based on a small set of mutations that have dramatic effects. Rather, some mutations have few or no pleiotropic effects but may be the main sources of morphological variation and evolution. Developmental studies of gene function provide a new way of conceptualizing and examining variation by identifying specific mutations that generate phenotypic variation and determining how they modify gene function, thus causally linking molecular and phenotypic variation.

Body plan evolution is an exemplar of the generation of variation. Zrzavy and Stys (1997) assessed the evolution of the arthropod body plan by examining morphological variation using cell lineage analysis, patterns of mitotic domains, and the expression zones of several genes. They found a number of relationships of body segments not previously expected, such as the structure of the mandibulate head having evolved several times. The incorporation of cell- and gene-based developmental features into analyses of basic body form is changing ideas about evolutionary trajectories of lineages and relationships between lineages.

Similarly, Raff's (1996) examination of the links between the evolution of animal body plans during and following the Cambrian radiation and the evolution of the developmental processes that acted on those plans elucidates a series of general principles. Raff considered the origins of animal phyla and the use of gene sequence data to examine questions of phylogenetic relationships and, therefore, patterns of evolution within and among lineages. He paid particular attention to the developmental problems posed by extinct and extant body forms and their evolution. His many examples illuminate the links among development, phylogenesis, and evolution in assessing the origins of new body forms.

Body form of other organisms can be similarly assessed. Andrews (1995) assessed evolutionary, physiological, and developmental determinants of the modification of growth form in fungi. His evolutionary determinants of form include phylogenetic and developmental constraint, adaptation, ecophenotypic factors, and chance; physiological factors relate to the morphogenesis of the cell wall, including the cytoskeleton and cytosis. Fungi respond to environmental signals by varying timing, extent, and mode of differentiation, decoupling sexual and asexual phases of the life cycle, and modifying growth form. The cellular totipotency, localized senescence, internal age structure, and indeterminate growth of fungi give rise to multicellular complexes that perform specific activities, such as dispersal, foraging, and reproduction. Andrews considered that the morphological plasticity of fungi surpasses that of other modular organisms such as benthic invertebrates and plants, yielding diverse, but still constrained, growth forms.

Evolution and Development of Morphology, Performance, and Fitness

The relationship among morphology, performance, and fitness is central to the issue of adaptive significance (Arnold, 1983). New tools, and the application of old tools to developmental systems, make possible the investigation of the functional morphology—the link between morphology and performance—of developmental processes and early ontogenetic stages of animals.

The mechanics of development profoundly influence early ontogeny and the consequent morphology. The rearrangements and movement of tissues that comprise embryogenesis must have their roots in either changing force regimes or changing material properties (Moore et al., 1995). Cell migration and the process of convergent extension are examples of the effects of forces that affect gastrulation (Davidson et al., 1995), neurulation (Dunnett et al., 1991), cell aggregation (Bretschneider et al., 1999), and patterning of the neural ectoderm (Elul and Keller, 2000). The forces of development can be measured directly (Moore, 1994). Significant progress has been made by mathematically modeling forces and testing the models against observed patterns of development (Odell et al., 1981; Davidson et al., 1995). Ontogenetic changes in solid properties such as stiffness (Adams et al., 1990; Moore et al., 1995), and viscoelastic properties such as stress relaxation (Moore et al., 1995), are linked to changes in form in the gastrula and later stage *Xenopus laevis* embryos. The difficulties of performing mechanical tests and measuring forces and material properties are daunting, but the potential payoff is high—understanding a different class of cell and tissue level signals and interactions.

At a larger scale, the relationship between morphology and performance is also important when organisms interact with their environments. Most animals go through a significant postembryonic larval period, a morphologically distinct form that metamorphoses into the adult (see Hall and Wake, 1999). Few individuals survive the larval stage, and even small differences in performance of a biologically vital task such as swimming or feeding could lead to higher survivorship. The functional morphology of developing larvae must be studied separately from that of the adult because (1) some features of larval morphology are simply not present in adults (Raff, 1996; Page, 2000); (2) larval growth and metamorphosis almost invariably involve allometric growth, so models of adult performance cannot be scaled down to larval sizes (e.g., Webb, 1999); (3) larvae often operate at a much smaller size scale than adults, in which different physical regimes apply (e.g., Hernandez, 2000); and (4) the material properties of both hard and soft tissues are different at extremely small size scales because of heterogeneity and anisotropic effects (Koehl, 1996).

At the same time, we recognize that, like larvae, fetuses are affected by environmental interactions. Van Der Muelen and Carter (1995) illustrate the ways that the mechanics of development influence the growth and shape of the long bones of mammals, the environment being that of the uterus. The literature abounds with examples of the relationship of morphology, performance, and fitness during ontogeny, but they are rarely presented in the context of development and evolution.

Phylogenetic Context

Examination of the development of morphology in a phylogenetic context is transforming our understanding of both morphogenesis and phylogenesis in many lineages of organisms. Gleissberg and Kadereit (1999) used ontogenetic comparisons of leaf development in one family, the Papaveraceae, and developed a series of potentially testable evolutionary scenarios to examine the evolution of variation in leaf morphology. A set of morphogenetic traits of leaves mapped onto a phylogenetic reconstruction of the family revealed extensive homoplasy, including both parallelisms and reversals.

Kerr and Kim (1999) used molecular and morphological characters to analyze phylogenetic relationships of the Holothuroidea, the least well known class of echinoderms. Their phylogenetic hypothesis is at great variance with previous assumptions of relationships of members of the class. Adult bilateral symmetry must have evolved three times, with one group regaining radial symmetry. Also, respiratory trees, unique to sea cucumbers, are derived from ectoderm, and the plates of the tests of some holothuroids are not homologous with those of other echinoderms and have evolved at least twice. Kerr and Kim also postulated that indirectly developing larvae evolved at least twice within the group. Phylogenetic reconstruction allows the identification of developmental scenarios that can be tested with additional data from ontogeny and from the genetic basis of the developmental patterns.

Raff et al. (1999) used a long-established technique, hybridization, to assess the basis for change in developmental mode. Two closely related species of sea urchins, one a direct developer and the other with a feeding larva, were crossed. Hybrids survived through metamorphosis and had structures characteristic of feeding larvae, with the restoration of paternal gene expression lost in the direct-developing maternal species. But the hybrids were not simply reversions to the paternal larval form; their ontogeny was distinct from that of either parental species. A number of features of the hybrid were those of the immediate common ancestor of the parental species, but they also had developmental features of distantly related echinoderms. The interaction of genomes that encode different developmental modes produces a novel but workable ontogeny in the hybrids that "recovers" both recent and ancient ancestral features of echinoderms.

Mabee's (2000) explicit analysis of the use of phylogenetic hypotheses to interpret patterns of evolution of developmental genes and gene processes illustrates a useful methodology for analyzing the developmental processes that underlie morphological variation. She mapped morphological and developmental features on a robust phylogeny of the vertebrate fore limb and found that, contrary to long-held hypotheses, teleost fishes and tetrapods may not share homologous skeletal elements. This implies that similarities and differences of Hox gene expression recently proposed must be interpreted differently. Specifically, the phase III Hox pattern of tetrapods may not be that of digits, but may be the normal expression pattern of the fish metapterygium. To test which hypothesis of Hox gene function is correct will require looking beyond established model systems to a phylogenetically appropriate taxon, either a shark or a basal bony fish.

Morphology, the phenotypic representation of development and evolution, is central to evolutionary developmental biology. These examples provide abundant evidence that the integration of development and evolution to investigate morphological variation yields new insights into the process that generates morphology and the patterns of structure and function that result.

SUSAN W. HERRING

Ontogenetic Integration of Form and Function

Most morphological features do not result from a straightforward translation of the genome, but rather are the product of complex interactions between the developing organism and its surroundings. Under normal conditions these epigenetic interactions result in typical somatic development; under unusual circumstances, they adjust the phenotype to accommodate altered environmental demands. Temperature variation, diet, population density, presence of parasites or predators, and environmental complexity are among the numerous factors that modulate the physical outcome of development.

Ontogenetic integration is the result of epigenetic matching of form to function. Other terms that have been used for this phenomenon or the processes that produce it are "dependent morphogenesis" (Schmalhausen, 1949), "developmental plasticity" (Gilbert, 2001), and "phenotypic plasticity" (Sultan, 2000; Pigliucci, 2001), but the last of these terms has also come to mean certain types of evolutionary change (Gordon, 1992). Furthermore, the word "plasticity" obscures the important role of ontogenetic integration in normalized development.

Ontogenetic integration reflects the existence of mechanisms that can produce different phenotypes from the same genotype (an element of plasticity) or stabilize the phenotype in the presence of variable genotypes and environments (canalization). An example is developmental matching of neurons to their end organs. Neurons are overproduced and survive only if they receive appropriate signals from their targets (Raff et al., 1993); thus, innervation ratios remain stable. Similarly, the differentiation, growth, and maturation of the muscular system are entwined with those of the skeletal system by mechanical interactions, ensuring that the levers and struts of the body are suited to the forces imposed on them (Carter et al., 1991; Herring, 1994).

Ontogenetic integration of form and function is the morphological equivalent of physiological acclimation (also called physiological adaptation), the extension of homeostatic feedback mechanisms to enable organisms to deal with varying internal or external environments. Like physiological acclimation, ontogenetic integration involves norms of reaction, which describe the response of the genome to various stimuli (Schmalhausen, 1949; Sultan, 2000; Gilbert, 2001).

In some cases physiological acclimation provides mechanisms for morphological change, thus producing integration of form and function. For example, activation of frog pituitary hormones in response to a dark background color (physiological acclimation) leads to a modification of skin color (Vazquez-Martinez et al., 2001). Similarly, the regulation of metabolism by dietary intake (physiology) can lead to obesity (morphology). Working out at the gym (physiology) enlarges the exercised muscles (morphology). Some authors (e.g., Futuyma, 1979, p. 53) would simply classify all these examples as physiological acclimation, because the changes occur during the postnatal life of an organism and are reversible. However, comparable phenomena take place during early ontogeny and are irreversible. The mammalian hard palate fails to form if an immobile tongue impedes it (Carey et al., 1982). Hatchling lizards raised on broad surfaces grow relatively longer hind limbs than those raised on narrow dowels (Losos et al., 2000).

Different Stages of Development and Levels of Examination

Some examples of ontogenetic integration are restricted to embryogenesis, while others indicate a lifelong flexibility. To a considerable extent, this reflects life history. Individuals with very short life spans or taxa with predictable environments have little to gain from the ability to alter their morphology. At the other extreme, long-lived organisms such as trees and large mammals must deal with changing weather cycles and changing community structure even as adults. In these individuals morphology is not static: it changes in response to extrinsic stimuli throughout life. Mechanisms to change morphology include not only differential tissue growth but also processes that recapitulate development. For example, mesenchymal stem cells, capable of differentiating in a variety of directions, persist in adult birds and mammals (Caplan, 1990). In other cases mechanisms to effect integration are different at different stages. Fetal muscles grow mainly by cell proliferation, adult muscles mainly by an increase in the size of existing cells. Bones can increase in diameter and density in adult mammals, but they cannot grow in length after ossification of the cartilaginous growth plate.

Ontogenetic integration of form and function is apparent at all levels of organization and is perhaps most significant at the cellular level. Cell birth and survival are often strongly influenced by milieu. In some strains of laboratory mice, the proliferation of neurons in the brain can be doubled by increasing the complexity of the environment; in other strains, improved cell survival leads to the same outcome (Kempermann et al., 1998). Endothelial cell survival depends on whether cell spreading takes place, which, in turn, is governed by the extracellular environment acting through integrin binding (Chen et al., 1997). Gene expression and cell phenotype are also modified by extrinsic signals. The mechanism for the above-mentioned color change of frogs on a black background involves not only up-regulation of pituitary hormones, but also the conversion of one cell subtype to another (Vazquez-Martinez et al., 2001). A variety of cell types, including endothelial and smooth muscle cells, modulate in response to mechanical stress using pathways that involve the cytoskeleton (García-Cardeña et al., 2001; Halayko and Solway, 2001).

Tissues and organs also show ontogenetic integration of form and function. Many cells die after a short life span and are replaced by younger cells, often formed from pluripotent progenitor cells. This process imparts considerable flexibility to accommodate new conditions or to effect repairs. The resulting changes in relative proportions of body parts can have striking effects on overall organism morphology, particularly in higher plants, in which environmentally caused variations in leaf shape, branching pattern, and even reproductive mode can make members of the same species difficult to recognize (Sultan, 2000). Such effects may be further exaggerated if the allocation of limited resources to one body part has repercussions for other organs (Emlen, 2001).

Although purely physical mechanisms may be involved in some aspects of morphogenesis and plasticity (Trussell, 2000), for the most part ontogenetic integration seems to be the outcome of selection for malleable and versatile gene networks (Greenspan, 2001). The resulting morphologies are usually adaptive (Sultan, 2000; Van Buskirk and Saxer, 2001).

Occasionally, regulation of morphology by function does not improve the phenotype. Traits may be constrained to change because of strong interactions with other processes (Gilbert et al., 2000; Trussell, 2000) or because the organism's capacity to accommodate has been exceeded. For example, vigorous exercise in roosters represses circumferential growth in lower limb bones, weakening structure (Matsuda et al., 1986). Similarly, poor supplies of nutrients result in small adult size and persistent effects in multiple organ systems (Hoefs and Nowlan, 1997; Ohlsson and Smith, 2001; Weinig and Delph, 2001). Although compensatory mechanisms exist, the resulting phenotype is rarely an improved one (Metcalfe and Monaghan, 2001). Nevertheless, an undesirable morphology still might qualify as adaptive if it were the best possible outcome under adverse circumstances; thus, short stature has been considered an adaptive response to chronic undernutrition (Balam and Gurri, 1994).

Because it can also lead to rapid phenotypic change, ontogenetic integration is a good adaptive strategy for ecological generalists and invasive species (Sultan, 2000). Indeed, distinguishing whether a phenotypic change is the consequence of ontogenetic integration or genotype evolution may be difficult (Sultan, 2000). Novel phenotypes caused by environmental changes may well have masqueraded as evolution in phylogenetic studies (Samadi et al., 2000; Pigliucci, 2001). However, ontogenetic integration contributes importantly to evolutionary change—for example, by stabilizing the phenotype. Furthermore, although somatic flexibility probably hinders the pace of genetic evolution (Behara and Nanjundiah, 1995), in combination with selective pressure, ontogenetic integration may accentuate and improve the eventual product (Behara and Nanjundiah, 1995; Pfennig and Murphy, 2000).

KENNETH M WEISS

Phenotype and Genotype

Phenotype and *genotype* are terms coined by Wilhelm Johannsen (Johannsen, 1909; Mayr, 1982). While various distinctions were not made then as they are today, the two terms have always related respectively to the characteristics of an organism and to the characteristics of inheritance.

Terms and Definitions

Phenotype refers generally to any trait measured on an organism, whether morphology, behavior, biochemistry, or any other character. The term usually refers to the functional elements of an organism, such as skin and organs, enzymes, hormones, antibodies, and so on. A phenotype need not vary to be measured; having four limbs is a phenotype. But biomedical or genetic interest often focuses implicitly on traits that vary, such as cholesterol levels, the number of bristles on a fly leg, or the number of legs on an arthropod.

Debate centers on what to measure for a given purpose, or how to measure it. For example, measuring blood pressure on calm, seated people serves a biomedical purpose but may not be the best approach to measuring the function or evolution of circulatory systems. Vague definition with precise measurement, or imprecise measurement with good biological definition, are problems common to the notion of phenotype. But once a phenotype has been defined, whatever it may be, individuals can be characterized by it, and let the user beware.

We usually do not include direct measures of genes under the term phenotype, but because we give them a special place of honor, genes are often lurking nearby. This is not surprising since the term was coined in a genetic context. However, aspects of DNA structure such as the packaging of chromosomes by histones or their methylation can be considered as phenotypes in the classical sense. When underlying genotypes are assessed indirectly by looking at the proteins for which they code, the resulting protein structures are referred to as phenotypes. Examples are the ABO blood types and the protein isoforms of the Apolipoprotein (APOE) gene product associated with cholesterol levels and Alzheimer's disease.

If phenotypes are definitionally arbitrary, the term *genotype* is more constrained, being restricted to specified DNA (or in some cases, RNA) sequences in an individual, usually in the context of inheritance. Usage is variable, and the term may refer to a named gene or to the entirety of the inherited DNA, including a set of a species' chromosomes, mtDNA, chloroplasts, and so on ("set" because some organisms inherit multiple copies of the genome). Genotype can also refer more vaguely to all genes known, unknown, or even just presumed to exist that are related to a trait of interest, such as the genes for diabetes or for bristle number in flies.

Genotype usually implies a referent "gene" or "locus," but these terms are not always clearly distinguished. Locus traditionally has referred to a position on a chromosome, while gene refers to logical or functional units, typically to protein-coding regions. Both terms may be used for regulatory sequences that control the expression of a gene, or even single nucleotides in a particular chromosomal location. Genotypes may have no phenotypic effect, as in the case of microsatellites (polymorphic repeats of short sequence elements) or single nucleotide polymorphisms (SNPs) that are used as markers of biological history or in genetic mapping to find co-inheritance correlations between such genotypes and phenotypes.

A few subtleties worth noting relate to contemporary species and to evolution (see Buss, 1987). In many species, reproduction is by tissue cloning or budding; new organisms develop from many cells shed by the parent. There may be sequence variation among these cells that accumulated in the life of the parent. Such an organism may thus not have a uniform inherited genotype as do organisms produced from single-celled zygotes. However, the distinction is somewhat of an illusion; because of somatic mutations, no multicellular organism is likely to have a single genotype after it has undergone at least one cell division. Multicellular organisms are almost inevitably genetically mosaic, with mutations at each division generating hierarchical variation among cell lineages, and hence tissues. Local mosaicism can also be actively produced, as in X-inactivation (Lyonization) in mammalian females. Common usage refers to the inherited genotype as the *constitutive genotype*. In cancer we may refer to the genotype of the tumor, which may evolve during its life history, relative to the constitutive genotype of the host. The distinction should probably be made more often, and more explicitly, because most studies of relationships between genotypes and phenotypes assume that the determinative relationship is with the constitutive genotype.

Genotypes can also be modified by the incorporation of exogenous DNA naturally, or experimentally by introducing a virus that will express a particular gene, or by transgenic experiments in which a context-specific enhancer is used to express a particular gene in a chosen tissue at a specified developmental stage. Such experiments alter the histologically local genotype, often just the local *expressed* genotype; transgenic procedures can be used to insert, replace, or remove a targeted gene in the inherited germ line. An experimental animal or plant is often denoted by its altered genotype, such as whether it has 0, 1, or 2 copies of an altered gene (e.g., +/+, +/−, and −/− genotypes, respectively).

Elements of Phenogenetic Relationships

In the modern view, the function of genes is to code for the biological nature of organisms, that is, for their phenotypes. The *phenogenetic* relationships that connect genotypes and their associated phenotypes are fundamental, and they are typically viewed in two basic ways. The first concerns proximate causation and hence *mechanism.* In developmental biology, phenotype usually refers to morphology. Morphology is the result of many steps of gene-product interactions occurring over developmental space and time. Thus, for a single genotype there can be many phenotypes, even within the same individual (or, alternatively, the entire developmental process or history can be viewed itself as a phenotype if it can be adequately defined). From a conceptus to the final stage under consideration in any given investigation, different genes can be identified whose expression has been activated or repressed. In current practice, biological traits are often treated as phenomena computable from the information contained in a set of necessary and/or sufficient gene expressions.

Experimental approaches to inferring genetic mechanisms are often typological. An experimental model like an inbred strain of "the" mouse or frog is selected often for practical reasons, such as a large existing knowledge base or a transparent manipulatable egg. The genotype is moot in that the organisms are selected to be genetically homogeneous. A common first approach is to document the expression of a candidate gene to identify the tissues, cells, or times during embryogenesis when it is expressed. This is followed by altering expression by inactivating, over-, or under-expressing the gene in embryological time or space to determine its effects. Inferences are then made about the tissues or structures for which the gene has necessary and/or sufficient effects. Genes that regulate, and those affected by, the expression of a candidate gene are also identified, to assemble a regulation-expression network. A physical morphology may be the final outcome, but the phenotype of interest may as well be any aspect of developmental history itself (Hall, 1999a), or homeostatic or metabolic patterns. Often, the discomfiting observation is made that the organism does not miss the candidate gene at all; rarely does the disappointed investigator disavow its hypothesized relationship to a putative phenotype, and in truth it is not easy to determine when to hold and when to fold.

A second major focus in developmental biology is not typological but concerns origins and evolution, and essentially involves variation (e.g., Hall, 1999a). In studies of natural variation, agricultural breeding, or human genetics, the issue has to do with the correlation between genotypic and phenotypic variation within a population or among strains in the same species. But phenogenetic relationships can be studied between species as well, by comparing the expression or effect of a given set of genes, relating sequence variation to phenotypic differences. Does a frog make legs the same way we do? The expression patterns of the Hox class of homeobox genes may be compared between species; mutations in regulatory elements, or changes in the number of members of the Hox genes or their chromosomal arrangement, may be considered as aspects of the genotype in this context. Phenogenetic relationships are frequently extrapolated hopefully across phylogenetic space. Indeed it is not unusual for similar effects to be identified in other species, as in Hox mutations leading to polydactyly in mouse and man, or even, more spectacularly, by demonstrating experimentally that a mammalian regulatory element can control the expression of the homologous gene in a fly in a nearly normal way.

Strictly speaking, a phenogenetic relationship pertains to an individual's ontogeny and life history. Each individual is a unique genotype and has a unique life history, so the phenogenetic relationship is also unique. Generally, however, we are interested in the replicable aspects of phenogenetic relationships, which can be characterized by the *penetrance functions,* that is, Pr(P | G), the probability of a specified phenotype for an individual with a given genotype. Phenogenetic relationships are statistical or population concepts. Some genotypes may have high penetrance and are strongly determinative, meaning that environmental and other factors have little effect on the phenotype (that is, Pr(P | G) is approximately equal to 1.0). But for many traits, like blood pressure or circulating lipid levels, penetrance is much lower: there can be large variance within genotypes relative to the variance among genotypes. Such genotypes have disappointingly weak associative relationships, not clearly in line with mechanistic causal concepts about genes. Thus some interpretation is needed to understand the meaning of observations such as "in a given sample a trait appears in 12% of the individuals with a given genotype." Such "incomplete" penetrance is completely compatible with substantial heritability (that is, correlation in the phenotypes of genetic relatives), for reasons to which we now turn.

Many-to-Many Nature of Phenogenetic Relationships

Inferences about both similarity and difference must take into account that species basically differ at all loci, and even individuals within species differ in most (showing how curious is our reliance on inbred animals for phenogenetic inference). Just as there are phenogenetic relationships among family members, there are broadly quantitative phenogenetic relationships among species who are, after all, distant relatives.

For complex traits affected by more than one gene, a number, sometimes a very large number, of different genotypes can be associated with the same phenotype and, conversely, many phenotypes are associated with the same genotype. These facts were known empirically at the turn of the twentieth century and impeded the acceptance of Darwinian evolution by natural selection until, among other things, it was shown that complex traits can be produced by multiple Mendelian (particulate) loci. Although observed by various biologists in the late nineteenth and early twentieth centuries (Provine, 1971), we generally associate the unifying understanding with Fisher's (1918) famous paper.

The unifying idea was combinatorial: Different combinations of alleles from different loci contribute to a trait in a kind of dose-response relationship. When this happens, alleles at even just 4–5 loci can yield roughly continuous variation in traits like blood pressure or stature, a distribution made even smoother by variation in environmental factors. For any complex trait—and most developmental traits appear to be complex in this sense—we should not be surprised to find variation and redundancy in phenogenetic relationships, including regulatory mechanisms.

Nomenclatural clarity. Genes are often given names that, at one time at least, were intended to impart phenogenetic clarity. Thus even before a gene itself has been identified and cloned, its presumed existence may be denoted by naming after a phenotype with which it is associated, such as by Mendelian-like inheritance patterns. For much of the twentieth century, when specific genes were hard to come by, this practice made sense. In the DNA age it has become a kind of terminological abusage, that may be cute but harmless in regard to genes in fruit flies (*"hedgehog"* or *"tinman"*), but can be misleading in fields like biomedical genetics, in which gene names have taken on the characteristics of brand names or even advertisements for investigators' work.

Gene misnomers can become awkwardly reified, as in the case of many diseases, for which genes have been named because in some particular sample a correlation was found between the frequency of the disease and marker variation in a particular chromosome region. Often the genes, once cloned, turn out to be widely or even ubiquitously expressed, and their wild type may have little (or nothing) to do with the trait they affect when mutated. Naming genes for traits in this way could also mislead the unwary into thinking the trait is directly inherited, which is usually far from accurate. Genes are sometimes given suggestive names even before they are found, such as many "diabetes" genes named for chromosomal regions implicated in a family study, but which subsequently cannot be replicated. Names, like sticks and stones, *can* hurt if they entrench notions that misguide public policy or research directions.

Functionally arbitrary aspects of phenotypes. The form of the protein is directly related to its function. Hemoglobin molecules must bind oxygen, collagen and chitin must have molecular rigidity. By contrast to these structural aspects of phenotypes, it is becoming increasingly apparent that many aspects of an organism are logically necessary (the organism must have limbs) but functionally arbitrary. Here we refer to the genetic mechanisms involved in the production of structural phenotypes—the cascades of regulation that lead to the appropriately coordinated expression of physiologically or morphologically specific genes. Regulatory mechanisms are genetic and constitute the developmental phenotype (Gerhart and Kirschner, 1997; Hall, 1999a; Gilbert, 2000a; Carroll et al., 2001; Davidson, 2001). But regulatory mechanisms themselves can be largely or even completely arbitrary relative to final structures.

We are familiar with the arbitrary nature of the protein-coding system: there is no correlation between a particular codon and the chemical properties of the amino acid for which it codes (whether there was such a correlation at the very beginning of life is debatable; see Maynard Smith and Szathmary, 1995). Gene regulation is similar. A gene is expressed when enhancer/promoter sequences in its flanking DNA are bound by regulatory proteins (transcription factors) that recognize those sequences. So long as this occurs, it does not really matter in principle what the specific regulatory factors are. And, like the coding system, there is nothing about the chemical properties of, for example, the *cdx1* gene or the enhancer sequence it binds to when regulating the caudal expression of *HoxC8* in mouse embryos (Shashikant et al., 1995) that has anything to do with the structure of somites, segments, or tails—or, for that matter, even mice, because the same elements are present in many other species.

Where the action is in the interaction. Complex organisms develop by context-specific combinatorial cascades of gene expression. Serially homologous, repetitive, or regionally patterned structures raise important phenogenetic (and phylogenetic) issues in this regard. The expressed combination of regulatory factors may be specific to a given structure;

for example, the digits in the hand or foot are produced in zones expressing unique combinations of such factors. But some repeated structures appear to result from the generation of spatially and temporally repetitive patterns of high and low expression levels of a small number of interacting signaling factors. In what are known generally as *reaction-diffusion processes,* the dynamics of the interaction of activating and inhibiting factors generates zones of expression in which structures develop, surrounded by zones of lateral inhibition in which they do not (e.g., Meinhardt, 1996; Murray, 1998; Meinhardt and Geirer, 2000). Under these conditions, each element of a periodic series, like each hair or tooth or leaf, can be specified by the *same* genes. In such periodic patterns, which are very common in complex life, one can meaningfully say that the phenogenetic relationship resides in the regulatory dynamics; there need be no genes "for" each feather or inter-feather, tooth or inter-tooth, and so on (e.g., Jung et al., 1998; Weiss et al., 1998; Jernvall and Jung, 2000; Jernvall et al., 2000). This is quite different from the standard one gene–one protein view of life and explains how genes can determine patterned traits, or how such traits can vary and evolve.

Phenogenetic drift and evolution. In the Western world, we like our causation to be clean—one cause, one effect. There is a natural tendency to think deterministically about phenogenetic relationships, and to assume that a similar structure means a similar underlying genotype. Highly conserved traits such as basic elements of the energy cycle, cell structure, and the DNA processing, transcription, and translation mechanisms often reinforce this view, as do comparable developmental mechanisms found in closely related (and some distant) species, such as in the use of the *Pax6* gene in photoreceptors or Hox genes in axial patterning. But there are exceptions, and we have learned in recent years that developmental genetic mechanisms can vary within as well as among species.

As discussed earlier, within a population or species, a complex trait can have a similar value (e.g., internode length in plant stems, insulin level) with different genotypes. Even if such a trait is continually maintained by strong natural selection, over time there can be *phenogenetic drift,* in which different lineages accrue different phenogenetic mechanism(s) (Wagner and Misof, 1993; Hall, 1999a; Weiss and Fullerton, 2000; True and Haag, 2001). Natural selection preserves the phenotype but cannot see and does not care what genotype is responsible—another aspect of functional arbitrariness in phenogenetic relationships. Phenogenetic drift is implied by the many-to-many relationships that are essential to complex traits. The rate at which phenogenetic drift occurs over evolutionary time is calibrated in part by the frequency of equivalent genetic mechanisms in a population at a given time, or the probability that acceptable alternatives arise by mutation, admixture, or other processes.

We are used to thinking of genotypes as the primary, and more permanent, biological entities. It may seem paradoxical in an age when we like our causation to be clean, but even though a phenotype may have to be generated anew from a fertilized egg cell each generation, if the trait is preserved by selection across generations, it may have more biological permanence (biological reality?) than the underlying genotypes. Phenogenetic drift introduces an element of uncertainty in the nature and assignment of homology; traits can be homologous in the classical evolutionary sense of shared ancestry, without being comparably homologous at the gene (or developmental process) level. Nor do homologous genes or regulatory mechanisms necessarily produce homologous traits.

An example of phenogenetic drift is the utilization of different, unrelated proteins in the formation of the eye lens in different taxa. The important feature of lens proteins is their light-transmitting properties. Regulatory examples include variation in the use of the *bicoid* gene in egg polarity among insect species (Schmidt-Ott, 2001), and the substitution of enhancer elements controlling the expression of the *eve* gene in corresponding stripes in the eggs of different fruit fly species (e.g., Ludwig et al., 2000). Sex determination and gonadal development are examples of fundamental, shared-derived traits whose phenogenetic mechanisms vary widely among vertebrates today (e.g., Renfree and Shaw, 2001).

Phenogenetic relationships are important in practical applications in agriculture, pharmacology, and public health. Complex relationships between genotype and environment, interactions among genes, and somatic and other aspects of intraorganismal genetic variation make phenogenetic relationships more elusive in nature than they may appear in the laboratory. The most tractable and manipulable phenogenetic relationships are those in which the penetrance of the genotype is high. Those were the relationships documented by Mendel in deliberately, artificially simplified experiments that successfully drove most genetic research in the twentieth century. But phenogenetic relationships are markedly less determinative for more complex traits.

Genotypes and phenotypes are subtle properties of organisms. Joined within each individual lifetime by the mechanistic role of genes as information storage units, the many-to-many phenogenetic relationships and phenogenetic drift mean that over evolutionary time, phenotypes can be liberated from genotypes so that each may have its own reality.

PAUL M. BRAKEFIELD AND PIETER J. WIJNGAARDEN

Phenotypic Plasticity

Phenotypic plasticity refers to variation in the phenotype—for example, morphology, life history, behavior, or physiology—of a given genotype when individuals complete their development in different environments. Some plasticity across environments is an almost ubiquitous feature of phenotypic variation, in some cases as a form of developmental constraint, an inability across environments to map phenotypes in a highly repeatable manner onto underlying genetic variation. In other cases, as exemplified by *Bicyclus* butterflies, phenotypic plasticity has evolved as an adaptive response to environmental heterogeneity (see Gotthard and Nylin, 1995). These butterflies, which live as active adults in the alternating dry and wet seasons in Africa, exhibit *seasonal polyphenism* in which alternative phenotypes are adapted to the particular seasonal environment in which they spend all or most of their adult lives (Shapiro, 1976). The alternative phenotypes are produced by control of potential variability in developmental pathways in response to an environmental cue perceived in an earlier stage of ontogeny.

To evolutionary biologists, phenotypic plasticity is a highly attractive concept because it provides a potential means of adaptation to divergent environments. The problem of an adaptive response to seasonally varying environments by genetic differentiation has been well stated by Shapiro (1984, p. 297): "An organism with two generations a year which faces two seasonal selective regimes a year will be permanently out of phase: the genetic makeup of the 'summer' generation reflects selection in a 'winter' regime and vice versa—such an organism is like a military general always planning for the last war. The shorter the generation time relative to environmental periodicity, the closer the organism may track its environment, and the less the load at any given time" (see also Levins, 1968).

The evolution of phenotypic plasticity can provide the solution to this genetic problem. Again, in Shapiro's words: "Polyphenism can reduce the load and the time lag of response to zero, if we assume that all individuals are equally competent to make correct developmental decisions and that the cue(s) they respond to in the environment is (are) trustworthy" (1984, p. 297).

To developmental biologists, phenotypic plasticity is a fascinating phenomenon because of its potential for understanding mechanisms involved in the control of development, and in the interactions between gene expression and the environment during ontogeny. From a genetic perspective, whether there are specific regulatory genes that determine plasticity ("genes for plasticity") and how they function is especially interesting. Finally, there is the contrast with genetic canalization in which a genotype yields similar phenotypes in different environments, and developmental properties tend to limit variation in the final phenotype. Integrating genetic studies of canalization and plasticity should enable novel insights about developmental stability.

Phenotypic plasticity is most clearly illustrated by variation among individual plants or animals of a clonal organism when a single clone is replicated and the individuals raised across different environments. Phenotypic plasticity provides information about the genetic variation for the expression of phenotypic plasticity and can provide ideal material for field experiments on natural selection.

Both field and laboratory experiments have been used to study the seasonal polyphenism expressed by *Bicyclus* butterflies of tropical Africa. The field work has addressed issues of population biology, migration, function, and performance. If the phenomenon has evolved through natural selection and thus reflects adaptive evolution, the alternative phenotypes will have a higher relative fitness in the environment in which they normally occur. The results demonstrate that seasonal polyphenism in *Bicyclus* butterflies is adaptive (see Brakefield and French, 1999). The insights from the field studies about relevant environmental gradients, in combination with an ability to rear one of the species, *B. anynana,* in the laboratory, has allowed us to analyze the genetic, developmental, and physiological mechanisms involved in generating the phenotypic variation. We applied an integrative approach, from genetic variation through to phenotypes, combined with analysis of the survival patterns of the phenotypes within natural populations to explore the extent to which the evolution of phenotypic plasticity may be constrained by genetic architecture and the internal organization of developmental pathways.

Norms of Reaction or Character States

The analytical framework used for much of the work on phenotypic plasticity has been the *norm of reaction,* which describes the set of phenotypes produced by a single genotype across a range of environments, and which is normally depicted graphically as the mean phenotypic value of a genotype plotted against the values along some abiotic or biotic environmental gradient. Reaction norms were first drawn by Woltereck (1909) and combine descriptions of environmental conditions experienced by individuals with information about the underlying physiological and developmental mechanisms in generating the phenotype. The approach has become widely used since Schmalhausen (1949; see van Noordwijk, 1989). Schlichting and Pigliucci (1998) have further elaborated the concept as the *developmental reaction norm* to include a more explicit role for developmental processes.

The mathematical analysis of norms of reaction is based on polynomial mapping of phenotypes onto genotypes and environments. An alternative is the character state approach, which focuses on the mean phenotypic values (the character states) expressed by genotypes across different environments. The expression of a character in two different environments can be studied as two genetically correlated characters. In contrast, polynomial descriptions of norms of reaction are based on the mean and genetic (co)variances of coefficients of these polynomials.

The debate about which approach is more productive in terms of yielding insights about evolutionary processes has centered around three interrelated issues of whether (1) plasticity should be considered a character in its own right; (2) there are indentifiable, so-called genes for plasticity; and (3) selection can act on plasticity itself. Via (1993a, b) has been a particularly forceful proponent of the character state approach and the view that plasticity evolves as a by-product of natural selection favoring different mean phenotypes in different environments. Hence she saw no need to treat plasticity as a trait or to invoke genes for plasticity. This work followed the highly influential review of Bradshaw

(1965), which concluded that phenotypic plasticity is specific for a character and a particular set of environmental influences. Plasticity has a specific direction, he claimed, and it is under genetic control that, while not necessarily related to genetic heterozygosity and polymorphism, can be radically altered by selection. Researchers including Scheiner (1993) and Schlichting and Pigliucci (1993, 1995, 1998) have emphasized experimental data that support the existence of plasticity genes and the ability to change plasticity through selection.

Perhaps a consensus is now appearing: although these approaches appear distinct, they are interchangeable in many circumstances, and many distinctions that have been made at one time or another at the environmental, phenotypic, and genetic levels of organization are not clear cut. A pluralistic approach is probably the most productive direction to take (Via et al., 1995), especially when applied in an integrative manner to those specific examples of phenotypic plasticity that can be studied at different levels of biological organization.

Before outlining the efforts made along these lines to understand the proximate mechanisms that underlie the evolution of phenotypic plasticity in *Bicyclus* butterflies, we introduce the population biology of their seasonal polyphenism.

Adaptive Phenotypic Plasticity

Bicyclus anynana of the wet season form have a pale medial band and conspicuous, submarginal eyespots on their ventral wings. In contrast, adults of the dry season form are uniformly brown in color with highly reduced eyespots. Compared with field surveys undertaken over successive seasons with climate changes, rearing experiments in the laboratory reveal that temperature provides the predictable cue for the adult environment (Brakefield and Reitsma, 1991; Windig et al., 1994; Brakefield and Mazotta, 1995). Temperature is high when larvae of the wet season form are developing; declining temperatures yield adults of the dry season form. Larvae are most sensitive to environmental cues in their final and penultimate instars (Kooi and Brakefield, 1999).

Field experiments in Malawi demonstrated that seasonal polyphenism in *Bicyclus* is an example of adaptive plasticity. Butterflies of the dry season form are inactive for much of their long adult life and are well camouflaged when at rest on dead brown leaf litter. Those of the wet season form reproduce quickly and are more active when among green foliage.

Their marginal eyespots are exposed when at rest and can function in deflecting the attacks of predators away from the vulnerable body (Brakefield and French, 1999). Seasonal polyphenism in the North American pierid butterfly, *Pontia occidentalis,* is an adaptive response to air temperatures, with associated changes in the amount of dark wing pigment and in thermoregulatory behavior (Kingsolver, 1995; Brakefield, 1996).

Physiological and Developmental Control

Developmental and physiological processes are often *canalized,* or quite resistant to genetic or environmental disruption. Phenotypic plasticity in which experience of specific environments during ontogeny leads to organized changes in development and variation in phenotype can therefore help us understand how developmental pathways can be mediated in response to environmental stimuli and therefore provide different phenotypic options (Nijhout, 1999; Evans and Wheeler, 2001). Examples in plants involve changes in growth form in response to such environmental stimuli as shading, salt contamination, and flooding. This type of plastic response is mediated via plant hormones (Van Hinsberg, 1996; Voesenek and Blom, 1996; Smekens, 1998). Ecdysteroid hormones regulate seasonal polyphenism in several species of butterfly, including *Araschnia levana* (Koch, 1987, 1992) and *Precis coenia* (Rountree and Nijhout, 1995a, b). However, other hormones can play a role, such as a summer-morph-producing hormone (SMPH) in *Polygonia c-aureum* (Endo et al., 1988).

Ecdysteroid hormones also mediate the development of seasonal forms of *B. anynana* (Koch et al., 1996). Thus microinjections of 20-hydroxyecdysone into young pupae of the dry season form yield adults with a shift in wing pattern toward the wet season. The increase in ecdysone titer after pupation occurs at a later stage in pupae of the dry season form than in those of the wet season form.

The development of eyespots is becoming comparatively well understood. The surface of a butterfly wing is covered by a layer of scale cells that contain color pigments. Each eyespot has concentric rings of numerous scale cells of different colors formed around a group of organizing cells known as a *focus* (reviewed in Brakefield and French, 1999). The number and position of eyespot foci are determined in developing wing disks in late larvae. Classical surgical experiments in early pupae demonstrated that at this stage, the foci each establish information gradients in the epithelial cell layer, presumably via the concentration of diffusible morphogens. Surrounding scale cells then respond in rings to this signal and become fated to synthesize a particular pigment just before adult eclosion. *Distal-less* and other genes of the *hedgehog* signaling pathway are involved in the early, larval establishment of the eyespot foci and in the processes leading to determination of the eyespot rings (Brakefield et al., 1996; Brunetti et al., 2001). In particular, the area of expression of several of these genes at different stages of eyespot formation appears to act as a marker for the activity and strength of the central organizing focus.

The eventual phenotypic differences between the seasonal forms, either when environmentally or genetically determined, are not apparent in divergent patterns of *Distal-less* gene expression in the wing disks at the larval stage, but were in one-day-old pupae. This immediately follows both the sensitive period for ecdysteroid hormone injection and the period of focal signaling prior to determination of the color pigments of scale cells in the adult eyespot (Brakefield et al., 1996). An exciting challenge is to understand precisely how the secretion of ecdysteroids in early pupae of the wet season form up-regulates the developmental genes involved in ventral eyespot formation.

Genetic Control and Evolution

Two types of genetic control of phenotypic plasticity have been distinguished (Schlichting and Pigliucci, 1993, 1995, 1998). *Control by allelic sensitivity* involves environmentally induced changes in the amount or activity of products of a gene or set of genes (and their alleles). Alternatively, in *regulatory control,* a plastic response might be controlled by the up- or down-regulation of specific genes in an environmentally dependent manner. Schlichting and Pigliucci speculate that allelic sensitivity is the likely candidate for the control of continuous reaction norms, whereas regulatory control may result in more discrete or threshold-like responses; they also coupled, respectively, the adjectives "passive" and "active" to these two kinds of genetic control. However, as Via et al. (1995) pointed out, both kinds of control could be involved in each extreme type of response. In regulatory control, the loci involved respond directly to a specific environmental stimulus and trigger a specific series of morphogenic changes. Their existence would appear to imply direct selection for the occurrence of a plastic response to different environments (Pigliucci, 1996).

Several researchers have used artificial selection to survey the availability of additive genetic variation for the evolution of phenotypic plasticity in laboratory stocks of a variety of plants and animals. All individuals in our stock of *B. anynana* from Malawi are competent to make a developmental decision in response to the temperature they experience as larvae to develop the adult wing pattern phenotype that enables them to best survive in the season in which they must primarily live. Although the phenotypic plasticity in field populations is in the form of classical seasonal polyphenism with rather discrete phenotypes (Windig et al., 1994), laboratory experiments that use several environments spanning the extremes characteristic of the field locality show that the underlying reaction norms are continuous in form (Brakefield and Mazzotta, 1995). The population-level reaction norm has a more or less linear relationship between eyespot size and rearing temperature. Apparently, intermediate phenotypes are rare in nature because intermediate temperature environments are usually too ephemeral to yield them; high and low temperatures are reliable cues for the alternative seasonal environments of the adult butterflies.

Bicyclus always reproduce sexually. The individuals within a family are the nearest we can effectively come to replicating individuals of a single (i.e., similar) genotype. A population consisting of different genotypes can be depicted as a bundle of reaction norms. The nature of genetic variation is then reflected in differences in shape among the reaction norms of different genotypes (van Noordwijk, 1989).

Plots of bundles of reaction norms for *Bicyclus* reared at different temperatures from split families clearly demonstrate quantitative variation among them. While only the wet season form is produced at the higher temperatures that prevail in the wet season in Malawi, there is quantitative variation among genotypes in the mean phenotype produced by individuals of different families. The situation is slightly more complex at low temperatures. Although all butterflies have the smaller eyespots of the dry season form, those reared in the laboratory at relevant temperatures for the beginning of the dry season in the field tend not to be as extreme as those collected in the field (which frequently have no hindwing eyespots at all). Because we now know that any factor that prolongs preadult development time leads to butterflies with smaller eyespots, we believe that variables other than temperature, such as host plant quality, play a role in accounting for the extreme dry season form with very small eyespots (Kooi et al., 1996). Whatever the precise mean phenotype produced at low rearing temperatures in the laboratory, there is also quantitative variation among families.

This variation has provided the material for several experiments that used artificial selection to change the developmental and phenotypic response of the ventral eyespots of *B. anynana* to rearing temperature (Holloway and Brakefield, 1993; Holloway et al., 1995). The particular target eyespot, as well as other eyespots, responded rapidly to selection. Responses occurred both at the temperature in which selection was performed and at other temperatures. In other words, heritabilities are high, and not only are there positive genetic covariances among eyespots, but also for the same eyespot across temperatures.

A later experiment to produce high and low lines for ventral eyespot size (Brakefield et al., 1996) used a much longer period of selection and progressively increased (low line) or decreased (high line) rearing temperatures over the generations. Eventually, the high line yielded only the wet season form across all rearing temperatures—although phenotypic plasticity remains, because butterflies reared at the higher temperatures have, on average, larger eyespots than those reared at lower temperatures. In sharp contrast, the low line produced only the dry season form at all temperatures. Analysis of phenotypic variation in crosses between these lines estimates the number of effective genetic factors that are fixed for alternative alleles across them as between about five and ten (Wijngaarden and Brakefield, 2000). Studies of the hormonal mediation of the eyespot phenotypes produced by these lines indicate that some of these genes are involved in specifying the timing of secretion of the ecdysteroid peak that follows pupation. Comparisons with the unselected stock also showed that the stock, when reared at high temperature, is similar in hormone dynamics and in phenotype to the high line, but at low temperatures it parallels material of the low line. Thus, the loss of the ability to produce both alternative seasonal forms in the selected lines appears to be an effect of changes in the pattern of hormone secretion following pupation across rearing temperatures (Brakefield et al., 1998).

Interestingly, the dorsal eyespots of *B. anynana* (which are not exposed to predators in butterflies at rest) do not express phenotypic plasticity, and lines selected for divergent size of these eyespots do not show changes in the dynamics of ecdysteroid secretion after pupation (Brakefield et al., 1998). The proximate mechanism of uncoupling of the eyespots on the dorsal and ventral wing surfaces is not clear. Because all eyespots are formed by essentially the same developmental pathway, understanding how such uncoupling is achieved may shed light on the evolution of genes for plasticity in this system.

The upward and downward pairs of lines established in the different selection experiments diverged, sometimes dramatically, in the elevation or height of the bundles of reaction norms. A response to selection for a change in phenotypic plasticity will, however, necessitate a change in reaction norm shape, which is only possible when reaction norms of individual genotypes (or families) cross (i.e., there is genotype x environment interaction). Although surveys in *B. anynana* have consistently described some "crossing of reaction norms," the selection experiments mentioned here have shown only minor changes in reaction norm shape and have not yielded novel shapes that are unrepresented in the base stock. Further studies using experiments specifically targeted at reaction norm shape and designed to change the form of the plasticity have also failed to yield populations with either substantially steeper or shallower reaction norms, or ones in which the overall shape had diverged from a more or less linear relationship between phenotype and temperature (Wijngaarden and Brakefield, 2001, 2002).

Thus while extreme changes in elevation can evolve rapidly, the same apparently is not true for changes in shape. This appears to be due to positive genetic covariances across environments; a response to selection in one environment is accompanied by a parellel response across other environments. Changes in shape may not be generated readily because the underlying hormones that mediate the developmental pathway of ventral eyespot size in early pupae are also involved in long-range and multimodal signaling within the organism (Wijngaarden and Brakefield, 2002).

Constraints on Evolution

These results from selection experiments suggest that in *B. anynana,* and perhaps in other species, there may be genetic constraints (at least in the short term) on responses to natural selection. Selection for changes in plasticity could arise because of climate change or through extensions of species ranges into regions with different relationships between climatic and ecological environments. Indeed, surveys of variation across species of *Bicyclus*, especially those adapted to seasonal environments north and south of the equator, suggest that given sufficient time and directional selection, evolution can yield parallel phenotypic changes in response to very different environmental gradients (Roskam and Brakefield, 1996, 1999).

Other forms of constraint have also been proposed to play a role in the evolution of adaptive phenotypic plasticity (Antonovics and Van Tienderen, 1991; Schlichting and Pigliucci, 1998). Plasticity might have costs that could act as a constraint; only recently has some effort been made to measure them experimentally (DeWitt, 1998; Scheiner and Berrigan, 1998; Smekens, 1998), or to incorporate such variables into models of the evolution of phenotypic plasticity (Van Tienderen, 1991, 1997; Moran, 1992). A cost of plasticity becomes apparent when a plastic organism has a lower fitness than a nonplastic one in an environment in which both organisms express the same mean phenotype. DeWitt et al. (1998) list five potential costs: energetic, production, information acquisition, developmental instability, and pleiotropy and epistasis. In addition to such costs, the benefits of plasticity may be limited if facultative development cannot produce a trait mean as near to the optimum as can fixed development. The same authors list four potential limits of this kind: information reliability, lag time, developmental range limits, and the epiphenotype problem.

Adaptive phenotypic plasticity provides many challenges for future research, not least in understanding the proximate mechanisms in sufficient detail to enable clearer insights about such evolutionary constraints and limits. It is also to be hoped that increasing numbers of developmental biologists and those working at the interface between development and evolution will take up the challenge of phenotypic plasticity as a means to understanding more about the control of developmental pathways. More complete descriptions of so-called genes for plasticity may take us closer to understanding control mechanisms that can act as developmental switches. In addition, the study of phenotypic plasticity that is not adaptive and that may represent an inability to eliminate developmental instability should take us closer to understanding the processes that underlie the evolution of developmental stability.

BRENT MISHLER

Phylogeny

Ever since the Darwinian revolution, *phylogeny*—reconstruction of the course of splitting of and change along lineages in the tree of life—has been one of the driving aims of evolutionary biology. The only figure in the *Origin of Species* is a hypothetical phylogenetic tree, and many other early evolutionists attempted to apply Darwin's ideas to postulated trees for real groups of organisms. Haeckel coined the term phylogeny and was its most influential early proponent. The concept of phylogeny was linked by Haeckel and others in various ways to the concept of *ontogeny* (study of the development of individual organisms). The desire to study phylogeny, and the temptation to compare it to ontogeny (another obvious temporal process of change in organisms) was understandable. However, rigorous methods for inferring phylogenetic trees were lacking; early trees were intuitively drawn without a clear connection back to the data. Haeckelian excesses appear to have led to a rather low esteem for phylogenetic tree building by the end of the nineteenth century. Phylogenetic thinking was eclipsed in biology in the early 1900s by genetic and natural history studies, which were melded together in the Modern Synthesis of the 1930s to form the new field of population biology.

This new field, which dominated evolutionary biology until at least the 1970s, was concerned with evolutionary change on very small spatial and temporal scales, the hope being that once processes at those scales were understood, it would be easy to scale up to the big tree of life. Genotypic features were given pride of place for study, while the phenotype was seen as an epiphenomenon. Systematics (the study of relationships of organisms) at any more inclusive level than species was relegated to a back seat during this whole period.

Beginning in the 1970s, several independent things happened to rectify the balance between genotype and phenotype, micro- and macroevolution, and population biology and systematics. The study of ontogeny, phylogeny, and broader space and time scales became fashionable again. Most of these new trends are detailed elsewhere in this book; this entry concentrates on what happened to systematics, in particular, and how an explicit methodology for systematics was finally derived, in what might be considered the culmination of the Darwinian Revolution.

The field of systematics underwent a conceptual upheaval throughout the 1970s and 80s; see Hull (1988) for an entertaining and insightful treatment of the "systematics wars." Many issues were at stake, foremost of which was the nature of taxa. Are taxa just convenient groupings of organisms with similar features, or are they lineages, marked by homologies? The consensus that emerged from this upheaval was that taxa are the latter, but why? Before dealing with that question, we need to address the topic of homology and how phylogenetic trees can be reconstructed.

The Hennigian Revolution

Many methods have been used to produce phylogenetic trees; most were subjective and intuitive, thus hard to evaluate. One promising approach, which became popular in the 1960s, was to use computer algorithms to evaluate the overall similarity of organisms in all describable features, a school of thought that became known as *phenetics*. Relationships were expressed in a branching diagram—a *phenogram*—sometimes (but not always) interpreted as a phylogeny. It was hoped that stable, repeatable classifications would result.

At about the same time, Willi Hennig (1966) was exploring a different approach, using what he called *special similarity*. Hennig's seminal insight was to note that in a system evolving via descent with modification and splitting of lineages, homologous similarities among organisms come in two basic kinds, *synapomorphies* due to immediate shared ancestry (i.e., a common ancestor at a specific phylogenetic level) and *symplesiomorphies* due to more distant ancestry (Figure 1). Only the former are useful for reconstructing the relative order of branching events in phylogeny; "special similarities" (synapomorphies) are the key to reconstructing truly natural relationships of organisms. This approach was made quantitative and objective with the development of computer algorithms by Kluge and Farris (1969; also Farris, 1970). This approach became known as *cladistics*.

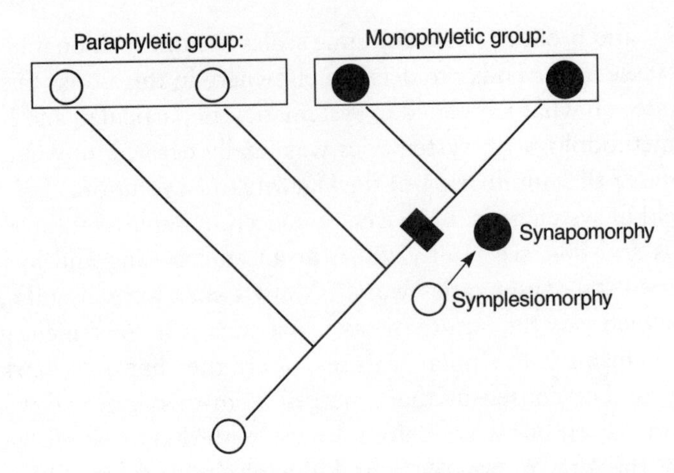

FIGURE 1. A hypothetical cladogram. Shown is a phylogeny of four species based on a number of characters. One character is mapped onto the cladogram; an evolutionary change has occurred in this character, giving it a two-state transformation series. Two possible higher-level groups are shown: the one on the right (supported by the synapomorphic state) would be monophyletic, and the one on the left (supported by the symplesiomorphic state) would be paraphyletic.

The phenetic and cladistic approaches thus shared some qualities; both were objective, quantitative, and repeatable. If both gave the same result, there would have been no point to a controversy over methods. But they don't. Farris (1983) and others have shown convincingly that the information content of the phylogenetic approach is greater. Overall similarity is an incoherent mixture of synapomorphy, symplesiomorphy, and nonhomology.

Thus, we now realize that only characters that changed state along a particular lineage can serve as evidence for the prior existence of that lineage after further splitting occurs. This fundamental idea driving recent advances in phylogenetics, known as the *Hennig principle,* is as elegant and fundamental in its way as is Darwin's principle of natural selection. Simple, yet profound in its implications, it is based on *homology,* one of the most important concepts in systematics, but also one of the most controversial. What does it mean to say that two organisms share the same characteristic? The modern concept is based on evidence for historical continuity of information (Roth, 1988); homology would then be defined as *a feature shared by two organisms because of descent from a common ancestor that had that feature.* There are two fundamental kinds of homology: *phylogenetic homology* (between organisms as defined above) and *iterative homology,* which is reserved for cases in which two or more parts within one organism are descended from a common ancestral part—serial homology in morphological characters or paralogy in molecular data.

This concept is clear in theory, but how do we recognize homology? The best early codification of recognition criteria was that of Remane (Wiley, 1981): detailed similarity in position, quality of resemblance, and continuance through intermediate forms. Also, an important contribution of cladists has been the explicit formulation of a phylogenetic criterion: a hypothesis of taxic homology of necessity is also a hypothesis for the existence of a monophyletic group. Therefore, congruence among all postulated homologies provides a test of any single character in question, which is the central epistemological advance of the cladistic approach. Individual hypotheses of putative homology are built up on a character-by-character basis, and then a congruence test (using a parsimony principle) is applied to distinguish homologies (i.e., those apparent homologies that are congruent with other characters) from homoplasies (i.e., apparent homologies that are not congruent with the plurality of characters).

Phylogenetic Systematics

To complete the Hennigian system, classifications are applied to the resulting branching diagram, or *cladogram.* A corollary of the Hennig principle is that classification should reflect reconstructed branching order; only *monophyletic groups* should be formally named. A strictly monophyletic group is one that contains all and only descendants of a common ancestor. A *paraphyletic group* is one that excludes some of the descendants of the common ancestor; see Figure 1 for the distinction between these two groups. This elegant correspondence between synapomorphy, homology, and monophyly is the basis of the cladistic revolution in systematics.

Returning to the question asked earlier, why is phylogenetic integrity necessary for taxa? Taxa could of course be whatever we want, because the whole nomenclature system is a human construct. Many kinds of nonphylogenetic biological groupings have been proposed that are unquestionably useful for special purposes—for example, predators, rain forests, succulent plants, and bacteria. However, phylogenetic systematists have settled on phylogeny as the best criterion for general-purpose classification, and they restrict the use of the formal Linnaean system to hypothesized monophyletic groups. The fundamental argument is that phylogenetic relationships are the best way to summarize known data about attributes of organisms and to predict unknown attributes (Mishler, 1995).

In addition, phylogenetic taxa are "natural" in the sense of being the result of the evolutionary process. Other processes that impinge on organisms are not so overriding in their effects. Natural selection might under some extreme conditions cause organisms to become very similar even though they are unrelated. Such similarity will not be across the board, however, only in the suite of attributes being influenced by convergent selection, e.g., a hummingbird pollination syndrome, thorns, or succulence. Across the board, detailed similarity is more likely to be due to descent (homology) than common environment (analogy). Thus phylogeny provides the best general-purpose classification for biology.

Methodology for Phylogenetic Reconstruction

The first issue to face in building a phylogenetic tree is development of a data matrix. The rows in the matrix are monophyletic taxa, but what are the columns, the characters? The central epistemological problem of systematic research is how to recognize, distinguish, and "define" taxonomic characters precisely. Going back to first principles, a taxonomic character—a putative taxic homology in the terms defined above—is a single, independent piece of evidence for the existence of a monophyletic group.

A good taxonomic character shows greater variation among observable taxonomic units than within. This variation must be heritable and independent of other characters, i.e., not genetically correlated with other characters in a specific evolutionary sense. There are other meanings of "correlation," some of which (such as phylogenetic congruence) do not disqualify characters from counting as independent. This view of taxonomic characters also requires that each be a system of at least two discrete transformational homologues, or character states. Note that this is a restricted usage of the term "character," derived from the ontology of phylogenetic systematics. For other purposes, as in functional/evolutionary studies, numerical phenetic comparisons, or identification, less strict usages are applied.

The second issue to be faced is how to turn the data matrix into a phylogenetic tree. The basic cladistic approach uses a parsimony criterion, as briefly outlined above. The data matrix is viewed as itself a refined result of character analysis, with each character an independent hypothesis of homology (Mishler, 1994). The parsimony algorithms basically test these independent hypotheses against each other, looking for the best-fitting joint hypothesis. The shortest tree is selected as the best supported by the data.

Supporters of the parsimony method argue that only the fewest and least controversial assumptions should be used: that characters are heritable and independent, and changes in state are relatively slow as compared to branching events in a lineage. When these assumptions hold, reconstructions for a character showing one change on one branch will be more likely than reconstructions showing two or more changes in that character on different branches. These amount to a joint assumption that an apparent taxic homology—a feature that has already passed strict observational and experimental tests of detailed similarity, heritability, and independence—is more likely to be due to true taxic homology than to homoplasy, unless evidence to the contrary exists, i.e., a majority of apparent taxic homologies showing a different pattern.

When can straight parsimony methods fail? In an important early paper, Felsenstein (1978) showed that branch-length asymmetries within a tree can cause parsimony reconstructions to be inconsistent. That is, if the probability of a parallel change to the same state in each of two long branches is greater than the probability of a single change in a short connecting branch, the two long branches will tend to falsely "attract" each other in parsimony reconstructions using a large number of characters (see also Sober, 1988). The region at which branch-length asymmetries will tend to cause such problems has been called the Felsenstein zone (Albert et al., 1992). The seriousness of this problem, measured by the size of the Felsenstein zone, is affected by several factors, the most important of which are the number of possible character states per character and the overall rate of change of characters.

To push back the boundaries of the Felsenstein zone, scientists can carefully select characters that show the appropriate amount of variation for the phylogenetic depth of the study. If they are available, more taxa can also be added to the study group to break up the long branches. If no additional characters or taxa are available, scientists can consider the probabilistic biases in character change that may be causing homoplasy. This can be done through character weighting in parsimony or through maximum-likelihood approaches.

What if there are valid reasons for not viewing all apparent taxic homologies as equal in the weight of evidence they bring to the analysis? In this case, character and character-state weighting can be applied (Albert and Mishler, 1992; Albert et al., 1992, 1993). If differential probability of change in different characters (or types of characters) can be postulated *a priori,* maximum-likelihood-based weights can be specified in parsimony analyses, essentially giving a larger "vote" to difficult changes. That is, a homoplasy in a character that is easier to change is preferable to one in a character more difficult to change. For example, weights could be applied that take into account differential probabilities of change at different codon positions in a protein-coding gene. This is a simple matter of introducing a multiplier representing the relative weight. The relative weight of a character is the negative natural log of its relative probability of change (so high probability of change = low weight).

Specifying differential probabilities of transformation among states *within* characters is a little more difficult algorithmically, but it can be done in a similar fashion (e.g., weights that take into account gains versus losses in restriction site data, or transition/transversion bias in sequence data). The method for applying such character-state weights is a *step matrix*. This specifies the "cost" of going from one state to another, and it can be very complex (even asymmetrical).

One important conclusion of attempts at modeling the major known transformational asymmetries is that the differential weights produced have little effect on parsimony reconstructions. With data having a reasonable rate of change, weighted parsimony topologies are usually a subset of the unweighted (or more properly, equally weighted) ones. Thus, paradoxically, our pursuit of well-supported weighting schemes has ended up convincing us of the broad applicability and robustness of equally weighted parsimony (Albert and Mishler, 1992; Albert et al., 1992, 1993).

Maximum-likelihood (M-L) approaches to phylogenetic estimation attempt to evaluate the probability of observing a particular set of data, given an underlying phylogenetic tree (assuming an evolutionary model). Among competing phylogenetic trees, the most believable (likeliest) tree is one that makes the observed data most probable. To make such a connection between data and trees, it is necessary to have auxiliary assumptions about such parameters as the rate of character change, the length of branches, the number of possible character states, and relative probabilities of change from one state to another. The likelihood function begins with the evaluation of each character, one at a time, considering the probabilities of all possible assignments of states to the internodes. The overall likelihood is the sum of the likelihoods of all the characters.

Maximum-likelihood approaches are sometimes presented as though they were an alternative to parsimony, but this is not actually the case. Instead, parsimony should be viewed as the simplest of a whole family of M-L approaches. Parsimony is based on the simple model of evolution already described, and more complicated M-L models can be developed if justified. Experience shows however, that with data carefully selected for the problem at hand (thus with appropriate rates of change), parsimony and more complicated M-L models give exactly the same solutions. Only when the boundary of the Felsenstein zone is being approached are the more complicated M-L models or character-weighting schemes possibly justified. But even there, care must be taken to be sure the models are accurate, or spurious results can occur (Pol and Siddall, 2001).

General Uses of Phylogenetic Trees

The area of comparative methods has seen explosive growth in recent years. This is an area of considerable controversy and rapid conceptual development. Virtually every issue of major journals and each new book on systematics and evolution contains something of interest on this subject. Many advances have been made in improving evolutionary model building as a route to understanding; "tree-thinking" is now central to all areas of systematics and evolution. The central importance of phylogeny reconstruction in systematics, ecology, development, and evolutionary biology has become widely realized in recent years (Donoghue, 1989; Funk and Brooks, 1990; Wanntorp et al., 1990; Brooks and McLennan, 1991; Harvey and Pagel, 1991; Miles and Dunham, 1993; Martins, 1996).

The large number of comparative methods can best be summarized by placing them into categories corresponding to the types of hypotheses meant to be tested:

1. *Cladogram comparisons.* Methods meant for comparing different

phylogenetic trees are necessary for the study of *coevolution.* Coevolution can be broadly defined as congruence between two or more systems undergoing tree-like evolution, i.e., evolution by descent with modification and with branching. This is a generalization of the phylogeny/homology relationship, or the "coevolution" of organism lineages and characters already discussed. Coevolution comes in many forms: vicariance biogeography (organism/earth coevolution), host/parasite relationships, and community evolution (e.g., symbionts, pollinator/plant coevolution, or other long-term ecological associations). Methodologies have diversified to compare cladograms in coevolutionary studies, including consensus techniques (Funk and Brooks, 1990), tree-to-tree distance metrics (Penny and Hendy, 1985), and parsimony techniques (such as Brooks parsimony; see Brooks, 1990; Brooks and McLennan, 1991).

More recently scientists have realized that the statistical nature of the M-L estimation process lends itself to testing evolutionary hypotheses (independent of controversies over its value as a reconstruction method). The likelihood of different trees can be compared using a likelihood ratio test statistic. Elegant tests have been developed for the presence of a molecular clock, comparing two trees (such as comparing a new tree with a classic tree from the literature), or comparing trees derived from different genes to see whether they are really different (e.g., Huelsenbeck and Bull, 1996; Huelsenbeck et al., 1997).

2. *Clade comparison within a cladogram.* Methods have also been developed to detect whether there are imbalances in symmetry between sister clades in the same cladogram, in order to address various questions in both micro- and macroevolution. What is the null expectation? Slowinski and Guyer (1989) showed that real trees should be expected to be quite asymmetrical even under a random model. Furthermore, even if trees are judged significantly asymmetric, how can we associate that judgment with some specific factor postulated to be the cause of that asymmetry? This leads to the hot topics of *key innovations* and *adaptive radiations.* There have been many, often conflicting definitions of adaptive radiations (Givnish and Sytsma, 1997). The postulated key innovation should cause rapid diversification of lineages in ecological or morphological features. Determining "rapidity" quantitatively has led to the development of number of methods meant to deal with the required time estimation problem, which involves two questions: Can we assume a molecular clock? If we can, how do we calibrate it (Sanderson

and Wojciechowski, 1996; Sanderson, 1997; Baldwin and Sanderson, 1998)?

3. *Discrete-state character comparisons on a cladogram.* Methods have also been developed to examine how discrete-state characters evolve on a tree individually and together. Such characters can be mapped onto cladograms using parsimony, so as to minimize the number of character-state changes. In this way, suites of characters are built up for hypothetical taxonomic units (HTUs). Specific types of hypotheses that can be tested include polarity of character-state changes in one character, and the association of state changes in two characters, either undirected (Ridley's test; Ridley, 1983) or directed (Maddison's test; Maddison and Maddison, 1992).

4. *Continuous character comparisons on a cladogram.* Other methods have been developed for examining how quantitatively varying characters are associated on phylogenies. The "bad old way" to compare two such characters was through direct correlations of species values (using species as data points). However, as pointed out by Felsenstein (1985) and others, this approach treats species as if they were equally related to each other. Quantitative comparative approaches were developed in an effort to remove the influence of history, for example, using ANOVA and ANCOVA (Harvey and Pagel, 1991), autocorrelation (Cheverud and Dow, 1985), independent contrasts (Felsenstein, 1985; Burt, 1989), and general linear model approaches to partition variance and subtract the phylogenetic effects (Martins, 1996). Conversely, other methods explicitly describe variation due to phylogeny by tracing the quantitative characters on a phylogenetic tree, reconstructing values for nodes, and looking at direction of change by comparing ancestors and descendants (e.g., Huey and Bennett, 1987).

The relationship between evolution and development cannot be studied except in the light of a rigorous phylogeny. Several of the comparative approaches described here should be used to study developmental evolution. For example, discrete-state or continuous changes in ontogeny must be mapped onto a cladogram to distinguish different modes of heterochrony. In other words, the only difference between paedomorphosis and peramorphosis is the polarity of ontogenetic characters on a cladogram (Fink, 1982).

Hypotheses about the effect of developmental constraints on evolution of phenotypic characters can be studied by comparing the relative rates of molecular and morphological evolution on the same branches of the tree. If developmental constraints are important in restricting the path of a lineage through phenotypic space over time, one should be able to demonstrate a slowdown of phenotypic change in the face of "normal" rates of molecular change. On the other hand, if a release of developmental constraints served as a key innovation making possible an adaptive radiation, then one should be able to demonstrate a speedup of phenotypic change in the face of "normal" rates of molecular change.

There is a reciprocal relationship as well: it is clear that developmental information can have a tremendous impact on the reconstruction of phylogenetic trees. Despite the abundance of molecular data now available for phylogenetic analysis, morphological characters remain essential evidence for branches, especially in "deeper" comparisons (Mishler, 2000). For such deep branches, quasi-clock-like characters such as DNA sequence data are poor markers, simply because they keep changing and thus can erase the evidence for a branch's existence. Characters that change in a major way, yet rarely, are the best markers for such ancient branches. Morphological characters often have much greater complexity in their structure than simple nucleotide matches in a DNA alignment, plus they have development that allows a temporal axis of comparison not available with DNA sequence data. Thus sounder initial hypotheses of homology can often be made in morphological comparisons.

The future of evolutionary biology lies in greater integration of data from different hierarchical levels in the organism, ranging from molecules through development to morphology and ecology. These data in turn should be compared carefully on phylogenies both to better reconstruct all the hierarchical levels in the tree of life and to better understand the causes of evolutionary change. Phylogeny will remain the key link between the two important biological hierarchies: the process hierarchy within each organism and the process hierarchy among nested lineages through evolutionary time.

JONATHAN M. W. SLACK

Phylotype and Zootype

The term *phylotypic stage* means a stage of development at which all members of a phylum, or other higher taxon, look the same. It was first used by Sander, following the German expression *Körpergrundgestalt* previously introduced by Seidel (1960). Sander was particularly concerned with insects and pointed out that although adult insects vary greatly in their morphology, they all look similar at the embryonic stage called the "extended germ band" (Sander, 1983). At this stage the segmental arrangement of the body is very obvious, including the gnathal segments that are hard to visualize when they later become the mouthparts of the adult. All insects have three gnathal, three thoracic, and a variable number of abdominal segments. The extended germ band is reached before dorsal closure of the embryo and consists of a plate composed of ectodermal and mesodermal components. There is a ventral nerve cord and an indication of anteroposterior tagmosis in terms of the appearance of the procephalon and the appendages on the gnathal and thoracic segments. Most details of the future epidermal structures, however, are yet to appear, and the endoderm, in terms of the anterior and posterior midgut invaginations, has yet to form. Sander's conception has been generally accepted by those interested in insects, although the more general acceptability of the concept has been a matter of some debate.

In Seidel's original *Körpergrundgestalt* article, the conception had been generalized to other phyla—in particular, to coelenterates, annelids, molluscs, echinoderms, and the subphylum of vertebrates (see Figure 1). Such a generalization involves a number of assumptions, for example, that the definition of a phylum implies the common possession of a particular and distinct body plan. Although elementary zoology textbooks often state that one phylum equals one body plan, this one-to-one correspondence is not obvious at all. Applying modern techniques for classifying morphological disparity might return interesting results regarding how uniform the phyla really are in their distinctive status.

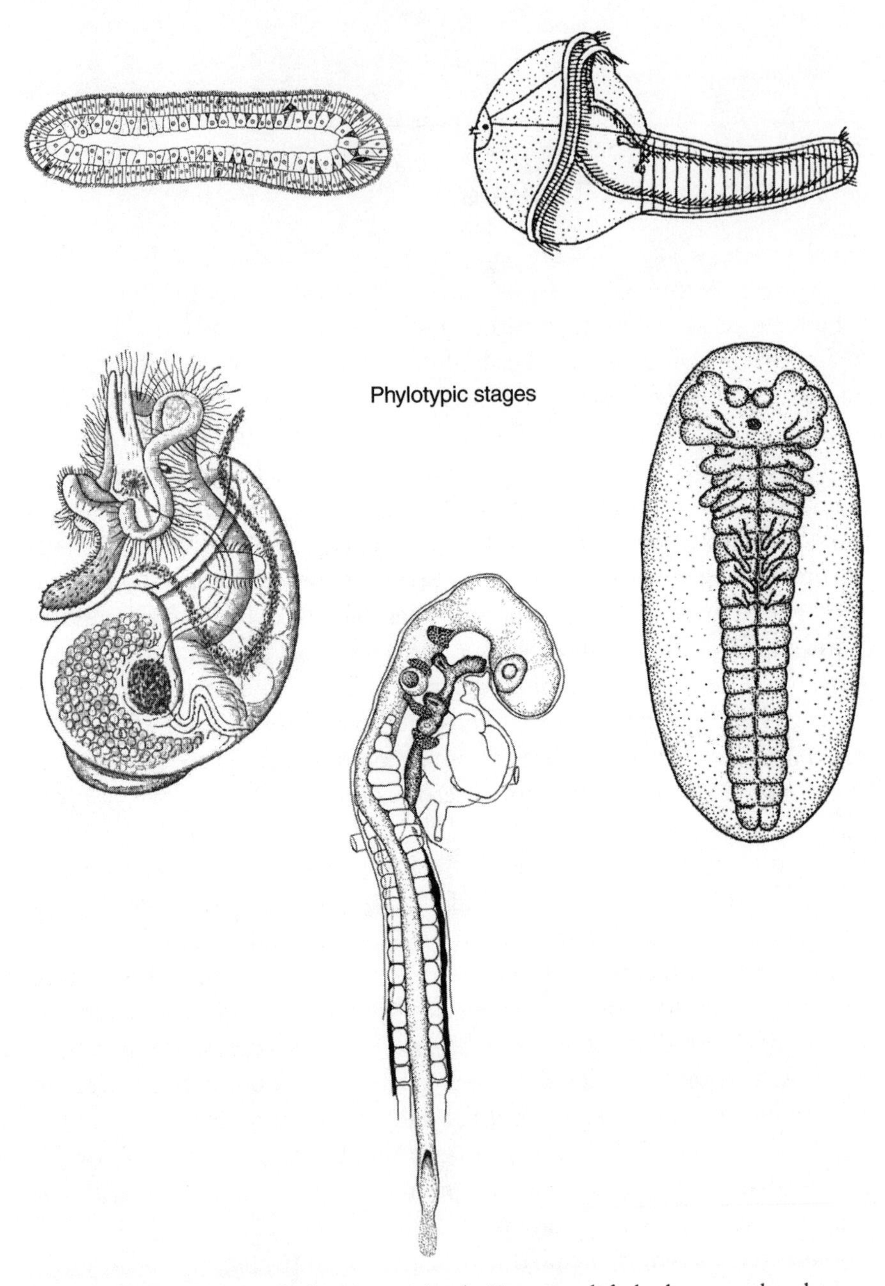

FIGURE 1. Phylotypic stages of various animals. From top left: hydrozoan planula, segmenting polychaete, insect at extended germ band stage, vertebrate at tailbud stage, gastropod at veliger larva stage.

Certainly to a lay person, the various classes of mollusc would appear more disparate than some of the minor vermiform phyla. But to some degree it is inevitable that higher taxa correspond to broad features of body organization, because the whole logic of taxonomy is to encode different degrees of difference in different orders of relatedness.

The next idea is that the body plan of a phylum is displayed in its entirety at one stage of development, normally the early to mid-phase of embryogenesis. There is some empirical evidence to support this. The second most widely accepted phylotypic stage is the tailbud stage of the vertebrates. Since the celebrated argument of von Baer and Haeckel that embryos of different vertebrate classes were almost identical, it has been widely accepted that the textbook vertebrate body plan approximates the appearance of an embryo at an early tailbud stage, which is after neural tube closure, the formation of a certain number of somites, and the appearance of branchial arch structures from the neural crest. Cohen and Massey (1982) have argued that the true phylotypic stage should be the neurula rather than the tailbud stage. And despite discussion of the considerable liberties taken by Haeckel to maximize the resemblance of vertebrates of different classes (Richardson, 1995; Richardson et al., 1997), the tailbud stage is still considered phylotypic.

Before considering exceptions and problems, we need to have some idea of why conserved body plan features should be displayed at one stage. Although it is often stated that the earliest stages of development cannot change in evolution because early mutational change would compromise later events (Arthur, 1988), it is nonetheless a familiar empirical fact that early embryonic morphology is very diverse. The main reason for this is the diversity of reproductive lifestyle and strategy. For example, the eggs of even quite closely related animals may differ markedly in yolk content. This is the result of evolutionary trade-offs: organisms come to occupy different niches in which they produce either a lot of small eggs with a poor chance of survival or a few large eggs with a good chance of survival. The presence of more yolk drives various changes in early development, including the disposition of cleavages (meroblastic rather than holoblastic) and the nature of gastrulation movements (epiboly rather than invagination) (Gerhart and Kirschner, 1997). Viviparity imposes even more drastic changes on early embryonic life and is necessarily accompanied by the early formation of a variety of extra-embryonic membranes and supporting structures from the zygote as well as the mother. So early development is necessarily diverse because reproductive behavior is diverse.

Late development is also diverse, but for quite different reasons. By late development, the embryo is becoming quite similar to the post-embryonic organism, whether larval or imaginal. Free-living organisms are subject to selection and must acquire distinct niches for their survival to the reproductive stage, so the late embryos have to be diverse because the organisms to which they give rise are diverse. The seeds of late events may be laid down early, as Richardson (1999) has claimed, but they are by definition manifested late in terms of morphology. It follows inexorably that the stage of maximum similarity within a large taxon is the early-middle period of embryonic development, after the constraints imposed by reproductive strategy and before the constraints imposed by the adaptation of the free-living organism. The extended germ band of insects and the tailbud stage of vertebrates are both early-middle stages that exist within egg cases, jelly layers, or a uterus, and so interact with the environment only minimally.

According to this way of thinking, there is no specific cause for the phylotypic stage; it is just the stage in the middle at which the selective pressures for change are minimized. As a result, it is the stage most likely to retain the features of the common ancestor. Is any further cause required? Perhaps it is a matter of efficiency: for example, a file of somites is the best way to make body muscles and vertebrae, and if they had not existed they would have been called forth independently in different vertebrate classes by natural selection. This is hard to disprove although the molecular homologies of early embryos make multiple independent origins of body plan features rather unlikely. Duboule (1994) has argued that the phylotypic stage is a consequence of the temporal colinearity of expression of Hox and maybe other gene families. In any case, these possible additional reasons for the phylotypic stage, while placing it in early-middle development, do not demand that all phylotypic features should necessarily be displayed simultaneously. Indeed they are not, as particularly emphasized by Richardson (1995) and Richardson et al. (1997), so the phylotypic stage is not really a stage at all. Instead, it is a more-or-less elongated phase of early-middle development within which the most highly conserved features of morphology are apparent.

So are there phylotypic stages for groups other than the insects and vertebrates, neither of which are actually true phyla? Among the molluscs, the veliger larva would seem to be a reasonable candidate. This has a head, a posterior, a through gut, a dorsal shell gland, and a ventral muscular foot—in other words, the pattern corresponds to the textbook idea of an archetypal mollusc. However, it is not strictly *phylo*typic, because one important class of molluscs, the cephalopods, do not have a veliger stage. A phylotypic annelid might be considered to be a stage at which a certain number of segments have arisen, and the organism has a head, a ventral nerve cord, a through gut, and segments involving all the germ layers. This description would apply to the leeches, which lack a trochophore, as well as the polychaetes and oligochaetes, which possess a trochophore. The cnidarians are a phylum whose body plan is generally considered to be a two-layered blind-ended sac. If so, then the planula larva is a phylotypic stage although, because the anterior end of the planula becomes the basal end of the polyp, there is the interesting question about which end, if either, might be considered the anterior (Yanze et al., 2001).

The echinoderms are a remarkable phylum in which metamorphosis from the larva to the adult is particularly dramatic. Here it would seem that the echinus rudiment, or its equivalent in the other classes, should be considered phylotypic because the earlier larvae are typically bilaterally rather than radially symmetrical and do not have echinoderm features such as a water vascular system or tube feet. All of these examples were suggested by Seidel in his original *Körpergrundgestalt* paper. A case has also been made for the nauplius larva of crustaceans, which has three pairs of head appendages, although this feature has been lost from some malacostracan taxa (Williams, 1994).

A few years ago, questions such as "are all animals monophyletic?" or "which end of the cnidarian is the anterior?" might have seemed academic to the point of transcendence, because no empirical evidence could possibly be brought to bear. But now we know the answers. The rise of molecular developmental biology has enabled the identification of a number of long-range molecular homologies in animal design that are truly compelling. The key to understanding the strength of this evidence is the arbitrariness of the molecular components used to build multicellular organisms. Some molecules are not at all arbitrary; they have to have certain features to exert their functions. For example, collagen needs to be fibrous and insoluble to fulfill its structural role. Myosin needs to have the motor protein activity to move along actin filaments.

Phosphofructokinase needs to retain its specific enzyme activity and hence to have particular amino acid residues in particular three-dimensional locations.

These constraints to molecular function do not apply to the main classes of molecule that control development. These are signaling molecules and transcription factors (reviewed in Slack, 2001). Signaling molecules are released from one type of cell and act at receptors on another type of cell to effect a change in gene activity. Transcription factors are the intracellular proteins that control gene activity and are often activated by the action of signaling molecules. While signaling molecules and transcription factors certainly need specific molecular features to do their biochemical jobs, their biological functions are informational so they are more or less interchangeable. In other words, there is no good reason why one signal should be used rather than another, or one transcription factor rather than another, in order to control the activity of a specific gene. If it is found that the same signals and transcription factors are used for the same biological purpose in different organisms, it is highly probable that the reason for this is homology: descent from a common ancestor in which the pathway was operative.

The first discovered and best-known long-range molecular homology is that of Hox gene expression. Hox genes are a family of genes that control anteroposterior pattern. They normally exist as a gene cluster whose members are expressed as a nested set in the early-middle stage embryo. To a first approximation, this means that all genes are on in the prospective posterior, and that successive members of the cluster have more anteriorly located anterior boundaries of expression. The evidence that Hox genes control anteroposterior pattern has been obtained from molecular-genetic experiments on model organisms such as mouse, *Xenopus,* and *Drosophila.* In general, loss of function leads to anteriorization, and gain of function (ectopic expression) leads to posteriorization of the axial body pattern (McGinnis and Krumlauf, 1992).

Doing these experiments in a wide range of animals from other phyla is not possible, but we can clone their Hox genes and examine the expression patterns in embryos using *in situ* hybridization. There are some variations—not all animals have the same number of Hox genes, and in vertebrates the cluster has been quadruplicated to give four similar clusters—but the general rule of nested anteroposterior expression has held up remarkably well across the whole animal kingdom, except for sponges (Kourakis and Martindale, 2000; Ferrier and Holland, 2001).

In fact the association of Hox expression with anteroposterior pattern is so well established that it has been possible to show that the true anteroposterior axis of the sea urchin becomes established in the echinus rudiment, with the anterior at the oral end and the successive coeloms representing more posterior levels (Peterson et al., 2000).

The Hox genes are not the only genes controlling anteroposterior pattern. A ParaHox cluster has been defined consisting of *gsx, lox,* and *cdx* type genes, also showing an anteroposterior expression, particularly in the gut (Brooke et al., 1998). Hox and ParaHox clusters are distantly related and probably arose through an ancient duplication of a primordial gene cluster. Both clusters are probably present in cnidarians as well as higher animals, but absent from sponges (Finnerty and Martindale, 1999; Ferrier and Holland, 2001). In the head, transcription factor genes of the *otx* and *emx* classes are expressed and are necessary for anterior development (Cecchi et al., 2000); at the posterior end, the Cdx drslt (Wu and Lengyel, 1998) and Evx (Sordino et al., 1996) family of transcription factors are required. All of these also show conserved expression patterns across a very wide phylogenetic range.

The next major aspect of the body plan that is highly conserved is the dorsoventral patterning system that operates in early development. In this case, there is only evidence from vertebrates and insects, and the homology requires the assumption that one of the lineages turned upside down at some stage in its history, such that its basal side became uppermost (Nübler-Jung and Arendt, 1994; DeRobertis and Sasai, 1996; Arendt and Nübler-Jung, 1997).

In vertebrate embryos, there is a ubiquitous synthesis of bone morphogenetic proteins that induce ventral type genes. On the dorsal side, the organizer (Spemann's organizer in amphibians, the embryonic shield in fish, Hensen's node in birds) emits inhibitors of BMP such that there is a gradient of BMP activity from ventral to dorsal. One of these inhibitors is chordin. In *Drosophila,* the homologue of BMP, called Decapentaplegic, is synthesized on the dorsal side and is antagonized by the homologue of chordin, called short gastrulation, from lateroventral levels. The net result is similar: an activity gradient of Decapentaplegic, with the high point dorsally instead of ventrally. This similar system, leading to an inverted pair of body plans, is consistent with the long-known difference of dorsoventral body arrangement in vertebrates and insects, with the insect having a ventral nerve cord and a dorsal heart, while the vertebrate has a dorsal neuraxis and a ventral heart.

The heart itself appears to have a long-range conservation. In *Drosophila* one of the key transcription factors required for heart development is called Tinman, after the character in "The Wizard of Oz" who had no heart. The vertebrate homologue is called, more prosaically, Nkx2.5, and this factor is also expressed in the prospective heart and is also necessary for heart development (Evans et al., 1995). Interestingly the *tinman* homologue in *C. elegans,* which does not possess a heart at all, seems to be involved in making the pharyngeal muscles (Haun et al., 1998).

Another remarkable long-range conservation is the transcription factor Pax6, required for vertebrate eye development. The homologue, Eyeless, is required for *Drosophila* eye development, and it is possible to force thoracic imaginal discs in *Drosophila* to become eye by introduction of the vertebrate *Pax6* gene (Quiring et al., 1994). Homologues of *Pax6* and *sine oculis* have also been found to be expressed in the developing eyes of planarians and cephalopods (Gehring and Ikeo, 1999), and there is some evidence of a functional requirement for these genes in eye regeneration (Pineda et al., 2001). It is a remarkable fact that optical sense organs that have formerly been textbook examples of parallel evolution are now thought by some to be genuinely homologous, based on a conserved role of these transcription factors.

The existence of this conserved molecular anatomy led us some years ago to propose the term *zootype* to represent the archetype of animal form (Slack et al., 1993). Although there are no actual body parts conserved across all animals, the functional domains of key developmental control genes are conserved. We pointed out that the stage of expression and action of these genes in different phyla corresponded to the previously defined "phylotypic stages" within each phylum. This is not surprising if the reason for the existence of phylotypic stages is considered to be the relative lack of selective pressure for change of anatomy at the early-middle developmental stage. Many more long-range conservations have been discovered since the zootype concept was first proposed and the argument has been correspondingly strengthened (see Figure 2).

Investigation of sponges has so far not shown evidence for a Hox cluster, so we can either claim that sponges are not animals or, if we wish them to be animals, define another name than zootype for the archetype of all diploblasts and triploblasts. The common ancestor of diploblasts and triploblasts must have had some anteroposterior polarity. We can

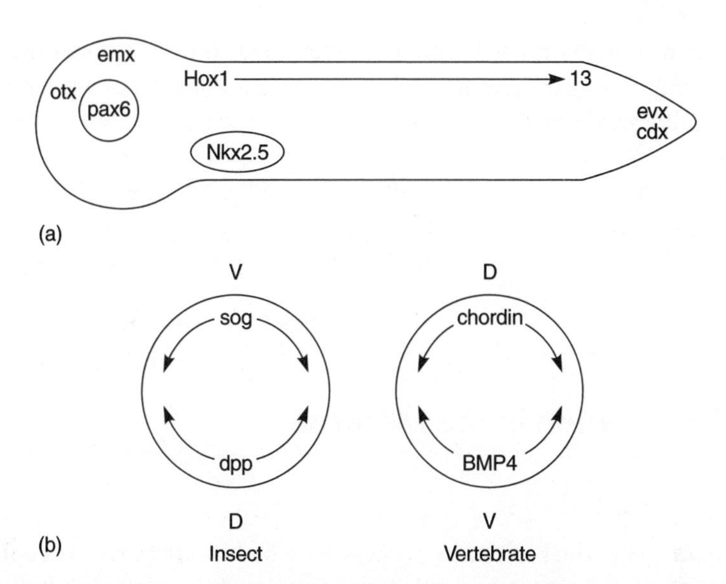

FIGURE 2. The new zootype. (a) indicates domains of transcription factor activity responsible for forming the anteroposterior pattern, the eye, and the heart. (b) indicates the homologous systems patterning the dorsoventral axis in insects and vertebrates.

further deduce that the common ancestor of the triploblasts possessed some anterior light-sensitive organ and that at least the common ancestor of insects and vertebrates possessed all these features plus a circulatory organ. The organism shown in Figure 2 may not be quite the common ancestor of all animals because evidence relating to dorsoventral pattern and the heart does not as yet extend to cnidarians or platyhelminths. But it is still fairly basal in animal evolution, corresponding to a protocoelomate or, if the classification of Aguinaldo et al. (1997) is preferred, a prototriploblast.

Any assessment of the current validity of the concepts of phylotypic stage and zootype must recognize that body plan definitions are subjective and not all key characters need to be displayed simultaneously. However, it must remain true that embryonic morphology is often a better guide to higher-grade taxonomic affinity than adult morphology, and that major molecular features of homology are displayed in early-middle stage embryos. Most remarkable of all, the recent study of the molecular biology of development has revealed some aspects of the anatomy of the long-lost common ancestor of triploblasts, for which we are most unlikely ever to find a fossil.

ALEJANDRO SÁNCHEZ ALVARADO

Regeneration in the Metazoa

Regeneration is the biological process by which body parts are replaced. Given our current ignorance about the molecular events guiding organ and tissue replacement in the Metazoa, this is an interim definition that will have to expand as knowledge grows. Unlike embryogenesis, in which regulatory sets of conserved molecular cascades are known to be shared by distantly related taxa, we do not know whether regeneration in the Metazoa is a homologous trait or the result of convergent evolution. Thus, comparative studies between various regenerating organisms are needed to understand the evolution of regeneration. This is not to say that categorizations of regenerative events have not been attempted (see Goss, 1956a; Needham, 1952). A classification that has withstood the test of time is Morgan's (1901), which cataloged regeneration as either requiring remodeling of preexisting cells *(morphallaxis),* cellular proliferation *(epimorphosis),* or a combination of the two (Sánchez Alvarado, 2000). Still, until more inter- and intra-phyletic molecular data are acquired, grouping or differentiating the various regenerative events that exist in many animals will be difficult.

In addition, we do not know the extent to which mechanisms involved in the replacement of cell types during physiological turnover are shared with regeneration. To what degree is maintenance of the differentiated state "regeneration"? In other words, how are aging, senescence, and longevity related to regenerative properties? Are the rates of these events or any other temporal transformation of differentiated tissues a reflection of the relative regenerative powers of a given cell, tissue, or organism? Only experimentation can answer these questions. However, the literature suggests that there may be a link between regeneration, aging, senescence, and longevity.

Measurements of mortality in *Hydra,* which has a high regenerative ability, suggest that a lack of senescence may exist in this organism (Martínez, 1998). Also, the tip of human digits, in particular in young children, are capable of epimorphic regeneration after amputation, a property that lessens and is lost as the individual ages (Illingworth, 1974). Another interesting example is the decrease of muscle mass (sarcopenia) that occurs as humans age (Lexell et al., 1988), apparently as the result of older individuals' inability to repair and regenerate myofibers. Intriguingly, cross-age muscle transplantation experiments in rats show that old muscle grafted into young hosts regenerates well, while young muscle grafted into old hosts does not (Carlson and Faulkner, 1996), suggesting that regenerative abilities are not entirely dependent on the tissue itself but also on its environment. Therefore, it may be conceptually and experimentally fruitful to keep an open mind, and consider events such as regeneration and the factors affecting physiological turnover (aging, senescence, and longevity, among others) as possible manifestations of the same phenomenon.

Regeneration: Rule or Exception?

Many, if not all, of the animal model systems that have propelled forward our molecular understanding of evolutionary, developmental, and biological processes display limited or no regenerative abilities. *Drosophila melanogaster* (fruit fly), *Caenorhabditis elegans* (roundworm), *Xenopus laevis* (frog), *Danio rerio* (zebrafish), *Gallus gallus* (chicken), and *Mus musculus* (mouse) represent a mere 0.0005% of the total number of species in the animal kingdom ($<1.2\times10^6$) (Brusca and Brusca, 1990)—which is hardly representative and therefore reflects poorly on the actual phylogenetic distribution of regenerative properties.

In fact, most, if not all, phyla possess organisms capable of undergoing regeneration as adults (Table 1). For instance, the simple diploblastic hydra regenerates its main body axis in both anterior and posterior directions after being medially transected. More complex animals such as the triploblastic planarian also display bidirectional axial regeneration, a rather remarkable property because anterior regeneration in the flatworms entails the *de novo* regeneration of a missing head. In contrast, unidirectional axial regeneration is observed in vertebrates such as sala-

TABLE 1. Distribution of regenerative properties in the metazoan phyla. Protostome and deuterostome classification of the phyla are indicated. Question marks refer to references in the literature about the occurrence of either morphallaxis or epimorphosis, but for which no confirmation or sufficient data are available

Phylum	Example	Morphallactic/ Epimorphic	Source
Cnidaria	*Hydra vulgaris*	"+/−"	(1)
Ctenophora	*Mnemiopsis mccradyi*	"?/+"	(2)
Protostomes			
Platyhelminthes	*Polycelis felina* (freshwater planaria)	"+/+"	(3)
Nemertea	*Lineus ruber*	"−/+"	(4)
Pseudocoelomates	*Stephanoceros*	"−/+"	(5)
Mollusca	*Octopus vulgaris*	"−/+"	(6)
Annelida	*Lumbriculus terrestris*	"−/+"	(7)
Coelomate Worms			
Sipuncula	*Sipunculus robustus*	"−/+"	(8)
Echiura		unknown	
Pogonophora		unknown	
Vestimentifera		unknown	
Arthropoda		"−/+"	(9)
Deuterostomes			
Echinodermata	*Antedon mediterranea*	"−/+"	(10)
Hemichordata	*Balanoglossus capensis* (enteropneust)	"−/+"	(11)
Chordata			
Urochordata	Ascidians	"−/+"	(12)
Cephalochordata	Amphioxus	"−/?"	(13)
Vertebrata	*Notophthalmus viridescens* (salamander)	"−/+"	(14)
	Homo sapiens	"−/+"	(15)

Sources: (1) Lenhoff, S. G. and Lenhoff, H. M. (1986). *Hydra and the Birth of Experimental Biology, 1744: Abraham Trembley's Memoirs Concerning the Natural History of a Type of Freshwater Polyp with Arms Shaped like Horns.* Pacific Grove, California: Boxwood Press. (2) Henry, J. Q., Martindale, M. Q., and Boyer, B. C. (2000). *Dev. Bio.* (220): 285–95. (3) Dalyell, J. G. (1814). *Observations on Some Interesting Phenomena in Animal Physiology, Exhibited by Several Species of Planariae.* Edinburgh. (4) Newth, D. R. (1958). *New Biology* (26): 47–62. (5) Jurczyk, C. (1926). *Zool. Anz.* (67): 333–336. (6) Lange, M. M. (1920). *J Exp. Zool.* (31): 1–57. (7) Randolph, H. (1892). *J. Morphol.* (7): 317–344. (8) Rice, M. E. (1970). *Science* (167): 1618–1620. (9) Rèaumur, R. A. F. (1712). *Mem. Acad. Roy. Sci.* 223–245. (10) Candia Carnevali, M. D., Bonasoro, F., Lucca, E. and Thorndyke, M. C. (1995). *J. Exp. Zool.*(272): 464–474. (11) Gilchrist, J. (1923). *Trans. South African Philos. Soc.* (17). (12) Nakauchi, M. (1982). *Am. Zool.* (22): 753–63. (13) Biberhofer, R. (1906). *Arch. Entwicklungsmech. Or.* (22): 15–17. (14) Brockes, J. (1997). *Science* (276): 81–7. (15) Illingworth, C. (1974). *J. Pediatr. Surg.* (9): 853–58.

manders, in which large, multitissue structures such as limbs, tails (including spinal cord), jaws, tissues of the eye, and even large portions of the heart regenerate with perfect patterning and restitution of structure (Eguchi and Okada, 1973; Oberpriller and Oberpriller, 1974).

The widespread distribution of regenerative properties suggests that regeneration is an ancient feature, maintained in some creatures but secondarily lost in others. Consequently, the ability to regenerate significant portions of the body should not be viewed as an unusual attribute of a few remarkable organisms. Instead, regeneration is a basic biological phenomenon that is widespread among organisms belonging to diverse branches of the metazoan phylogenetic tree.

Cellular Totipotentiality

A key characteristic of most metazoan somatic cells is that their nuclei contain all the necessary genetic information to produce whole organisms. This has been demonstrated by adult, somatic cell nuclei transplantations into enucleated eggs, which then produce cells capable of developing into complete animals (Gurdon, 1962; Campbell et al., 1996). To regenerate missing body parts, therefore, an injured organism simply requires access to the inherent totipotentiality of its uninjured cells. However, access without selection would be meaningless and potentially harmful. If an organism were to regenerate a limb by resorting to its uninjured cells' totipotentiality, such cells, if completely unbridled, would not only produce the missing limb, but also other parts that were not needed. Therefore, access to totipotentiality and precise regulation of the identity of the structure(s) being restored appear to be prerequisites for regeneration to occur.

Accessing and deploying only the specific developmental programs involved in the ontogeny of the missing body part(s) can work only if such programs are themselves autonomous from all other genetic and epigenetic events involved in creating an embryo. For instance, if only an adult eye needed to be regenerated but the eye program depended on the developmental programs driving head formation, a fully developed animal could not regenerate a missing eye without first regenerating its head. This lack of autonomy is not observed during eye regeneration in salamanders (Reyer, 1954); some differences must exist between embryonic and regenerative molecular choreographies. In fact, body part regeneration as illustrated by hydra, planarians, and salamanders displays a unique developmental autonomy or modularity (Santos, 1929, 1931; French, 1986; Hobmayer et al., 2000). Consequently, a fundamental difference in the functional organization of the genome must exist between the adult cells of regenerating and nonregenerating animals. This difference must allow for a high degree of developmental plasticity, such that specific morphogenetic cascades that are activated only during embryogenesis can remain accessible even after normal spatial and temporal embryonic deployment has occurred.

Dedifferentiation

Understanding how the stability of the differentiated state is regulated in mammals, salamanders, planarians, and hydra, for example, should shed much needed light on regeneration. The regenerative strategies used by these and other animals vary. Some recruit differentiated cells into regeneration, while others resort to undifferentiated reserve or stem cells. In salamanders, appendage regeneration occurs by dedifferentiation of adult somatic cells at the site of amputation (Echeverri et al., 2001). Unlike wound healing in mammals, the amphibian epithelium establishes a direct interaction with the underlying tissues, without forming a basal lamina or scar tissue. The direct interaction of epithelium and the underlying cells triggers dedifferentiation, in which terminally differentiated cells lose their morphology and become capable of proliferating again. Progenitor cells for the regenerating structure are derived from this new cell population. In essence, wound healing in salamanders represents the establishment and maintenance of developmental processes within the context of an adult organism by creating *in situ* a population of short-term stem cells.

Recent research on amphibian muscle cells indicates that a factor in vertebrate sera can trigger their dedifferentiation. The multinucleated, skeletal muscle cell fibers (myotubes) of newts can be cultured and maintained indefinitely *in vitro* using a minimum medium containing salts, amino acids, and a source of carbon. If bovine, ovine, or chicken serum is added, the salamander myotubes dedifferentiate from multinucleated to mononucleated cells, re-enter the cell cycle, and proliferate (Tanaka and Brockes, 1998; Tanaka et al., 1999).

This response contrasts with that of mammalian myotubes, which fail to change their morphology or enter mitosis when placed in serum-containing media (Lassar et al., 1994). This result has led to the assertion that in mammals, myotubes are terminally differentiated, i.e., that muscle cells are in a state of post-mitotic arrest that renders them incapable of proliferation. Yet the observation that the sera of many vertebrates that are incapable of regenerating their appendages contain a factor(s) that can trigger the dedifferentiation and concomitant re-entry into the cell cycle of newt myotubes is intriguing. It also demonstrates that, even though mammalian myotubes are naturally exposed to the circulating dedifferentiation factor(s), the cells remain incapable of responding to it.

Three recent lines of evidence suggest that the "terminally" differentiated mammalian myotubes can in fact be driven to dedifferentiate. The first is pharmacological and involves the discovery of a synthetic purine derivative known as myoseverin (Rosania et al., 2000), which, when added to minimum medium, triggers the dedifferentiation of multinucleated mouse myotubes to mononucleated cells. A similar cellularization process occurs in newt myotubes (Kumar et al., 2000; Velloso et al., 2000). It is not entirely clear how a purine derivative could so drastically modify the differentiated state of mammalian myotubes; myoseverin may carry out its destabilizing effect by binding to tubulin, which may disrupt microtubule integrity and consequently cellular architecture (Rosania et al., 2000).

The second line of evidence is genetic and involves the homeobox transcription factor msx-2. When *msx-2* is artificially introduced into mouse myotubes and its expression is induced, a fragmentation of 5–10% of the mouse myotubes into individual, mononucleated cells occurs. Moreover, the clonal progeny of these myotube-derived mononucleated cells are pluripotential, because their culture in various tissue-specific growth media gives rise not only to muscle but also to bone and fat cells (Odelberg et al., 2000).

The third and final observation is cellular in nature. Newt/mouse hybrid myotubes can be produced in culture by mixing their mononucleated myoblasts. If the hybrid myotubes are exposed to serum-containing media, the mouse nuclei exit their post-mitotic arrest and re-enter the cell cycle (Velloso et al., 2001). Failure of mouse myotubes to dedifferentiate when exposed to vertebrate sera lies not with the nucleus. Rather, it is the inability of the cell to respond to the extracellular factor(s) found in vertebrate sera that prevents the intracellular changes that must be responsible for modulating genomic totipotentiality. Taken together, the pharmacological, genetic, and cellular data clearly demonstrate that inducing dedifferentiation in so-called terminally differentiated mammalian cells is possible. Overcoming the differentiated state may be the first step to inducing regeneration of organs and appendages in mammals.

Stem Cells

Dedifferentiation is not the sole cellular mechanism for regenerating missing body parts. A strategy found frequently in vertebrates and invertebrates is the use of stem cells, which have been described in mammals (McKay, 1997; Prockop, 1997; Cheshier et al., 1999), even in tissues traditionally believed to be incapable of growth through cellular proliferation such as the brain (Palmer et al., 1997). Regulation of stem cells is under both genetic and epigenetic control. The gene responsible for the neurodegenerative disorder Ataxia-telangiectasia *(Atm)* is required for neural stem cell development in mice. Interestingly, in addition to this genetic component, exercise boosted brain cell survival in mice afflicted with this disorder (Allen et al., 2001). Therefore, understanding how genetic and epigenetic processes interact with each other to modulate stem cell populations would help us comprehend how developmental programs are coordinately deployed by stem cells to give rise to form and function.

A good experimental system in which to study the genetic and epigenetic components of stem cell biology is provided by the platyhelminths. In free-living members of this phylum, stem cells can encompass 20–30% of the entire cell population. In freshwater planaria, stem cells or *neoblasts* (Randolph, 1892) are randomly scattered through most of the mesenchyme (Newmark and Sánchez, 2000). Neoblasts are small (5–10 μm in diameter), undifferentiated, and the only mitotically active cells in planarians (Figure 1c). Neoblasts are also known to give rise to all the different cell types, including the germ line (Brøndsted, 1969; Baguñà et al., 1989). After amputation, these cells are locally signaled to proliferate, and eventually differentiate to restore missing body parts.

In addition to their ability to regenerate and restore proper proportions to the newly formed tissues, planarians show a high degree of plasticity in their ability to either grow or de-grow, i.e., proportionally shrink in size, depending on environmental conditions. During periods of prolonged starvation, planarians shrink: a 10 mm-long worm can be reduced to less than 1 mm over several months (Abeloos, 1930). Consequently, mechanisms must exist to regulate the number of neoblasts, such that the fine balance between cell death and cell proliferation is maintained to prevent a disruption of the animal's form and function.

Neoblasts, therefore, play a key role in the regenerative powers of planarians. However, little is known about the true nature of these cells or how their numbers are controlled as the organism grows and degrows. Recently, it has been shown that neoblasts can be labeled with DNA precursor analogs and that mitotic activity can be assayed using mitosis-specific antibodies (Newmark and Sánchez, 2000). The ability to follow the neoblast lineage, combined with gene expression abrogation methods such as RNA interference (Sánchez Alvarado and Newmark, 1999), lays the foundation for studies on the genetic and epigenetic processes responsible for the management of stem cells in adults and the mechanisms by which these cells are recruited to regenerate missing body parts.

In the eighteenth century, Voltaire, pondering on the ability of various invertebrates to regenerate their heads, wrote to Madame du Deffand (who was blind), expressing his hope that humans might one day be able to harness this regenerative ability (Rostand, 1951). Although we are nowhere near being able to implement medical applications for the regeneration of human missing body parts, we are closer than ever to deciphering the molecular principles behind this phenomenon. What is known so far clearly indicates that the functional organization of the genome in terminally differentiated cells can be modified to allow access to the cell's inherent totipotentiality. Also, a clear understanding of how certain organisms such as planarians can maintain long-term stem cells and regulate their rate of proliferation according to physiological needs will help us understand how totipotentiality is perpetuated, regulated, and repressed in different cell types generation after generation. The advent of genomics and the rapid maturation of bioinformatics will expedite the identification and characterization of the molecular choreographies executed during regeneration, and will allow us a deeper understanding of this fascinating biological property.

GRACE PANGANIBAN

Segmentation

Segmentation is a developmental process about which we understand a great deal, largely due to experiments in the model arthropod *Drosophila melanogaster,* the fruit fly. Because of the fruit fly's tractability, the *segmentation cascade,* or suite of genes that regulate segmentation, has been characterized using classical genetics. A large number of the genes have been cloned and their products studied using molecular and biochemical techniques. We now know how fruit fly segmentation proceeds from oogenesis through late embryogenesis, including how maternal transcripts are differentially localized at the poles of the embryo, how their translation is regulated, and how their products directly or indirectly control downstream segmentation genes of the gap, pair-rule, and segment polarity classes (Figure 1; reviewed in Sanson, 2001.

However, segmentation does not proceed in the same way in other organisms, raising the possibility that segmentation evolved multiple times. Just how many times—one of the more intriguing questions in evolutionary biology—is proving to be difficult to answer. Animals in several major phyla are either overtly segmented as both embryos and adults (e.g., arthropods and annelids) or more covertly segmented during some portion of their development, typically embryogenesis (e.g., chordates). This entry defines segments, evaluates the existing information regarding the evolution of segmentation, discusses why this information is insufficient to draw a firm conclusion regarding whether segments evolved once or more than once, and proposes experiments designed to address this issue more satisfactorily.

Segments are recognizably reiterated although frequently modified units found along the anteroposterior (AP) axis of a body plan. Reiterated segments may be identical, or *homonomous.* Myriapod arthropods such as centipedes and millipedes, for example, possess many leg-bearing homonomous segments along their trunks. Both homonomous segments and reiterated segments that are similar but not identical are termed *serially homologous.* Dipteran insects such as fruit flies possess

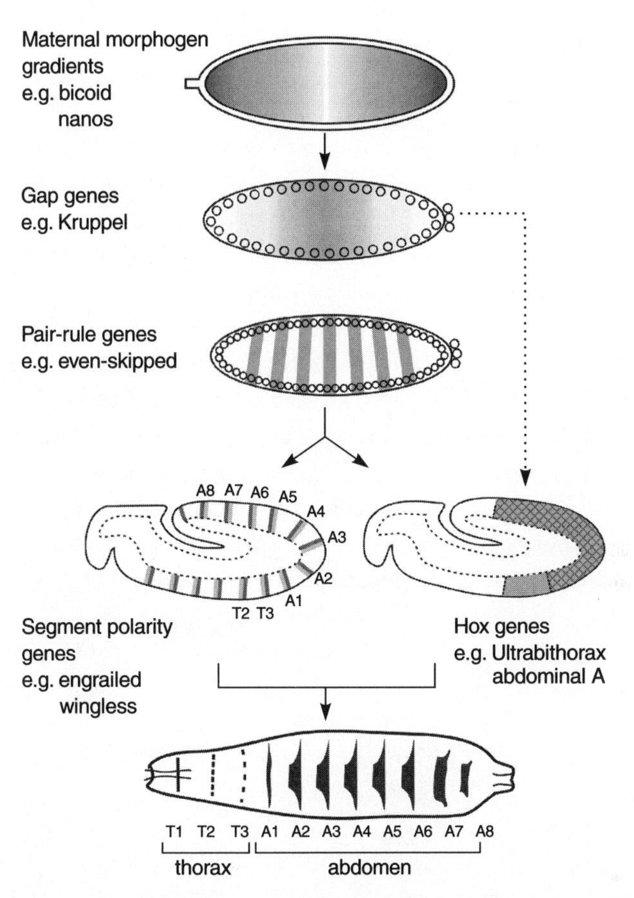

FIGURE 1. The *Drosophila* segmentation cascade. Patterning along the AP axis of *Drosophila* embryos. A cascade of maternal and zygotic genes is activated in the syncytial blastoderm, subdividing the ectoderm into smaller domains. After activation of the pair-rule genes, the embryo cellularizes and undergoes gastrulation. The pair-rule genes activate the segment polarity and Hox genes; however, a subset of gap genes also directly regulate the Hox genes. Segment polarity and Hox genes are thought to act in concert to control the differentiation of each segment. Reproduced with permission from Sanson (2001).

several types of serially homologous segments. For example, the second thoracic segment has a pair of wings, while the third thoracic segment carries a pair of reduced and modified wings called halteres. Each segment is composed of recognizable subdivisions along its AP axis. These subdivisions, termed *compartments,* exhibit unique gene expression patterns that are shared among the segments of an animal. For instance, the posterior compartment of each fruit fly segment expresses the segment polarity genes *engrailed* and *hedgehog* (reviewed in Ingham, 1991; DiNardo et al., 1994).

We do not know what the last common ancestor of the chordates, arthropods, and annelids—the Urbilaterian—looked like or whether it was segmented (DeRobertis and Sasai, 1996; Valentine and Collins, 2000). Based on ancient fossil tracks from the late Vendian/early Cambrian (ca. 600 million years ago), it has been proposed that the Urbilaterian was vermiform or worm-like (Valentine, 1994). If Urbilaterians were segmented, segmentation would need have arisen only once during animal evolution (Kimmel, 1996), and the segments of modern chordates, arthropods, and annelids would have a common origin (Figure 2A). If, on the other hand, Urbilaterians were not segmented, segmentation may have arisen twice—once in the chordate lineage and once in the arthropod/annelid lineage (Figure 2B; Brusca and Brusca, 1990), or even three times, once in the chordate lineage, and separately in the arthropod and annelid lineages (Figure 2C; Willmer, 1990).

In the absence of evidence to the contrary, the most parsimonious explanation for the existence of segmentation in multiple and distantly related animal phyla is that segments arose only once. Nonetheless, there are valid reasons for thinking that segments might have arisen multiple times.

First, segments represent a form of modularity, and modularity is incredibly useful. Modular building materials facilitate both rapid assembly and changes in function. Building a house with bricks is more straightforward than using mixed materials of random sizes and shapes. And although the bricks may be the same size and shape, the structures built with them need not be. Plants and animals both possess modular

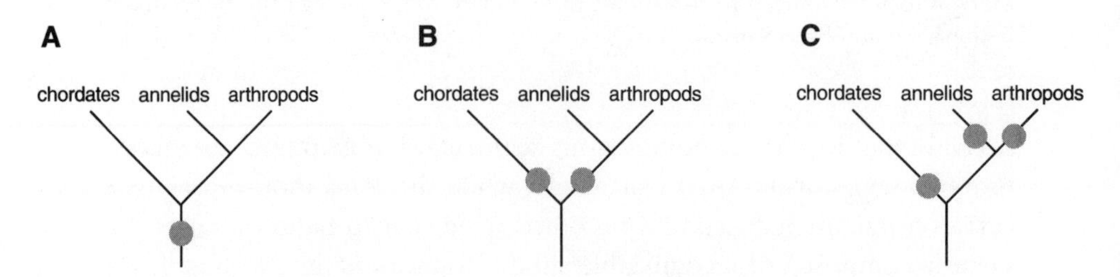

FIGURE 2. Possible scenarios for the evolution of segments. Segmentation (circles) may have arisen once (A), twice (B), or three times (C) during animal evolution.

body plans and modular organs (e.g., arthropod and vertebrate limbs and multifaceted arthropod eyes), yet are thought to have evolved from a unicellular common ancestor (reviewed by Raff, 1996). Thus other types of modularity have evolved multiple times. Segmentation could have, too.

Second, some unsegmented animals are more closely related to annelids or arthropods than annelids and arthropods are to one another (reviewed in Davis, 1999). This lack of overt segmentation could have resulted from loss in one lineage versus its independent acquisition in another, so by itself it does not support any particular model of segment evolution.

Third—and perhaps the most compelling reason for thinking segmentation arose multiple times—is the apparent lack of developmental genetic parallels among segmentation mechanisms in different phyla. For instance, the annelid homologue of the essential *Drosophila* segment polarity gene *engrailed* is expressed too late to play a conserved role in segmentation (Shain et al., 1998). *engrailed* function also does not appear to be conserved during zebrafish segmentation (Patel et al., 1989). Because parallels do exist between arthropods and chordates in fundamental developmental processes ranging from dorsoventral patterning to eye differentiation (reviewed in Halder et al., 1995; Holley et al., 1995; DeRobertis and Sasai, 1996; Ferguson, 1996), one might reasonably expect that parallels would have been observed by now in segmentation, if segmentation in these taxa does share a common origin. Similar arguments have been used to support the view that segmentation arose multiple times (discussed in Davis, 1999).

Nonetheless, the existing lack of developmental genetic parallels is neither surprising nor sufficient for us to conclude that segmentation evolved more than once. The modern developmental genetic mechanisms underlying segmentation are so different even among the arthropods that we should not expect that paradigms obtained primarily from studies of a single arthropod, such as the fruit fly, would hold in the distantly related chordates. Moreover, because insect body plans are highly derived (modified) relative to those of other arthropods, insect segmentation mechanisms are unlikely to be representative of ancestral ones.

Determining whether chordate, annelid, and arthropod segmentation share a common origin would be easier if we knew how segmentation proceeded in the earliest segmented animals of each of these taxa. With this information, we could assess whether the segmentation mechanisms were similar and therefore likely to share a history. However, we cannot (yet) do developmental genetics on animals that have been deceased for ~600 million years. We therefore are forced to look for alternatives. With some caveats, the logical choice is to study segmentation in the modern representatives of each taxon whose body plans have changed the least since the chordate/annelid/arthropod divergences. Candidate chordates include amphioxus and ascidians (see Holland and Holland, 1999). Candidate annelids include polychaetes (Shankland and Seaver, 2000). Candidate arthropods are less obvious, but would not include the fruit fly.

The extant (living) arthropod groups include insects, crustaceans, myriapods, and chelicerates, all of which are segmented. Within the arthropods, insects are most closely related to the crustaceans, from which they probably arose (Averof and Akam, 1995b). Insect segments form during embryogenesis in either of two ways. In long germ band insects, all segments form more or less simultaneously from cells present in the blastoderm embryo. In short and intermediate germ band insects, segmentation is sequential, with a few head segments being allocated from blastoderm cells and the remaining body segments forming in an anterior to posterior series from a growth zone located at the posterior of the embryo (see Nagy, 1994). Short or intermediate germ band development is more common and probably represents the ancestral mode.

It is therefore unfortunate—and somewhat ironic from the evolutionary biologist's perspective—that we understand most about segmentation mechanisms in the fruit fly, whose mode of segmentation is relatively unusual and significantly modified from that of its crustacean ancestors. Not only do fruit flies employ a long germ band mode of segmentation, but also many important steps in the fruit fly segmentation cascade occur while the embryo is a syncytium. Once sperm and egg nuclei fuse in *Drosophila,* the zygotic nuclei divide 13 times prior to establishing individual cell membranes. Thus, until there are approximately 4,000 nuclei, the fly embryo has only a single cytoplasm. Most segmentation can be and is regulated by cytoplasmic gradients of transcription factors, and cell-cell communication via surface receptors and their transduction machinery is unnecessary.

Based on the morphological aspects of short to intermediate germ band segmentation alone—for example, initial segment formation in a cellularized blastoderm and sequential addition of segments—we would predict that segmentation mechanisms would differ substantially between long and short to intermediate germ band animals. Indeed, the emerging evidence (primarily in the form of spatial and temporal expression data) from other insects and from crustaceans indicates that genes such as *even-skipped (eve)*, *fushi-tarazu (ftz)*, *wingless (wg)*, and *engrailed (eng)* that play critical roles in *Drosophila* segmentation may not have identical functions in these other animals (Nagy, 1994; Davis, 1999; French, 2001). Short to intermediate germ band segmentation is widespread among the arthropods and is used by crustaceans, from which insects are derived. It therefore is likely to represent the ancestral mode of arthropod segmentation. Detailed molecular genetic information about crustacean segmentation therefore would be extremely valuable for assessing whether arthropod segmentation shares a common origin with annelid or chordate segmentation. Parallel information from representative annelids and chordates also will be essential.

At least two potential caveats need to be considered as we evaluate emerging information regarding segmentation mechanisms in different animals. The first is that once a mechanism for segmentation evolved, it probably was possible to modify that mechanism in a variety of places while still producing a similarly segmented organism. The known differences between insects and crustaceans already make this clear. If we accept the well-substantiated premise that insects and crustaceans are monophyletic (Averof and Akam, 1995b), the observed differences among the insects, or between insects and crustaceans, in the genetic programs regulating segmentation must be secondary to segmentation itself. Once a genetic pathway evolves, it is possible to modify its steps or components without modifying the outcome; scenarios by which these types of modifications may occur have been described (Salazar-Ciudad et al., 2001b).

A second potential caveat is that directly orthologous genes do not necessarily have identical functions between species. For instance, even within the chordates, different *Wnt* family members are used in limb patterning in the chick than in the mouse (reviewed in Tabin et al., 1999), and orthologous *Dlx* genes are differentially expressed between chick and mouse and are likely to have divergent functions (Dhawan et al., 1997). Looking for shared functions between orthologous genes from one phylum to another is thus likely to be a hit-or-miss proposition, and we should not be surprised to find differences, even when there is a common evolutionary history. Regardless of these complicating factors, I remain optimistic that we will soon learn how many times segments arose and have a far clearer image of the Urbilaterian body plan.

MANFRED D. LAUBICHLER

Selection: Units and Levels in Developing Systems

Natural selection is arguably the central element of the modern theory of evolution. Perhaps not unexpectedly, it has also been one of the most contested concepts in modern biology. Ever since Darwin (1859) introduced the notion of natural selection in strict analogy to the well-established concept of artificial selection known to have produced the remarkable diversity of domesticated animals and plants, the efficacy and range of natural selection has been disputed. Indeed, Darwin's two chief propagandists, Thomas Henry Huxley, Darwin's bulldog, and Ernst Haeckel, his German counterpart, both voiced serious concerns about the ability of natural selection to account for major transformations of morphology.

At the core of Huxley's and Haeckel's objections were two separate, yet related issues: the nature of variation and the role of morphology within a theory of evolution. The latter had an illustrious history in the seven decades between Goethe's christening of the discipline in 1795 and the publication of the *Origin of Species,* giving rise to such important concepts as

Goethe's notion of the *Urpflanze* and his theory of metamorphosis
Cuvier's introduction of four distinct animal body plans and their corresponding systematic categories *(embranchments)*
Geoffroy Saint-Hilaire's opposing notion of universal analogy and his famous dispute with Cuvier in front of the French Academy, resolved only recently with the discovery of the homology between *Drosophila sog (short gastrulation)* and *Xenopus chd (chordin)*

and their respective domains of expression in Protostomia and Deuterostomia (De Robertis and Sasai, 1996)
Oken's system of classification based on the linear addition of organs
Meckel and Serres's concept of recapitulation
von Baer's laws of embryology
Owen's definition of homology and his introduction of the vertebrate archetype (see Zimmermann, 1953; Gould, 1977; Appel, 1987; Hall, 1992b; Nyhart, 1995).

This activity in the fields of morphology and embryology was an important resource for Darwin (Richards, 1992), yet it also gave rise to some of the most persistent objections that continued to challenge the explanatory power of his proposed mechanism of natural selection (Bowler, 1983).

Fundamentally, these objections rested on a different perspective with regard to the scale of variation and the scope of evolutionarily significant comparisons. Morphologists looked across a wide range of taxa and different forms in search of common structures, building blocks, and formative principles, whereas Darwin and his followers focused on small variations within populations and argued that the diversity of forms could be explained by extension from the gradual changes observed in the context of domestication and artificial selection. Darwin's population thinking thus clashed with the typological tradition of morphology (Mayr, 1982).

Morphologists also tried to explain, and not just establish, the connections between different types. This task became even more pressing in the decades following the publication of the *Origin of Species,* when the program of evolutionary morphology began to organize the diversity of life according to the principle of homology, which after Darwin was interpreted historically (Spemann, 1915; Nyhart, 1996). In the context of his theory of recapitulation, Ernst Haeckel discussed the role of heterochrony and additional mechanical causes in explanations of morphological transformations, but his attempt to interpret phylogeny as the direct mechanical cause of ontogeny was soon eclipsed by Roux's program of *Entwicklungsmechanik.* Following Roux, a young cohort of biologists approached development from a mechanistic perspective, searching for general principles of development. As a rule, they did not pay much attention to natural selection, even though Roux himself had earlier in his career attempted to employ the notion of struggle between the parts of an organism in the context of a theory of biological organization and development (Roux, 1881). Over the years, this tendency to ignore the principle of natural selection only grew stronger among developmental biologists, especially in Germany.

Part of the problem was that nobody had any idea how small phenotypic variations in more or less terminal traits (from a developmental perspective) could add up to the major morphological transformations that were the domain of evolutionary morphology. Consequently, as soon as genetics established itself around the paradigm of particulate inheritance after the rediscovery of Mendel's laws in 1900, the debate was re-enacted in the arguments between Mendelians and biometricians (Provine, 1971). In the absence of any clear evidence about the nature of the genetic material, the participants in these debates were free to project their own ideas about what constitutes a character (trait) into the new domain of genetics. Hugo de Vries's *Mutationstheorie* was just one attempt to fit genetics into the explanatory framework of major evolutionary transformations.

Throughout most of the mid-twentieth century, the majority of developmental and evolutionary biologists pursued their research independently of each other. Of those that continued to integrate both development and evolution, Richard Goldschmidt stands out, both because of the wide range of his interests and because of the criticism he drew from evolutionary biologists, newly consolidated behind the Modern Synthesis (Mayr and Provine, 1980). Goldschmidt's argument for the importance of macromutations, which he referred to as "hopeful monsters" (Goldschmidt, 1940), best illustrates the conceptual differences between selectionists and developmentalists with regard to the nature of genetic variation. For Goldschmidt and others like him, major transformations of morphology required so many correlated changes that they had to be accomplished with one bold stroke; simply adding small individual changes would never be able to produce the harmonious integration of different parts.

It is probably one of the great ironies in the history of modern biology that Goldschmidt's best illustrations for his theory were so-called homeotic mutations, a class of mutations such as *Antennapedia,* in which a fully developed leg grows instead of an antenna. The molecular study of this class of mutations led to major new insights into the genetic architecture of development, such as the existence of so-called master control genes that can regulate a whole battery of genes needed to produce complete morphological structures, revealing that different classes of mutations do exist, at least with regard to their phenotypic effects. These insights, in turn, enabled the newly emerging synthesis of evolutionary developmental biology.

Despite recent attempts at détente, present-day developmental and evolutionary biologists, as part of their intellectual heritage, are frequently trapped within different disciplinary paradigms. These are best characterized by the respective explanatory structures of these disciplines. Evolutionary biologists, in the end, refer to a framework of ultimate causality (Mayr, 1961), most commonly to one that involves natural selection (and therefore whole populations) as part of the explanation. Developmental biologists focus instead on proximate causes that concur with a mechanistic explanation of individual events. As a consequence, seemingly similar concepts, such as variation or modularity, can sometimes have radically different meanings in those disciplines. Anyone who attempts to integrate evolution and development within one explanatory framework needs to pay special attention to the different conceptual implications of these separate traditions. In that sense, it does matter whether we talk about evo-devo or devo-evo (Hall, 2000b).

The Unit of Selection Debate

In the context of the modern theory of evolution, natural selection is considered to be one of several processes that can account for observable changes in the distribution of types (genotypes or phenotypes) over time; random genetic drift would be another such process. The fact that natural selection can act cumulatively, as long as all of its requirements are fulfilled, explains why it is often seen as the major factor of evolutionary change (e.g., Dawkins, 1982). Furthermore, the formal properties for natural selection to occur are well understood. Whenever there is variation between individuals within a population for some trait, and this variation is correlated with a corresponding variation in fitness (the ability of individuals to survive and/or reproduce) and is also heritable—i.e., a positive correlation that is at least partially independent of environment exists between the parent's value of the trait and that of their offspring—natural selection occurs and the distribution of the trait within the population changes (Lewontin, 1970; Endler, 1992).

While these criteria are straightforward and have given way to a well-established mathematical theory of natural selection, the issue of the units on which natural selection acts has been debated ever since Darwin and Wallace first formulated the principle of natural selection. Objects from all levels of the biological hierarchy—clades, species, groups, individuals, cells, chromosomes, genes, and selfish DNA—have been suggested as potential units of selection.

The debate was further refined with the introduction of the canonical distinction between replicators and interactors (vehicles) by Dawkins (1976) and Hull (1980). *Replicators* are defined as objects that make an accurate copy of themselves; *interactors* are those biological objects that interact with the environment in such a way that reproduction is differential. These two entities thus represent two different dimensions of the process of evolution by means of natural selection.

Even before Dawkins and Hull introduced the replicator/interactor distinction, Lewontin (1970) analyzed the formal requirements for an entity to be a potential unit of selection and concluded that natural selection can act at any level of biological organization at which heritable variation that causes variation in fitness exists. Wimsatt further refined Lewontin's analysis and based his definition of a unit of selection on the additivity or context-independence of the genetic variance: "A *unit of selection* is any entity for which there is heritable *context-independent* variance in fitness among entities at that level which does not appear as heritable context independent variance in fitness (and thus, for which the variance in fitness is *context-dependent*) at any lower level of organization" (Wimsatt, 1980, p. 144).

This definition has since been modified (Lloyd, 1988), and its implications have been analyzed by Sober (1984), Maynard Smith (1987), Sterelny and Kitcher (1988), Waters (1991), Godfrey-Smith (1992), and Sarkar (1994). The main problem with Wimsatt's notion of context-independent variation is that it is too vague and can lead to counterintuitive interpretations even of simple cases such as heterosis (Godfrey-Smith, 1992). However, it is now well established that epistatic interactions will lead to context-independent variance in fitness at levels of organization other than the individual genes, and that even with epistasis, a certain amount of the additive genetic variance can always be attributed to the individual genes and will therefore be context dependent at any higher level of organization (Sarkar, 1994).

Wimsatt's analysis also led to another insight, namely that the *unit* of selection problem has to be distinguished from the question of the *levels* of selection. Brandon (1982) introduced the level of selection question by asking, "At what level does the causal machinery of selection really act?" This is another way of separating the problem of interaction with the environment, which in the case of organismic selection is a property of individual organisms, from the question of which biological entities contribute to the additive genetic variance and thus to heritable differences in fitness. Even if selection acts on the level of the individual organism, different units of selection (genes, gene groups, chromosomes, genotypes) can contribute to the additive genetic variance.

The analysis of epistatic interactions between individual genes and their consequences also yielded important new insights into how the details of the genetic architecture can modify the received notion of a unit of selection. While most contributions to the unit of selection debate have focused on a traditional level of the biological hierarchy (selfish DNA, genes, chromosomes, cells, individuals, groups, or clades), the detailed analysis of epistatic interactions between individual loci reveals that functionally (epistatically) connected networks of genes can act as units of selection, in that their contributions to the total additive genetic variance of a specific trait cannot be reduced to any other level of organization (Wagner and Laubichler, 2000). These networks of interacting genes are therefore functional units of selection and, as such, are closely related to the modular organization of developmental systems that has recently become the focus of attention of both evolutionary and developmental biologists (Wagner, 1996, 2001; Winther, 2001).

Fundamentally, the unit of selection problem is one of identifying the relevant biological units relative to a specific biological process. It is therefore a special instance of the more general problem of identifying functionally relevant biological units or characters in the context of evolutionary biology (Wagner and Laubichler, 2000; Wagner, 2001). In the case of selection, this process and its formal properties are relatively well understood. Recent work on this issue has led to several new insights and proposals, such as the notion of a unit of inheritance defined by its contributions to the additive genetic variance (the fraction of the total genetic variance that is relevant for selection) and the recognition that conflicts between different levels of selection can play an important role in explanations of major transitions in evolution (Buss, 1987; Maynard Smith and Szathmary, 1995; Wagner and Laubichler, 2000).

Specifically with regard to the latter issue, Leo Buss suggested that the patterns of early development and of the metazoan life cycle can be explained as a result of a conflict between the "selfish" interests of individual cells to maintain their own potential for indefinite replication and the "overriding" interest of the whole organism to tightly control the fate of individual cells. Under this view, the basic features of metazoan development, such as the early sequestration of the germ line and the conservation of many features in early development, are a consequence of interactions between different levels of selection. Individuality in the sense of the distinct biological (physiological, developmental) integration of organisms is therefore a derived evolutionary outcome of this conflict between different levels of selection.

Maynard Smith and Szathmary (1995) further expanded the idea of conflicts between different levels of organization in their in-depth analysis of major transitions in evolution. They suggest that, in the course of life's history on earth, several major transitions, defined as a dramatic change in the organization of biological units, occurred and that conflicts between different levels of selection played a major role in these transitions. Like Buss, they interpret the specific features of development as a response to the inescapable conflict between the interest of individual independent cells and their differentiation and specialization within the context of a multicellular organism. The unit of selection debate, which originated in the context of attempts to explain the evolution of social behavior, thus provides some important insights into the evolution of development.

Logic of Natural Selection and the Integration of Body Plans

In the ongoing conflict between proponents of the exclusivity of natural selection as the main mechanism of evolution and proponents of some additional morphological and developmental principles of organization, the existence of stable and conserved body plans *(Baupläne)* was one of the most contentious issues. How to explain the stability and conservation of these organizations of animal design in light of continuous variation and selection was one of the most challenging problems for selectionists. Indeed, many of the early contributions to evolutionary developmental biology after its re-emergence in the early 1970s were dedicated to this problem. The notion of developmental constraints as limiting factors to the expression of phenotypic variation was one of the first attempts to focus systematically on how development might be able to affect the phenotypic outcome of genetic variation (Alberch and Gale, 1985; Maynard Smith et al., 1985). The importance of other long-recognized developmental mechanisms, such as heterochrony, or the role of epigenetic mechanisms of development were also discussed, and several old debates about the "tempo and mode of evolution" (Simpson, 1944) were revisited (Eldredge and Gould, 1972).

Among the many contributions during that time, Rupert Riedl's (1975, 1977) synthesis of the logic of natural selection with older conceptions of developmental and theoretical biology, such as general systems theory, stands out. Riedl further developed the view of organisms and development as interdependent hierarchical systems and processes and introduced the notion of *burden* to account for the enormous selective pressure that exists on those elements in the developmental hierarchy that are the foundation of all downstream processes. In this interpretation of developmental systems, natural selection accounts for the stability and conservation of major body plans by simply assuring that any variation in those central elements of animal design would immediately be selected against.

In the early 1970s Riedl's conceptual synthesis had an enormous intuitive appeal, but the genetic architecture of developing systems was still largely unknown. The isolation of homeotic genes and the subsequent discoveries of detailed developmental pathways now provide the specific knowledge about the genetic mechanisms of development that is needed for a synthesis of the distinct traditions of quantitative and population genetics with developmental biology. The core of this synthesis is devoted to the problem of the genotype-phenotype map. Developmental mechanisms mediate between genetic variation and its phenotypic correlation, and the lack of a clear understanding of this relationship was one of the major drawbacks of the adaptationist program (Mayr, 1983). In the absence of any evidence about the details of this map, it has often been assumed that continuous genetic variation is correlated with equally continuous phenotypic effects. This view can no longer be upheld.

There is now experimental evidence for the specific genetic basis of species differences, such as the role variation in *cis*-regulatory sequences of *Ubx* (Ultrabithorax) plays in the different patterns of bristles on the third leg of *Drosophila* (Stern, 1998), as well as for the role of specific molecular changes in the Ubx and Abdominal A Hox proteins that are implicated in some of the most dramatic macroevolutionary changes of the insect body plan: the transition from a crustacean-like ancestor with multiple limbs to the clade-specific hexapod body plan characteristic of all insects (Galant and Carroll, 2002; Ronshaugen et al., 2002). These studies also suggest a possible scheme for the evolution of *Ubx* function, thus directly linking analysis of the developmental pathways to evolutionary models. In insects, *Ubx* has acquired one or more additional functions (i.e., constitutive repression domains); modeling the evolutionary dynamics of these systems will thus involve the analysis of epistatic and/or pleiotropic interactions. Such effects are, however, generally correlated with the existence of higher-level units of selection, such as networks of interacting genes.

As we begin to understand the structure of the genetic toolkit of development (Carroll et al., 2001), we start to see that these results support the view that parts of these developmental systems can act as units of selection. Such a perspective might also contribute to a synthesis of diverging interpretations of modularity, despite a general agreement that biological systems are modular in nature. In particular, it might unite the traditionally different perspectives of developmental and evolutionary biologists on the emergence of higher-level units—the individualization problem (Winther, 2001)—by demonstrating how selection on variations between epistactically and pleiotropically connected loci can eventually lead to the mechanistic integration and conservation of these developmental systems.

Today, in the context of evolutionary developmental biology, the long conflict between natural selection and developmental mechanisms as competing explanations for morphological evolution might finally be resolved. New empirical evidence that uncovers the details of the basic genetic toolkit of development rapidly changes our understanding of the nature of variation. It is now clear that simple mutations of regulatory genes can have dramatic phenotypic effects; no longer is there a need to evoke a separate class of macromutations to account for dramatic phenotypic changes.

Furthermore, as our understanding of both the role and the consequences of epistatic interactions between individual loci—as well as of the details of the molecular mechanisms of development, including the recognition of several layers of stable epigenetic systems and developmental processes—grows, we are now in a good position to integrate these diverse fields. The notion of modularity, recently the focus of much attention, already provides a conceptual framework for the integration of different lines of research. However, no unified theory of modularity that captures the various usages of the term in evolutionary and developmental biology currently exists. Nor do we have all the formal tools for the analysis of such systems. But compared to earlier decades of irreconcilable differences, we now have a well-established research agenda in evolutionary developmental biology and both evolutionary and developmental biologists willing to contribute to these efforts.

MIRIAM ZELDITCH

Space, Time, and Repatterning

Virtually every evolutionary change in development could be construed, regardless of cause or effect, as involving a change (repatterning) in either space or time. Construed so broadly, repatterning is uninteresting and uninformative. However, it becomes interesting when repatterning is taken more narrowly, not as a change *in* space or time but rather as a change *of* spacing or timing. This narrowed class is still broad but excludes changes in the processes themselves, even if they could somehow be reduced to changes in time, as well as fundamental reorganizations of epigenetic systems generating novel structures or interactions. Such dramatic reorganizations have been termed "ontogenetic repatterning" (Roth and Wake, 1985; Wake, 1985) but *epigenetic reorganization* might better capture the meaning of the concept. The narrowed class includes only modifications in the spacing or timing of otherwise conservative developmental processes, in other words, those that do not alter modularity, integration, or processes, and that use the same developmental mechanisms.

Spatial and temporal repatterning includes heterochrony, the most familiar kind of evolutionary change in development and the one most often studied, and heterotopy. As evident from the etymology of the words, heterochrony is an evolutionary change in developmental timing and heterotopy is an evolutionary change in developmental spacing. Determining whether a change involves space or time can be difficult because space and time, unlike heterochrony and heterotopy, are not categorically distinct. Even if it is sometimes useful to think about space and time separately, such as when one concept is of greater theoretical importance, evolutionary changes in development might typically involve *spatiotemporal* rather than spatial or temporal repatterning.

This entry treats the categories of temporal and spatial repatterning separately, as they are usually considered, and then discusses the relationship between them. Spatiotemporal repatterning may include some well-known phenomena that have been construed as purely temporal.

Temporal Repatterning

Temporal repatterning includes heterochrony, a concept sometimes defined broadly and vaguely, but also sometimes defined narrowly and precisely. Narrowly defined, *heterochrony* is the subset of evolutionary changes in developmental rate or timing that result in parallels between ontogeny and phylogeny (Gould, 1977; Alberch et al., 1979). Alternatively, heterochrony can also be narrowly defined as a temporal dissociation among individual processes (Raff and Wray, 1989). More broadly defined, heterochrony includes both these and other changes in relative rates of development of individual events or anatomical parts (Smith, 2001). Even more broadly, it can include all these changes along with any change in development that is manifested by a change in some trait (or measurement thereof), process, or part relative to time (e.g., McKinney and McNamara, 1991). This last definition is sometimes considered so all-encompassing as to be uninformative, and it has been dismissed for that reason (Wake, 1996; Zelditch and Fink, 1996; Rice, 1997, Zelditch, 2001), although some scientists regard it as the definition conventionally accepted by evolutionary biologists (Klingenberg, 1998).

To some extent, this semantic confusion is a natural result of the evolution of language; the theoretical meanings of concepts change as the theories do. But part of the confusion results from different perceptions of developmental rate and timing. These concepts can be understood from (1) an organismal perspective, in which developmental rate and timing are properties of the whole organism; (2) a more analytical and anatomical perspective, in which developmental rates and timings are properties of anatomical parts or events; or (3) a mechanistic perspective, in which individual processes have their own rates and timings.

Clearly, these are not mutually exclusive perspectives; indeed, it is unlikely that we will ever achieve a satisfactory theory of temporal repatterning without reconciling all three. Nevertheless, these perspectives can lead to different (and sometimes conflicting) interpretations of data, imply different causes and effects, and play different theoretical roles.

An organismal perspective on temporal repatterning. The organismal perspective on development and growth is typical of life history theory, which views rates of development and growth as fundamental parameters. Size and age at maturity, birth, or weaning have important consequences for fitness (Millar, 1977; Case, 1978; Glazier, 1990; Ricklefs et al., 1994). Morphology and morphogenesis are rarely of primary theoretical concern; to the extent that morphology enters into life history analyses, it is to provide a temporal sequence of developmental events that indicate degree of developmental maturity. However, many morphological studies adopt this organismal conception of growth and development, especially when morphology is thought to evolve as an indirect effect of selection on life history parameters (e.g., Gould, 1977). This view is particularly prevalent in studies of allometry, which test the hypothesis that evolutionary changes in shape result from evolutionary changes in body size (e.g., Huxley, 1932; Shea, 1983; Falsetti and Cole, 1992). Extending this theory to cases in which growth rates are dissociated from developmental rates, or when both are dissociated from time, broadens the analysis of allometric extrapolation to encompass heterochrony.

Two conceptual schemes for analyzing heterochrony were formulated by Alberch et al. (1979), both of which represent the ontogenetic trajectory as a path through three-dimensional space. One scheme uses the axes size(s) as a measure of growth, shape (σ) as a measure of development, and age (a) (Figure 1A). The other replaces shape with an axis of developmental events ordered from first to last (Figure 1B). Comparisons are made among taxa with respect to four parameters: (1) age at onset of development, α; (2) age at offset of development, β; (3) rate of development, k_σ; and (4) rate of growth, k_s (Figure 2). There are eight possible modifications of these parameters, taken separately, each of which has a name (Table 1). Those that involve development (i.e., α, β, k_σ) produce one of two morphological outcomes: a descendant resembling the ancestor either at a younger age (paedomorphosis) or at an older age (peramorphosis). For a more realistic phylogenetic approach to these comparisons, see Fink (1982).

TABLE 1. Definitions of heterochronic perturbations and their morphological expression, as defined by Alberch et al. (1979)

Control Parameter	Increment at Change	Process	Morphological Expression
α	$-\ \delta\alpha$	Predisplacement	Peramorphosis
	$+\ \delta\alpha$	Postdisplacement	Paedomorphosis
β	$-\ \delta\beta$	Progenesis	Paedomorphosis
	$+\ \delta\beta$	Hypermorphosis	Peramorphosis
k_σ	$-\ \delta k_\sigma$	Neoteny	Paedomorphosis
	$+\ \delta k_\sigma$	Acceleration	Peramorphosis
k_s	$-\ \delta k_s$	Proportional giantism	
	$+\delta k_s$	Proportional dwarfism	

Both schemes presuppose that all taxa under comparison share either the ontogeny of shape or the sequence of developmental events. That is neither a flaw in the schemes nor a questionable assumption; the schemes were specifically and explicitly designed for cases in which that assumption is met. They were not devised for cases in which morphogenesis also evolves. But modifications of morphogenesis are clearly of interest to evolutionary developmental biologists, and some might result from temporal repatterning—just not from the temporal repatterning of development as understood from this organismal perspective.

An anatomical perspective on temporal repatterning. One kind of temporal repatterning that can alter morphogenesis is a shift in relative rate of timing among individual parts. Studies that focus on timing, without considering changes in overall (organismal) rate, take an anatomical perspective on developmental rates. Not surprisingly, this perspective is most often found in the morphological literature. A particularly explicit discussion of this approach is in Smith (2001): she argues for comparing sequences of events so as to standardize timing across taxa, rather than using standards based on absolute (organismal) time. This distinct perspective views rates of growth and development as properties of individual parts or events. It differs from the one discussed next, which is also analytical rather than holistic, in that dissection of the organism is into parts rather than mechanisms.

Changes in relative timing or rate among parts, like changes in timing or rate of organismal growth and development, is called heterochrony. This definition is consistent with the term's original usage (Haeckel

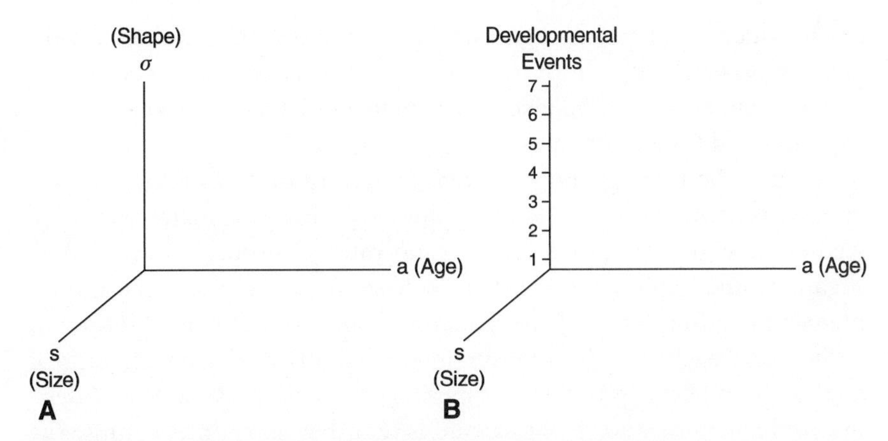

FIGURE 1. Two versions of the Alberch et al. (1979) formalism. A, comparing rates of growth and development. B, comparing timings of discrete events.

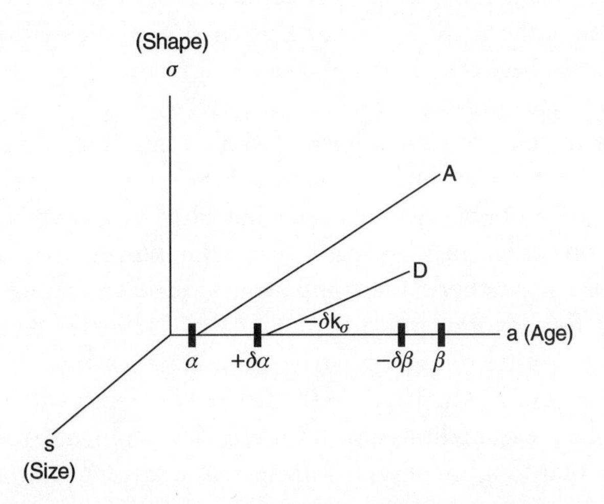

FIGURE 2. Comparing ancestral (A) and descendant (D) parameters. α, onset of a developmental process; β, offset of a developmental process, k_σ, rate of development.

1875). It is also sometimes called *dissociated heterochrony* because the rates and timings of different parts are dissociated from each other in their evolution. *Global heterochrony,* on the other hand, is not equivalent to a change in organismal rate and time. When contrasted to dissociated heterochrony, global heterochrony refers to probably rare cases in which the ontogenetic trajectory as a whole is extrapolated or truncated, i.e., in which the ancestral organismal rates and timings are altered without any other changes in morphogenesis. Dissociated heterochrony refers to changes in morphogenesis that involve dissociations among relative rates and timings of parts.

Because allometry is about relative growth rates, the anatomical perspective is often found in studies of allometry. However, many allometric studies view growth and developmental rates as properties of a whole organism and explicitly test the hypothesis that changes in those rates predict the adult shape of descendants. Some studies (e.g., McKinney, 1986) treat each part (or measurement thereof) as a separate unit of analysis, comparing growth rates measurement by measurement. A serious problem with studies of this kind is that they sometimes confuse the axes of bivariate allometry plots with those of the Alberch et al. (1979) scheme, and some mistake changes in relative growth rates with $\pm \delta k_{\sigma}$, thereby producing considerable conceptual and semantic confusion; see Godfrey and Sutherland (1996) for a discussion of the mathematical incompatibilities between these approaches and their impact on debates over human heterochrony.

Changes in relative developmental timing can alter ontogenetic sequences and extend or contract the time between developmental events. Such changes probably demonstrate that later events are not causally dependent on earlier ones because a reversal in sequence of cause and effect is unlikely. Methods for comparing sequences require a different conceptual framework, and different analytic schemes than those applied to conservative ontogenetic trajectories (see discussions of methods in Mabee, 1993; Smith, 1997; Velhagen, 1997; Hufford, 2001; Smith, 2001). While these methods are still under development—they seem designed for analyzing merely rank orders of events, not more complex networks—sequence approaches promise insight into causality and may provide characters for phylogenetic inference (Hufford, 2001).

The primary benefit of this perspective is that it takes morphology seriously, analyzing changes in morphogenesis neglected by studies of organismal rates and timings. Also, it is clear from numerous empirical studies that evolutionary changes in organismal rates of growth and development do not alter morphology in a uniform and entirely integrated fashion. The major limitation of this approach is that the decomposition of an organism into parts (or of its ontogeny into a linear sequence of events) is potentially arbitrary. It is possible to decompose morphology into units, such as a measurement of length, that have no characteristics other than rate or time. No biological insight is gained by showing that such one-dimensional features differ in rate or timing because no other possibility is even mathematically possible.

A mechanistic perspective on temporal repatterning. This perspective construes developmental rate and time as properties of individual *processes,* not of whole organisms or anatomical parts. Because of the variant interpretations of process, this approach is highly heterogeneous. Processes can include DNA translation, the reactions initiated by expression of a gene, and more complex inductive interactions among cells and tissues. All of these qualify as processes, but changes in the temporal patterning of one might be causally unrelated to the temporal patterning of another. A particularly critical perspective on temporal repatterning of mechanisms shows that the evolution of pigment patterns might be caused by temporal repatterning of processes (such as cell migration) but could also be caused by other sorts of changes, such as the sorting of differentially adhesive cell types (Parichy, 2001). Raff and Wray (1989), Raff (1996), and Hall (1998) also offer critical assessments of temporal repatterning of processes, although these discussions are directed partly at demonstrating that heterochronies inferred from morphology need not arise by temporal repatterning of processes.

Raff and Wray (1989) provide an analytic scheme complementary to that of Alberch et al. for comparing rates and timings of processes. Fortunately, they do not transfer the terminology from Table 1 of Alberch et al. (1979) for the very different kind of phenomena analyzed by their scheme. In addition to comparisons among rates and timings, sequence of processes can be analyzed using methods discussed earlier (Smith, 2001). But just as it is possible to decompose morphology into units that have no characteristics other than rate or timing, it is possible to decompose processes into units of pure time, such as the time at which a particular gene is expressed.

Process-based studies have an obvious benefit—they provide insight into the evolution of developmental processes. Clearly we wish to know whether development often evolves by temporal repatterning, or whether processes more often evolve by other kinds of changes, such as in location or causal organization. But it is important to note that the processes at issue are specifically developmental; if processes are considered from an evolutionary perspective, these developmental studies are purely descriptive, telling us *what* evolves, but not *how.* The evolutionary mechanisms responsible for fixing the novelty in a species, such as selection or random genetic drift, are not the processes revealed by these analyses.

Spatial Repatterning

Spatial repatterning is rarely studied, so it is a conceptually less heterogeneous category. However, it too has been viewed from potentially conflicting perspectives—the morphological and mechanistic. For example, a change in spatial patterning from the perspective of morphology would be a positional change in an organ, whereas from the perspective of process it would be a positional change in where processes occur. Certainly these are not mutually exclusive views, but there is no necessary relationship between them—a change in the spatial distribution of a process need not alter the location of a structure on an organism.

Several kinds of spatial repatternings can be distinguished, but an exhaustive classification should await further studies of the phenomena. At present, it is unlikely that all types have been characterized, and some that now seem distinct may be found to have underlying similarities. *Heterotopy,* another term introduced by Haeckel (1875), has been used both for changes in the location of organs (e.g., Frohlich and Parker, 2000; Frohlich, 2001) and for changes in the spatial distributions of processes (Zelditch et al., 2000). In some cases, it is both; a change in the location of a process results in a change in location of a structure, such as in the origin of internal cheek pouches of geomyoid rodents (Brylski and Hall, 1988).

Another kind of spatial repatterning is *homeosis,* which also involves a change in the location of an organ but differs from heterotopy in that one structure replaces another, e.g., a leg is located in place of an antenna, and many examples in plant evolution (Li and Johnston, 2000). A third kind of spatial repatterning is a change in pattern formation, such as in the mammalian limb, affecting both where limbs are located and also what kind of limbs develop (Lovejoy et al., 1999). Finally, *allometric repatterning* can arise by a change in spatial patterning of growth rates and by dissociated heterochrony, such as by shifts in location of fields of bone deposition and resorption (Lieberman, 2000), changes in the spatial structure of growth gradients, or shifts in where localized accelerations and decelerations of rates occur (Zelditch et al., 2000, 2001).

Attempts to determine whether a change is either temporal or spatial may be based on a false dichotomy, i.e., on the idea that we can think of space versus time. When only one of these dimensions evolves, we can distinguish between them; but in many cases, the modification is *spatio-temporal* rather than spatial or temporal. For example, a change in the spatial organization of growth rates is equally a change in rate and space; while it is a change in spatial patterning, it is also a change in relative rates of processes in particular places. Similarly, a change in the sequence of ossification could be viewed as an anatomical temporal repatterning, or as a change in where ossifications occur over time. Such ambiguous cases are problematic if space and time are viewed in dichotomous terms, especially if efforts are made to decide whether spatial or temporal repatterning is primary. These efforts seem misdirected because neither space nor time is either logically or biologically primary. Rather than attempting such a separation, it seems both simpler and biologically more realistic to recognize that spatiotemporal repatterning is a matter of both space *and* time.

JUKKA JERNVALL

Speciation

Superficially, speciation might appear almost a peripheral problem for studies and issues addressing evolutionary developmental biology. The apparent lack of integration between speciation and development is because most work on evolutionary developmental biology has been largely focused on the generation of large morphological differences among species, i.e., on macroevolution and, by extension, on megaevolution. Conversely, studies of speciation mechanisms lie firmly in the domain of microevolution. Thus, the Modern Synthesis has focused on linking microevolution to speciation while evolutionary developmental biology has focused on morphological changes in species once species are formed. One further reason for the few studies on speciation and development is that even large morphological changes do not necessarily imply a speciation event, while speciation itself may involve hardly detectable changes in morphology.

The problem of speciation, however, has been a central subject in evolutionary biology, both as the outcome and the promoter of changes in genetic composition of populations. Much of the research effort has centered on the extent and mechanisms by which new species arise in isolation via *allopatric speciation,* or whether populations also can give rise to new species via *sympatric speciation.* Of these, allopatric speciation has traditionally been considered the most prevalent mechanism (Dobzhansky, 1937; Müller, 1942; Mayr, 1963), usually involving some sort of geographical barrier to interbreeding. New species may arise in isolation without concomitant changes in their resource use or morphology. Thus allopatric lineage separation may not have to be accompanied, or followed, by morphological divergence. The emphasis on allopatric speciation may have contributed to the lack of literature linking evolutionary developmental biology with speciation.

In the case of *sympatric speciation,* development has a much more tangible role. While the frequency and mode of sympatric speciation remain controversial issues, speciation can and does take place *without* external barriers on interbreeding. Theories on mechanisms promoting sympatric speciation include ecological (Schluter and McPhail, 1993; Schluter, 1996, 1998, 2000) and competitive speciation (Rosenzweig, 1995).

When environmental variance (environmental heterogeneity) exceeds that of population variance, ecological-phenotypic divergence can drive speciation, followed by reproductive isolation. With enough niche vacancy (Rosenzweig, 1995), divergent selection can be presumed to produce speciation even within a single population. Obviously, the degree of isolation required for speciation depends on multiple factors such as environmental heterogeneity and species composition. Relatively weak selection or subtle differences in environment may be required when adjacent subpopulations belonging to a metapopulation become distinct species, a process called *parapatric speciation.* Because delineating sympatric, parapatric, and allopatric speciation into distinct processes is a quixotic endeavor, sympatric speciation is considered here in the broadest sense—basically anything but the standard allopatric mode of speciation.

Sympatric speciation involving concomitant morphological change must have a connection with mechanisms of evolution of development. The most direct way that development may be involved in sympatric speciation is when characters on which natural selection acts allow exploitation of new resources and hence invasion of a new adaptive zone. These characters are often called *key innovations*. While their evolution often involves little initial modification of morphology, key innovations may promote successive morphological and taxonomic divergence (Hunter, 1998).

One characteristic feature of key innovations that can be linked to high species diversity, and thus presumably to speciation, appears to be their frequent evolutionary origin (but see Hunter, 1998, on speciation versus origination in the fossil record). Evolution of nasal emission of echolocation pulses in bats (Teeling et al., 2002) and evolution of the hypocone, a fourth cusp in mammalian upper molars related to the evolution of new dietary specializations (Hunter and Jernvall, 1995), have been inferred to have occurred multiple times. Furthermore, extensive modification of morphology may follow the acquisition of a key character, such as in diverse mammalian tooth shapes derived from teeth with the hypocone and the modification of cichlid fish (labroid) pharyngeal jaw apparatus into different dietary specializations (Liem, 1974). The multiple evolutionary origins and subsequent modification of some key innovations raises the possibility that the underlying changes in development may have been relatively simple, requiring little modification of the genome. This should be the case particularly if speciation events are sympatric; both the "ancestral" and "descendant" phenotypes are retained in a population and the invasion of a new adaptive zone would probably require multiple attempts.

We can also consider the argument from the point of speciation studies. In a search for cases of sympatric speciation, we may find the best examples among key innovations that have evolved multiple times. Incipient forms of these kinds of morphological changes may also be detectable at the population level as normal population level variation (Meyer et al., 1990; Hunter, 1998). A related observation is the outstripping of morphological change compared to speciation at the bases of evolutionary radiations. The most dramatic ecological and morphological divergence in mammals occurred in the Eocene during the Cenozoic, after the extinctions of large portions of the Mesozoic fauna and flora, including dinosaurs. Mammalian radiations are characterized by both increase in the range of anatomical design and speciation. Available evidence supports the general notion that evolution of new morphologies, hence increase in disparity, begins first, followed by increase in taxonomic diversity (Foote, 1993, 1997; Jernvall et al., 1996). This fits with the ideas of sympatric speciation in which invasion of new adaptive zones, facilitated by evolution of new morphologies, promotes consequent speciation (e.g., Rosenzweig, 1995). The dietary divergence of cichlid fishes may also be an example of morphological change preceding further speciation (Albertson et al., 1999).

In combination with different ecological adaptations, sexual selection by assortative mating is thought to play a significant role in facilitating sympatric speciation (Lande, 1981; Turner and Burrows, 1995; Dieckmann and Doebeli, 1999; Higashi et al., 1999; Takimoto et al., 2000). Changes in characters affecting assortative mating, such as sexual size dimorphism, coloration, and display structures, while not necessarily ecologically adaptive, require changes in development during the speciation process.

Multiple examples of evidence linking assortative mating and speciation are found in cichlid fishes in East Africa and the Neotropics. Lake populations of extant cichlids are characterized by as many as hundreds of species that differ in feeding specializations, behavior, and coloration. Molecular and morphological reconstructions of cichlids have shown that individual lake populations originated from one or a small number of closely related species (Meyer et al., 1990; Kornfield and Smith, 2000; Schliewen et al., 2001) and that there is considerable convergence of morphology (Stiassny, 1981; Kocher et al., 1993). Furthermore, some of the lake radiations are very recent (within 50,000 years), which is a challenge for molecular reconstructions of phylogenies; species have retained their ancestral genetic polymorphisms (Kornfield and Smith, 2000). Sexual selection is thought to function in cichlids by female mating choice of male breeding coloration (Seehausen et al., 1997, 1999). Strong assortative mating has been inferred to promote speciation concomitantly with adaptive divergence (Schliewen et al., 2001) or earlier, before ecological differences emerge (Wilson et al., 2000).

The cichlid radiations are particularly interesting from the point of view of development and speciation because the cichlid pharyngeal jaw apparatus has been proposed to be a key innovation allowing diversification of dietary specializations (Liem, 1974). The evolution of functional versatility of the pharyngeal jaw apparatus appears to result from relatively small changes in muscle attachment that decouple the upper and lower pharyngeal jaws (Liem, 1974; Galis and Drucker, 1996). From the point of view of morphological evolution, assortative mating may promote speciation before ecological divergence (Wilson et al., 2000). This does not necessarily conflict with inferences in which morphological change outstrips speciation (Foote, 1993, 1997; Jernvall et al., 1996); speciation driven by assortative mating may be considered to be a relatively constant phenomenon through time in that sexual selection should be influenced by ecological opportunity indirectly. On the other hand, the specific developmental changes giving rise to the cichlid pharyngeal jaw apparatus may have been a more opportunistic shift in developmental processes, promoting morphological divergence and further speciation, only after ecological opportunity.

Initially, even subtle ecological opportunities may have been relatively easily exploited if aspects of development are phenotypically plastic. For example, the replacement tooth size and number in cichlids is affected by the hardness of their diet (Huysseune, 1995). In general, developmental processes involved in sexual selection may be hypothesized to form a more ancient set of mechanisms facilitating mate recognition and constantly producing incipient speciation. It will be interesting to uncover how conserved the aspects of developmental mechanisms involved in sexual selection are.

Neither initial stages of speciation caused by sexual selection nor adaptation to new ecological regime necessarily produce reproductive isolation. Hence, incipient species, such as many cichlids, may qualify as species according to the *ecological species concept* but not qualify as *biological species* because of a lack of hybrid sterility. Evolution of hybrid sterility, and by extension reduction of hybrid fitness, is another classic problem in evolutionary biology dating back to Darwin (1859). Accumulation of deleterious genetic combinations is thought to lead to hybrid sterility, an explanation known commonly as the *Dobzhansky-Müller hypothesis* (but see Orr, 1996; Wilkins, 2002). Incipient species that are effectively allopatric can accumulate mutations in a complementary fashion, resulting in genetic incompatibilities when they have an opportunity to hybridize.

Two central issues for developmental biology can be brought into the realm of hybrid sterility. First, a greater number of genes in a developmental pathway, or a greater number of gene interactions, have been modeled to increase the likelihood for reproductive isolation (Johnson and Porter, 2000, 2001). Second, gene duplication—the time-honored prime engine for evolution of new genes (Haldane, 1932)—has been proposed to be an efficient mechanism allowing Dobzhansky-Müller incompatibilities to emerge; this is the "divergent resolution" hypothesis (Werth and Windham, 1991; Lynch and Force, 2000; Taylor et al., 2001). Taken together, potential for reproductive isolation can be considered to increase with a greater number of genes in developmental pathways and with a greater degree of genomic redundancies due to duplication events. Because gene duplications and mutations causing Dobzhansky-Müller incompatibilities accumulate through time (Lynch and Force, 2000; Orr and Turelli, 2001; Taylor et al., 2001), any two allopatric populations will diverge at the genomic level even though morphologically (depending on selection regime) there may or may not be a change. Within this context of allopatric speciation, it remains to be seen how allopatric *(sensu stricto)* populations have to be for true reproductive isolation to occur. One may ask, for example, whether strong assortative mating (Takimoto et al., 2000), coupled with incipient niche separation, could provide sufficient isolation for the generation of hybrid sterility.

Considering speciation and morphological change, regardless of whether sympatric or allopatric mechanisms are splitting populations, we can hypothesize that developmental processes diverge due to genomic duplications and the preservation and co-option of duplicated genes. Because rates of gene duplication and loss of their copies are likely to be dynamically balanced (Lynch et al., 2001), speciation should initiate different patterns of genomic turnover that might facilitate morphological divergence. We can presume three principal outcomes for newly duplicated genes. Genes may be co-opted to a new developmental context, or remain active in their current developmental pathways and thus provide redundancy, presumably by buffering development. These two options appear not to be mutually exclusive; changes in *cis*-acting control regions can diversify developmental interactions of gene duplicates.

A third outcome, presumably the most frequent fate of gene duplications, is that one of the copies is silenced. Indirect estimates on rates of gene silencing suggest that there can be "flickering" for a period of time when a gene may regain functionality (Marshall et al., 1994).

The issue of gene duplication and speciation allows us to link the biological and evolutionary species concepts. When we consider evolving species lineages, the turnover of the genome—and presumably evolution of developmental processes—can be hypothesized to affect individual lineages differently and perhaps even promote their morphological divergence. One relevant issue to this hypothesis is that when we examine evolving lineages and compare periods of morphological divergence with separation of lineages, we commonly observe a lag between speciation and major morphological change.

For example, most mammalian lineages appear in the fossil record at the base of their radiation. However, estimations of lineage separations based on the molecular clock hypothesis have been used to suggest that lineages of many modern orders separated in the Mesozoic, up to 64 million years before their first appearance in the fossil record (Kumar and Hedges, 1998; Madsen et al., 2001; Murphy et al., 2001). These extreme estimates have been used to suggest major temporal or geographical gaps in the fossil record. However, at the very least, estimated fossil preservational rates do not support temporal gaps in the fossil record (Foote et al., 1999). Nevertheless, while the extremely deep origins of modern mammalian orders in the Mesozoic remain controversial (Archibald et al., 2001; Cifelli, 2001), some of the lineage splitting may well have occurred by the early Paleocene, some ten million years before the major period of morphological divergence in the Eocene.

While we currently lack estimates of whether the temporal gap between lineage separation and morphological divergence had any discernible effects on the nature of morphological change in mammals, the divergence lag hypothesis suggests that speciation history may have a central role in the evolution of development. In other words, time to previous speciation can affect the course of evolution *via* development. Regardless of the exact length of the divergence lag in evolution, any temporal gap between lineage separation and morphological divergence allows independence in genome duplication events. By the time mammals experienced availability of new adaptive opportunities in the Eocene, some of the lineages probably had attained unique features in their genome and developmental pathways, despite the lack of extensive morphological divergence. One possibility is that the increase in morphological disparity (different morphological solutions) may have been promoted by previously insignificant alterations of developmental pathways.

These kinds of differences in the genotype may have been cryptic at the level of phenotype (Lauter and Doeble, 2002) or could have involved minor changes in variational properties of morphological structures. However, even cryptic variation may have produced slightly different morphological solutions to similar ecological regimes after shifts from stabilizing selection to directional selection. A putative example is the Cenozoic evolution of herbivorous mammals in the Northern Hemisphere. These mammals evolved different details of cheek tooth morphology despite the overall increase in tooth shearing crest efficiency (Jernvall et al., 1996). In general, in the beginning of an adaptive radiation, morphologically similar lineages that have been separated for a long time may respond differently to an environmental change at the level of morphology, but not at the level of ecology. On the other hand, during an adaptive radiation, further speciation of evolving lineages would be likely to produce variations on a set of morphological themes because the genome and developmental processes have less time to evolve.

Any change in development that allowed or facilitated the evolution of key innovations may have allowed some lineages to diversify both ecologically and taxonomically, while other lineages that lacked appropriate developmental change may have remained relatively depauperate taxonomically and morphologically. Developmental changes involved in these key innovations may be related to the decoupling of developmental pathways, thereby allowing an increase in evolutionary versatility (Vermeij, 1973)—which is another process that may benefit from gene duplication and promote sympatric speciation. This process does not necessarily require that any fundamental differences in development produce small or large morphological changes. Rather, differences in developmental processes and their effect on morphological change may only influence variational properties of individual species and their responses to changes in adaptive regimes. At the genome level, this influence could be mediated by changes in the protein-binding properties of *cis*-acting enhancers. Such changes are more likely in genes that have recently duplicated. Large morphological changes during adaptive radiations could still result from more extensive modification of the genome, such as a higher proportion of evolution of new enhancer regions (Wilkins, 2002, p. 464). Perhaps extremely long divergence lags between speciation and adaptive radiation allows modification of development to a degree that facilitates abrupt morphological changes reminiscent of saltation *sensu* Goldschmidt (1940).

While it is very likely that all aspects of the genome can be affected in morphological change, different scenarios of genomic and morphological change may provide slightly different predictions on how conserved genetic networks should remain in the course of evolution. A classic example is the rise of animal body plans during the Cambrian explosion, an intense period of morphological change (Gould, 1989a). Again, molecular clock studies provide estimates of lineage separations that predate, sometimes by hundreds of millions of years, the documented morphological divergence (Wray et al., 1996; Lynch, 1999). While molecular clock estimates are, strictly speaking, estimates of gene trees, these results still raise the possibility that divergence lag, presumably imposed by environment, may have contributed to the apparently high disparity among organisms during the Cambrian explosion. The general problem with deep time inferences, however, is that evolutionary changes in the genome and development may erase the genetic past of character evolution (Leroi, 2000). Thus regardless of the quality of the fossil record, the linking of speciation, development, and morphological change may be best done on lineages that are relatively recent.

In conclusion, speciation can provide many insights into evolutionary developmental biology. Most notably, by considering developmental changes in the context of speciation, we are compelled to include issues such as variation, evolutionary history, and ecology in our theories on developmental evolution. This integration will also help us to identify and resolve the issues that may still be beyond the explanatory power of the Modern Synthesis, contribute to understanding problems like sympatric speciation, and demonstrate that evolutionary developmental biology forms an integral part of evolutionary biology.

JOHN O. REISS

Time

The discovery of time as a linear, measurable phenomenon was essential for the development of modern science because it provided a scale against which rates of various processes could be measured (Toulmin and Goodfield, 1990). In the present context, the dimension of time is the critical axis of both development and evolution; both involve processes whose most fundamental feature is that they take place (usually irreversibly) through time. Nevertheless, there are some obvious differences between these two phenomena in their relation to time, both in terms of the phenomena themselves and the mode by which we can investigate them.

In studying development, we trace changes that occur within a single organism and are thus physically continuous. In studying evolution, we trace change through time in populations of life forms that are recreated each generation. Development occurs repeatedly, on a time scale that is short relative to the human life span. Evolution occurred only once, over a long time scale. Time thus enters the study of evolution primarily as an external factor or scale against which events can be measured. In studying development, by contrast, not only can we ask how development plays out in time, but we can also investigate the mechanisms by which timing is controlled. Through the integration of these approaches, we can analyze how developmental timing, and its underlying mechanisms, have changed through evolutionary time.

From the outset we must distinguish between two aspects of time: *sequence* and *duration*. Time as sequence involves only the ordering of events; time as duration also requires that one can define an interval between successive events. The history of the geological time scale makes this distinction very clear. The discovery of the antiquity of the earth—what Martin Rudwick (1995) calls "Deep Time"—helped lay the intellectual foundations for the discovery of evolution. Today we can put dates on the geological time scale. We know, for example, that the Cambrian period began about 580 million years ago, and that the earth is about 4 billion years old. When it was first developed, however, the geological time scale measured only the sequence of events.

Much debate in the late nineteenth century came from the attempt to associate geological ages with absolute (astronomical) time. Some idea of the duration of different periods could be obtained by comparing the thickness of various formations with modern rates of sedimentation. There were, however, wide variations in the estimates. It is well known that Lord Kelvin caused Darwin no end of trouble by his insistence that the earth could be no more than 100 million years old, based on its rate of cooling from an initial molten state. The discovery of radioactivity provided the extra heat (and time) that Darwin needed. Radiometric dating made the geological time scale datable in absolute time. That does not mean that the timing of major evolutionary events (such as lineage divergence, gene mutation or duplication, or the appearance of particular morphological novelties) is always determinable. In fact, much of our knowledge of such timing will always remain inferential, relying on phylogenetic reasoning and assumptions such as the "molecular clock," based on constant rates of nucleotide substitution (Nei and Kumar, 2000).

In moving from evolutionary to developmental time, the situation is much more complicated. Even within a single species there are four distinct aspects of developmental timing to consider, two involving aspects of sequence and two duration. Each provides different ways to describe patterns of development and to set the framework within which mechanistic studies are conducted. These four aspects are most easily explained through comparison with possible mutant phenotypes.

Within the general notion of developmental sequence, we can first examine the particular transformation sequence (Velhagen, 1997) of states or events in the development of a specific feature (such as a morphological element or cell lineage). In this case mutant phenotypes are ones in which a qualitative difference in the sequence is observed (Figure 1A). Second, we can examine the relative sequence of events between distinct transformation sequences; these relative sequences have been called *event sequences* (Velhagen, 1997). Mutant phenotypes are ones in which distinct transformation sequences alter sequence relative to each other (Figure 1B).

Within the general notion of developmental duration, we can first examine the relative timing of the events of transformation sequences with respect to some intrinsic time scale of development. Even if sequence is

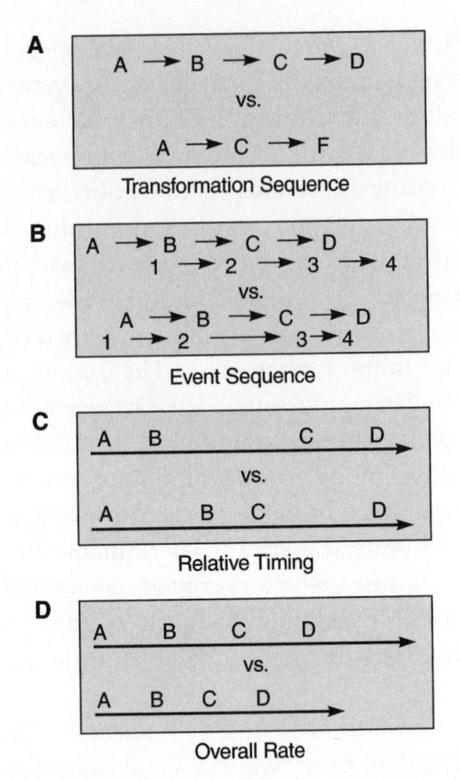

FIGURE 1. The four aspects of developmental timing.

conserved, mutant phenotypes may show events that occur earlier or later with respect to such a time scale (Figure 1C). In practice, this situation is almost always also a change in event sequence, because the duration of an interval in one transformation or event sequence (e.g., the period between standard developmental stages) is used as an index of the intrinsic time scale with respect to which the relative timing of another has changed. Last, we can examine the overall rate of development with respect to absolute time. In this case mutant phenotypes will show the same sequence and relative timing of events, but the absolute length of development will be scaled uniformly longer or shorter (Figure 1D).

Evolution of Developmental Timing

To study the evolution of developmental timing, one must first measure developmental time in a way that allows comparisons among species, and then analyze the variation among taxa. As with studies in individual species, we might find four distinct types of changes in comparisons between taxa (see Figure 1). Each of these includes changes that have been termed "heterochrony" (see Zelditch, 2001). Each has its own associated problems of measurement and evolutionary analysis.

In studying the evolution of transformation sequences (see Figure 1A), the problem of measurement consists largely of the problem of homology: Can individual temporal states of transformation sequences be homologized between taxa? Once this decision is made, analysis of variation in sequence is relatively straightforward. For phylogenetic analysis, the entire sequence of temporal states can be treated as a character state that is either present or absent; alternatively, each temporal state can be so treated (Mabee and Humphries, 1993). There have been surprisingly few formal phylogenetic analyses of this kind (e.g., Mabee, 1993; Hufford, 2001), though many informal scenarios have examined the evolution of transformation sequences (e.g., the evolution of the aortic arches in vertebrates). The classic heterochronic results of paedomorphosis and peramorphosis (Alberch et al., 1979) represent terminal deletion or addition of temporal states to a transformation sequence within this analytical framework.

In studying the evolution of event sequences (Figure 1B)—another type of heterochrony—we typically have much greater problems. As with transformation sequences, we must deal with the issue of homology among temporal states. In addition, extensive intrapopulation heterochrony may make it difficult even to define a single modal event sequence for comparison with other taxa, though methods for dealing with this problem have been suggested (Colbert and Rowe, 2001). Finally, how to correlate event sequences among different species is not at all clear. Unless there is a reason to choose some particular transformation sequence as a privileged standard against which to compare others, everything shifts simultaneously. However, this problem is not necessarily intractable. Velhagen (1997) suggested a method in which the need for correlation is eliminated by treating pairs of events as the characters; character states are defined by the sequence of the two events. More recently, Schlosser (2001) suggested a method in which event sequences of two taxa are plotted against each other on x,y coordinates, and Jeffery et al. (2001) outlined one in which a sort of average sequence is constructed as a basis for comparison.

In studying the evolution of relative timing (see Figure 1C), scientists must first decide on a metric for developmental duration. Two phenomena make absolute time inadequate (Reiss, 1989). The first is the temperature dependence of developmental rate. In almost all organisms, higher temperatures result in faster development. Though many model organisms have well-defined timing at given temperatures, making it possible to use absolute time as a rough metric of development (e.g., the 33 hr chick, the 72 hr *Drosophila*), how to compare these sorts of times across species is not obvious: which temperature should be chosen? The second phenomenon that causes problems for cross-species comparisons is the size dependence of developmental rate: bigger organisms take longer to develop. Variation in timing due to variation in adult size may be so great as to obscure any other reasons for variation, and so ideally should be factored out.

Detlaff and Detlaff (1961) suggested a method by which absolute times of development at a single constant temperature could be scaled to the duration of a particular interval (e.g., the time for the first cleavage division, or time to neurulation) to produce a dimensionless time that could be compared among species. They showed that this scale was roughly constant for a given species across a wide range of temperatures. Some recent work in ontogenetic evolution (Chipman et al., 2000) used this method in cross-species comparisons, with promising results.

I proposed an alternative metric for developmental time based on metabolic rate (Reiss, 1989). Starting from the fact that adult mass-specific metabolic rate times life span is approximately constant across broad taxonomic groups (Calder, 1984), I suggested that physiological time units (PTUs) might be defined based on mass-specific metabolic rate of embryos. This metric does not depend on the assumption that a particular reference interval has not changed with respect to the intrinsic time scale. Literature data supported the possible usefulness of this approach, but it has not been tested further. Given that a metric can be defined, it is a relatively simple matter to analyze the evolution of developmental timing using phylogenetic methods developed to deal with other quantitative data (Harvey and Pagel, 1991).

Finally, in studying the evolution of overall developmental rate (see Figure 1D), we enter into the well-developed field of life history theory (e.g., Stearns, 1992). Rather strangely, this field has traditionally had little interaction with studies of the evolutionary changes in development that are of interest here, a notable exception being the pioneering work of Gould (1977). Instead, life history theory typically focuses on such temporal parameters as birth rates, death rates, age-specific reproduction rates, and timing of life cycle transitions. Here the problem of measurement largely consists of the problem of dealing with population variability. Assuming that this can be done, standard phylogenetic methods again can be applied to examine evolutionary patterns.

General Model for Temporal Control of Development

Understanding the underlying timing mechanisms is essential if we are to move beyond the descriptive stage in our study of the evolution of developmental timing. Much criticism of studies of heterochrony has focused on their purely descriptive nature: how can heterochrony be a "mechanism" of evolution if it is only a result of underlying mechanistic changes? In particular, an understanding of mechanism should allow us to understand the connections among the distinct aspects of developmental time outlined above.

The concept of "positional information" (Wolpert, 1969) has been extremely useful in thinking about the mechanisms that cells might use in the spatial patterning of development. Similarly, we might postulate the existence of "temporal information," conveyed from an internal timekeeper, that acts as an intrinsic time scale used in the temporal patterning of development (Figure 2; see Banerjee and Slack, 2002). Cells would respond by carrying out the proper stage-specific program, resulting in the observed transformation sequence of developmental events. The internal timekeeper would be related to absolute time by what might be called a "developmental throttle" that sets the overall developmental rate. Event sequences do not appear explicitly in this model; they are epiphenomena of how distinct transformation sequences relate to the intrinsic time scale.

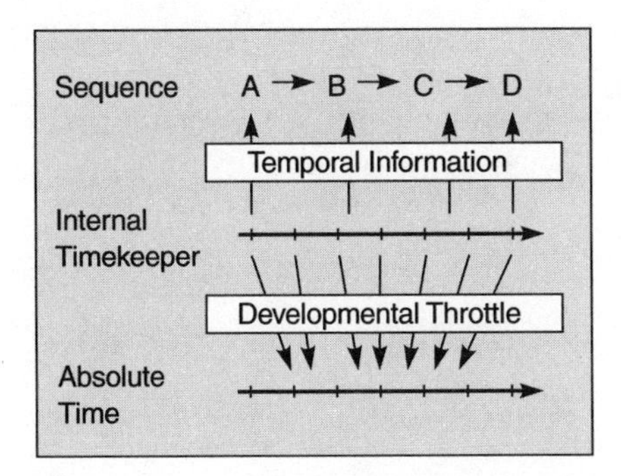

FIGURE 2. A general model for the temporal control of development.

This model ignores rather obvious possibilities for complications. For example, it does not incorporate any feedback between the internal timekeeper and either the sequence or absolute time. The potential presence of multiple, partially redundant internal timekeepers is ignored, as is the likely hierarchy of temporal control, involving both local timekeepers and timekeepers acting through global cues. Finally, lateral interactions among developmental sequences of distinct features—which might provide a mechanistic basis for event sequences—are also ignored. Nevertheless, the model is useful at least as a working hypothesis, providing a framework within which to examine the data.

Clearly, in this model the most important issue is the nature of the internal timekeeper. In a recent review, Johnson and Day (2000) proposed a useful distinction between two types of mechanisms that might be used to time development: *hourglass* and *clock* mechanisms. In hourglass mechanisms a constant decay or buildup of products from an initial time point is used to control timing, with threshold levels of the product being used as cues to initiate developmental events (Figure 3A). By contrast,

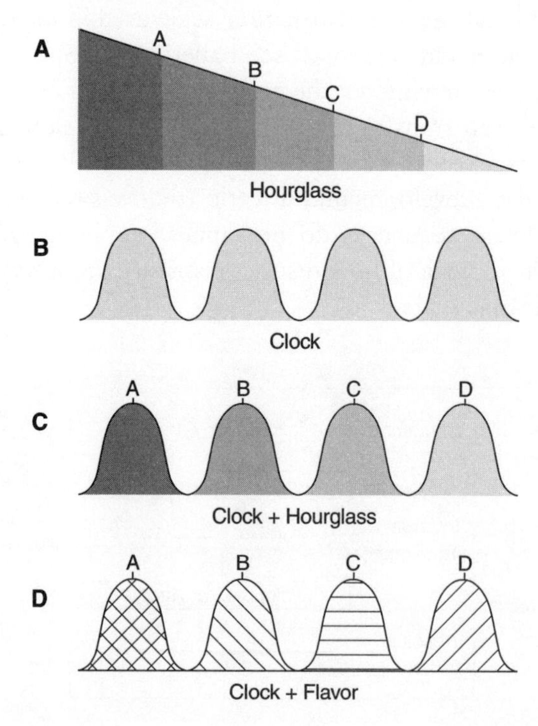

FIGURE 3. Potential mechanisms for an internal timekeeper.

clock mechanisms use periodic oscillations in levels of some substance to time development (Figure 3B); the oscillations in cyclins that control the cell cycle are a well-studied example. Because a pure clock mechanism repeats itself each cycle, it cannot provide unambiguous temporal information beyond a single cycle. However, many potential modifications of a clock mechanism can. For example, it is possible to combine a clock and hourglass system (Figure 3C). In such a system, each cycle of the clock results in breakdown (or buildup) of a product beyond the level experienced in the previous cycle. Alternatively, each cycle of the clock may initiate a qualitatively new state ("signature"), which combined with the clock, would provide a precise temporal identity to each time point (Figure 3D). Once an internal timekeeper is established, it is easy to imagine how temporal information could be interpreted to give rise to a specific sequence of developmental events, and how the rate of progress in absolute time could be modulated.

Experimental Approaches and Evidence

Let us now consider some of the evidence bearing on the control of timing, with respect to the general model developed earlier (see Figure 2). The first and best studied issue is the nature of the temporal information that cells "read" to carry out temporally appropriate behaviors: is it cell-autonomous or more global? A classical approach to examine this issue is *heterochronic transplantation* experiments. When cells are moved from an embryo at one stage to one at another stage, they may continue to develop on their original schedule or regulate and develop on the host schedule. The former result is good evidence for a cell-autonomous time scale, and the latter for global temporal cues. Both results, of course, have been obtained. For example, in a recent study involving neuroblast induction in *Drosophila,* Berger et al. (2001) found evidence for sequential, stage-specific inducers. Another well-known example of global, stage-specific temporal cues is the hormonal control of developmental transitions (e.g., arthropod molts, amphibian metamorphosis). On the other hand, in an experiment using heterochronic co-culture and transplantation of retinal precursors in *Xenopus,* Rapaport et al. (2001) found evidence for a cell-autonomous clock governing competence to respond to photoreceptor inducers. That both results have been obtained suggests that both cell-autonomous and global signals are involved.

Another approach for looking at this issue is through mutant screens for animals with abnormal developmental timing. To date, these have been most successfully conducted in the nematode worm *Caenorhabditis elegans,* in which a number of *heterochronic mutants* have been isolated (reviewed by Ambros, 2000). In *C. elegans,* wild-type hatchlings progress through a series of four larval stages (L1-L4) before the definitive adult stage. In the heterochronic mutants, developmental events appropriate to one larval stage are shifted so that they occur in a different larval stage, but the timing of development is otherwise unaltered. In terms of the distinctions developed earlier (see Figure 1), these shifts involve a change in both event sequences and relative timing.

Interestingly, two of these heterochronic genes, *lin-4* and *let-7,* encode small ($\approx$ 22 nt) antisense RNAs, which bind to the 3′ UTR of target mRNAs and repress their translation. Use of such small temporal RNAs (stRNAs) to control the timing of developmental events may be widespread. The *let-7* stRNA is evolutionarily conserved in a diversity of metazoans and appears to be temporally regulated in all of them (Pasquinelli et al., 2000). Moreover, screens to identify similarly sized regulatory RNAs in *C. elegans* and *Drosophila* have shown them to be present in a greater diversity than previously suspected (reviewed by Ruvkun, 2001, and by Bannerjee and Slack, 2002). The use of stRNAs as temporal cues, and their newly uncovered diversity, suggests that stage-specific developmental programs may be regulated by a combinatorial code of upstream regulatory genes, analogous to how segment identity in *Drosophila* and vertebrates is regulated by a homeobox gene code (Bannerjee and Slack, 2002).

As exciting as these results are, it is important to remember that from the standpoint of the model developed earlier, such stRNAs are acting as signaling molecules providing temporal information from the internal timekeeper; they are not the timekeeper themselves. Nevertheless, it is likely that as understanding of the role of stRNAs in development expands, we will acquire a much better understanding of the proximate mechanisms for temporal control of development.

As for the internal timekeeper, it is still unclear what its mechanisms are, or whether it even exists in a form that we can recognize. Mutant screens for alterations in developmental timing should uncover the elements of such a timekeeper, but no mutant so far identified seems to fit that role. A tantalizing hint, however, comes from the fact that one of the *C. elegans* heterochronic genes, *lin-42,* encodes a protein with sequence homology to the *Drosophila period* gene (Jeon et al., 1999). This gene is a central component of the circadian clock mechanism of *Drosophila* and mammals. *lin-42* mRNA levels oscillate in synchrony with the larval molt cycle, so that molts always occur during troughs in *lin-42* mRNA levels. It appears possible that LIN-42 protein functions to control the within-stage timing of developmental events, just as circadian rhythm proteins control the diel timing of various behaviors. As suggested by Johnson and Day (2000), perhaps circadian cycles are only a special, temperature-insensitive case of a more general cyclical mechanism that is also used as part of an internal timekeeper.

Let us now consider evidence bearing on the "developmental throttle" that controls overall developmental rate in our model. As noted above, developmental rate is generally temperature sensitive, with faster development at warmer temperatures. The observation that metabolism is also temperature sensitive, together with observations correlating mass-specific metabolic rate with life span both within and across animal species, has led to a model in which overall mass-specific metabolic rate controls developmental rate. In the field of aging, this is commonly known as the "rate of living" theory, and as such it goes back at least to Loeb and Northrop (1916) and Pearl (1928); the competing theory is the "oxidative stress" theory; see, for example, Perez-Campo et al. (1998).

Some of the most intriguing evidence bearing on the "rate of living" theory again comes from work on *C. elegans*. In particular, mutants in the *clk-1* gene of *C. elegans* show correlated lengthening of life span, cell cycles, pharyngeal pumping cycles, defecation cycles, and decreased development and locomotion rates (Branicky et al., 2000). Metabolic rate is also reduced, though not enough to explain by itself the slowing of development. While it is still unclear what the function of the CLK-1 protein is, Branicky et al. (2000) suggest that it may be involved in cross-talk between the mitochondria and the nucleus. Moreover, the mutant phenotype doesn't exclude the possibility that the *clk-1* gene is functioning as a component of an internal clock, rather than as part of the developmental throttle.

As our knowledge of the mechanisms involved in the temporal patterning of development grows, we can expect that comparative analysis of these mechanisms will provide a better understanding of the relation between the four modes of altering developmental timing. We may again look to the analogy of spatial patterning by Hox genes: molecular signatures of segment identity have provided a much clearer understanding of the developmental significance of morphological variation (e.g., Burke et al., 1995). Similarly, understanding of temporal control mechanisms, particularly the nature of the intrinsic time scale, should help us better understand the significance of the abundant variation in developmental timing seen at the descriptive level.

BENEDIKT HALLGRÍMSSON

Variation

Phenotypic variation is the central concept linking studies of evolution, development, and population genetics. The integration of evolutionary and developmental biology has been guided by two central questions about variation: (1) is the variation that leads to evolutionary change continuous or discrete; and (2) how does the structure of development bias or constrain the production of variation and hence the course of evolution? Both questions have deep roots. The former can be traced from the correspondence between Huxley and Darwin, through Bateson (1894), and Goldschmidt (1940) to recent approaches grounded in developmental genetics (Stern, 2000; Haag and True, 2001). The latter can be traced to von Baer's and Haeckel's interests in the relationship between ontogeny and phylogeny in the early nineteenth century to questions about the role of developmental timing in evolutionary change (Gould, 1977) and interest in the developmental basis for evolutionary change.

Interestingly, the study of variation as a subject in itself has remained peripheral to the ongoing synthesis of evolutionary and developmental biology. By variation as a subject, I mean the question of how variability—or the tendency to vary (Wagner et al., 1997)—is patterned among traits, populations, species, and higher taxa. Perhaps this is because, as Van Valen (1974) points out, the additional ingredients of ecology and morphometrics must be added to the mix of evolution, development, and genetics to study and appreciate the significance of patterns of variability. Alternatively, as argued by Hall (1999a), the lack of interest in patterns of phenotypic variability may reflect the focus of evolutionary developmental biologists on a few model organisms reared under controlled laboratory conditions.

Components of Phenotypic Variation

The variance of any aspect of the phenotype is the result of genetic and environmental effects. The genetic component of phenotypic variance can be broken down into the additive effects of genes and the deviations from simple additive effects due to dominance and epistatic interactions among genes. Although only additive effects contribute to heritability in the narrow sense, dominance and epistasis probably contribute significantly to genetic variance in most cases; see Falconer and Mackay (1996) or Roff (1997) for fuller discussions.

The environmental component of phenotypic variance can be partitioned in various ways depending on research aims. In human genetic studies, for instance, it is often useful to divide the environmental variance into familial and nonfamilial effects (Rice and Borecki, 1999). A useful distinction in evolutionary contexts is between environmental factors that contribute to differences among individuals and those that contribute to variation within individuals. This distinction defines the boundary between canalization and developmental stability. Canalization refers to the reduction of variation among individuals, whereas developmental stability refers to the minimization of variation within individuals (Clarke, 1998).

A final variance component is true developmental noise. Waddington (1957) regarded developmental noise as imprecision at the molecular level of developmental processes. Such variation is not of environmental origin as it is intrinsic to the material substrate of development. Unfortunately, true developmental noise is very difficult to measure. Within-individual variation, such as that measured by subtle departures from symmetry (fluctuating asymmetry), is caused by both environmental effects operating within individuals and by developmental noise.

Determinants of Variability

Soulé (1982) lamented nearly 20 years ago that there was no general theory of morphological variation. Sadly, this is almost as true today. Such a theory would allow us to predict patterns of variability in nature from knowledge of its determinants. The following is a brief synopsis of what is known about the potential determinants of variability among characters, populations, and species.

Inbreeding. The relationship between inbreeding and the genetic component of phenotypic variance is well worked out in population genetics (Wright, 1969). Inbreeding reduces genetic variance and is complicated mainly by a dependency on initial gene frequencies when dominance is present (Falconer and Mackay, 1996). In natural populations, however, the relationship between inbreeding and phenotypic variance is less simple. Frequently, genotypes differ in their sensitivity to the environment. Lerner (1954) argued that more homozygous individuals are more sensitive to environmental perturbations and presented several examples of variability increasing with inbreeding as a result of increased environmental variance. Schmalhausen (1949) presented a somewhat different argument that predicted the same outcome. Inbreeding results in a loss of genetic variability that is exposed to stabilizing selection, and the weakening of stabilizing selection results in an increase in variation of environmental origin. He points out that in laboratory-reared inbred lines, the reduction in stabilizing selection is extreme, and a greater increase in environmental variance can be expected.

A third and related possibility is that inbreeding results in increased expression of deleterious alleles (Charlesworth and Charlesworth, 1999). Deleterious alleles may be associated with higher environmental variances. Falconer and Mackay (1996) list several examples including fruit flies, mice, rats, and maize, in which inbred strains exhibit higher phenotypic variances than do hybrids. How often inbreeding increases variability and why is not clear. It is interesting to note, however, that a recent meta-analysis of the relationship between heterozygosity as determined by allozyme electrophoresis and morphological variability found only a very weak relationship (Reed and Frankham, 2001).

Selection. The effect of selection on genetic variance has and continues to be an active area of research in population genetics. Directional selection reduces genetic variance at a rate that depends on dominance, pleiotropic effects, and the strength of selection. Stabilizing selection and, less obviously, disruptive selection also reduce additive genetic variance (Roff, 1997). This expected reduction in phenotypic variance as a result of selection has been used to measure the intensity of selection in natural populations (Lande, 1979; Cheverud, 1989). There are, however, situations in which selection is expected to increase phenotypic variance. Slatkin and Lande (1976) showed that density-independent selection produces a larger variance in a fluctuating environment and a smaller variance in a constant environment. Similarly, frequency-dependent selection can increase phenotypic variance under certain conditions (Roff, 1997). Finally, as argued by Schmalhausen (1949), if strong directional or stabilizing selection produces a very significant loss of genetic variability, environmental variance might increase due to increasing homozygosity.

Mutation. Mutations are the ultimate source of variability, so it is obvious that mutations increase phenotypic variance. Less obvious is the possibility that mutations can also affect the variance through disruptive effects on development. Schmalhausen's (1949, p. 10) definition of a mutation as a "change in the reaction norm" encompasses the potential of a mutation to affect the gene-environment interaction. Both Schmalhausen and Waddington (1957) noted that mutant phenotypes are not just different but also more variable; Scharloo (1991) presents a list of examples of this phenomenon. In their view, the mutant represents a developmental configuration that has not undergone selection for canalization and is thus more sensitive to environmental perturbations. A fair amount of experimental and anecdotal evidence exists to support this claim. The increase in variability is thus an important variable to bear in mind when using mutations with significant phenotypic effects to study developmental processes.

Malformations. Malformations are basically extreme and usually dysfunctional variants. Such variation figured prominently in early discussions of variability from Vesalius in the sixteenth century (Straus and Temkin, 1943) to Bateson in the nineteenth (1894). Malformation, however, is not only a category of variation but also a source of variability. This can occur in two ways. First, it follows from the arguments of Schmalhausen and Waddington that malformed phenotypes should be less canalized and, therefore, more sensitive to environmental variation. Malformations fall at the extremes of phenotypic distributions and represent developmental configurations that have not been subjected to canalizing selection. This conjecture is supported by the extensive evidence for asymmetry in the expression of various malformations such as cleft lip and cleft palate in humans (Cronin and Hunter, 1980), and by evidence for dependency of expression on the genetic background or environmental variables (Rutherford, 2000). Malformations are also a source of additional variability in that they are often accompanied by secondary disruptions in developmentally related structures. The frequent coincident occurrence of multiple related malformations is well known in the teratological literature (Hagberg et al., 1998).

Stressful environments. There is considerable evidence that phenotypic variance is increased in stressful environments (Burla and Taylor, 1982). In fact, developmental geneticists of the 1950s and 1960s often used environmental stress such as heat or ether to reveal genetic variants. The reason for this relationship, however, is not well understood. Both Schmalhausen (1949) and Waddington (1957) argued that environmental stress would increase variability. Canalizing selection, they argued, should reduce variability within the most frequently encountered environmental contexts. In unusual environments, its effects would not necessarily hold. Hence, unusual environments can reveal genetic variation that remains hidden in the more highly canalized phenotype that is expressed under more usual circumstances. Note that because species adapt to the environmental conditions to which its individuals are most commonly exposed, most changes in those conditions would be stressful.

Recently, Rutherford and Lindquist (1998) pointed to one developmental genetic mechanism that may underlie the relationship between environmental stress and phenotypic variability. When they interfered with the function of the *Drosophila* heat shock protein Hsp90 through either mutation or an administered drug, the incidence of phenotypic abnormalities increased dramatically. Hsp90 is a molecular chaperone that stabilizes a variety of signaling proteins. They argue that under conditions of environmental stress, such as temperature extremes, available Hsp90 levels could fall as the chaperone protein is used up by stress-damaged proteins. This, in turn, will result in increased morphological variability. Rutherford and Lindquist's study suggests an important role for molecular chaperones as buffering mechanisms during development.

Function. There are several reasons to expect that the functional importance of a trait, as measured by its relation to fitness, should be related to its phenotypic variability. First, intense selection reduces genetic variance. Traits that are most closely related to fitness, therefore, should have lower genetic variances (Roff and Mousseau, 1987). Based mostly on anecdotal evidence, Simpson (1944) argued that while vestigial or nonfunctional characters are highly variable, functionally important characters are tightly integrated (highly correlated with other characters) and exhibit low variability. Aside from examples provided by Yablokov (1966), this issue has not been systematically addressed.

Second, more functionally important traits should be more highly canalized and exhibit greater developmental stability. Gummer and Brigham (1995) found that hind limbs have higher fluctuating asymmetry than fore limbs in little brown bats *(Myotis lucifugus)*. Barring some developmental reason, their study provides one example in which the more functionally important traits are more developmentally stable.

Third, the variance of a trait can itself be adaptive. Schmalhausen (1949) argued that the increase in variability in stressful environments can be of evolutionary importance by exposing hidden genetic variation to selection; see also Rutherford and Lindquist (1998). Variability could also be selected for at macroevolutionary scales, as lineages with higher variability exhibit greater evolutionary potential. This argument was developed by Dobzhansky (1937) and revisited more recently by Lloyd and Gould (1993). The data to support their argument are scanty. While Simpson (1944) argued that there is no association between variability of mammalian dental characters and the evolutionary success of lineages, genetic mechanisms enhancing variability are known to be an important factor in microbial evolution (Morschhauser et al., 2000). Phenotypic plasticity, or adaptive environmental deviations, can also be selected for under a variety of circumstances, and this has been an active area of research (Stearns, 1989; Callahan et al., 1997).

Increasing phenotypic plasticity is one of the major trends of evolution (Schmalhausen, 1949; Rensch, 1959). Within-individual variability may even be selected for under very specific circumstances. Simons and Johnston (1997) argue, for example, that fluctuating asymmetry can be adaptive as a bet-hedging strategy for plants in variable environments. These possibilities have barely been explored. Given its central importance in the evolutionary process, it is surprising that we know so little about the relationship between function and variability in nature.

Population-level factors. Several studies demonstrate a positive association between niche width and morphological variability (Van Valen, 1965; Rothstein, 1973) and genetics (Nevo, 1977; Steiner, 1977; Lacy, 1982). Niche width refers to some measurable aspect of the "proportion of the total multidimensional space of limiting resources used by a species" (Van Valen, 1965, p. 378). The niche-variation hypothesis holds that individuals vary in how adapted they are to different aspects of the niche occupied by the entire species. This hypothesis occupies an important place in evolutionary theory because it is one of the alternative explanations for the maintenance of genetic variation under selection (Mayr, 1963).

While changes in population size certainly affect variability (Schmalhausen, 1949; Roonwal, 1953; Hutchinson and MacArthur, 1959; Yablokov, 1966), the existence of a general relationship between population size and variability is arguable. Darwin (1859) predicted that more abundant species should be more variable, and Fisher later presented some quantitative evidence to support this claim (Fisher and Ford, 1928; Fisher, 1937). It is not known how general this relationship is. Van Valen (1965) suggests that a relationship between population size and variability is explained by the niche-variation hypothesis.

Widely distributed species should be more variable because of adaptation to varying local conditions. A great deal of anecdotal evidence supports this claim, while there are also examples to the contrary, such as humans versus chimpanzees. Analysis of this relationship is complicated by body size and dispersal patterns. Habitat variation is more significant at smaller scales for smaller organisms (Mayr, 1963), and the distances that individuals travel within that range can have profound effects on gene flow.

Gene flow can affect variability through its effect on the magnitude of among-population or geographic variation and also within-population heterozygosity levels (Mayr, 1963). Home range and group size are important determinants of gene flow because they affect the number of potential mates. The existence of a general relationship between variability, dispersal distances and patterns, and their scaling to geographic range is unknown. A limited test of this hypothesis found no relationship between home range size and variability in forty-one species of mammals (Hallgrímsson and Maiorana, 2000).

Character size and allomeric variation. Character size and allomeric variation. For similar traits within an organ system, such as individual bones, size-relative variability is negatively correlated with the mean (Yablokov, 1966). Lande (1977) proposed an explanation for this observation, showing that for a structure composed of a series of imperfectly correlated components, the variance of the structure as a whole will be lower than that of each component. The more components in the structure, the greater this difference in variability is between the whole and its parts. Soulé (1982) argued that this phenomenon, which he termed allomeric variation, generally produces a negative relationship between the size and size-relative variability of homologous structures. Allomeric effects have also been documented for within-individual variation (Lajus, 2001).

Dimensionality of traits. As originally noted by Schmalhausen (1935, cited in Yablokov 1966), variability is not comparable across traits of different dimensions. Both Lande (1977) and Van Valen (1978) have treated this problem, and both of these papers are required reading for anyone interested in the analysis of variation. Van Valen's solution is simpler and more elegant. He argues that coefficients of variation should be corrected by dividing them by an estimate of the effective dimensionality of a trait *(D)*. *D* is estimated from the correlations among measures of the individual dimensions of the trait.

Body size. Using a large sample of mammals (65,000 individuals in 351 species) and birds (237 species from published sources), Hallgrímsson and Maiorana (2000) found that size-relative variation for measures of body size is negatively correlated with the mean across species. The correlation increases when variability values are averaged across species into higher taxonomic categories (see Figure 1) and holds true after analysis of phylogenetically independent contrasts that take into account how the taxa in the sample are phylogenetically related. After considering a variety of alternative explanations, we concluded that the most likely explanation is that larger mammals and birds are composed to a relatively greater extent of intrinsically more variable body mass components such as bone, fat, and muscle. Smaller mammals and birds are composed to a relatively greater extent of components such as the viscera and the nervous system, for which size variation is more highly constrained by energetic factors. This fairly large scale regularity in variability has not been examined for other taxonomic groups.

The study of variability brings together diverse fields in biology, ranging from population genetics to developmental biology and ecology. The diversity of topics covered in this entry also clearly illustrates that a century and a half after Darwin, we still do not have a comprehensive theory of phenotypic variation. Instead, in Soulé's (1982, p. 751) words, we have "a collection of hypotheses" and "a lot of undigested pheno-

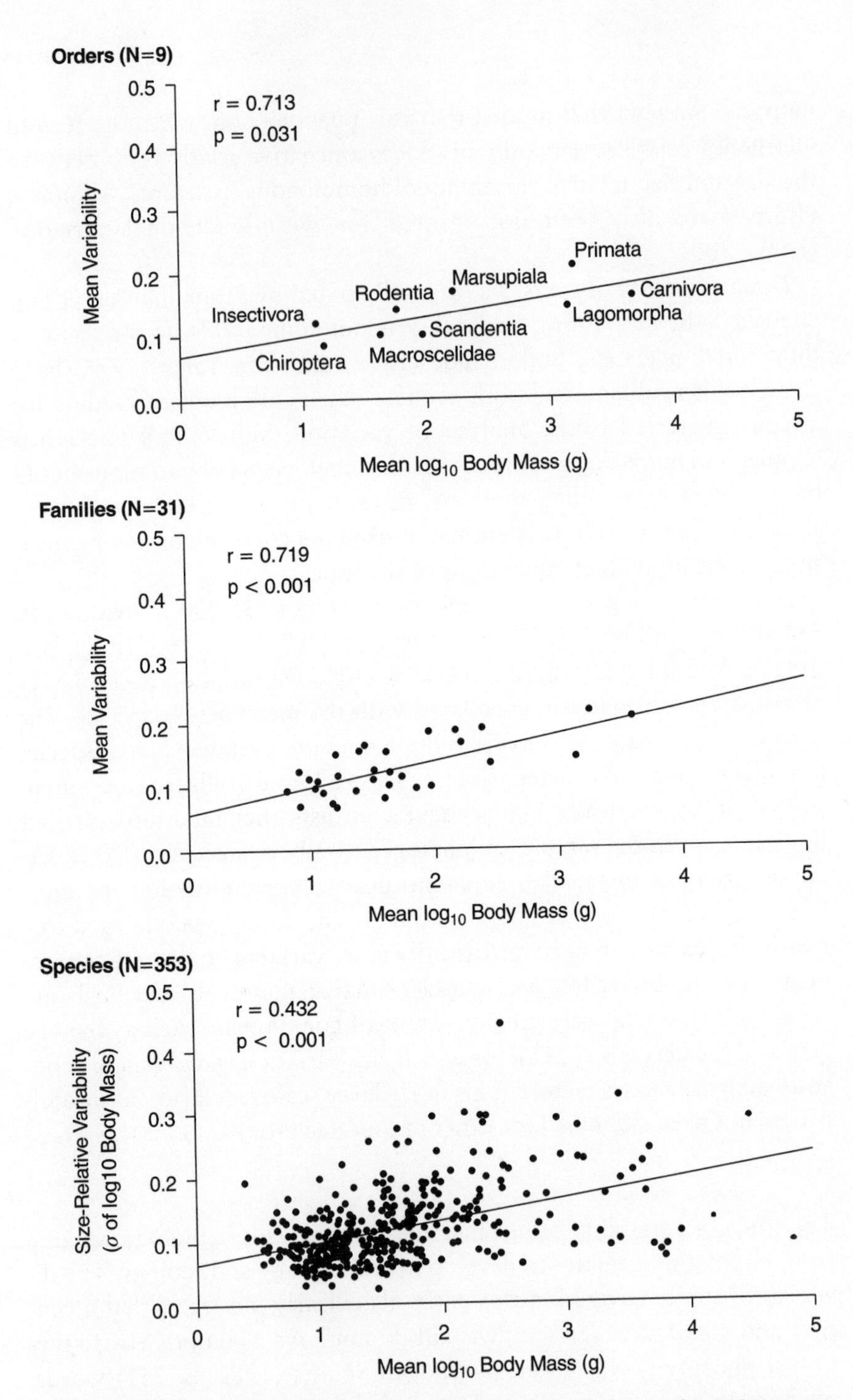

FIGURE 1. Relative variability of body mass against mean body mass for 353 mammalian species. The top graph A plots orders represented by ≥3 families and the next plots families represented by ≥3 species. At the specific level, size explains 21 percent of the among-species variance in variability. The slopes do not differ significantly among taxonomic levels (from Hallgrímsson and Mairoana, 2000).

mology." A successful theory of phenotypic variation will correctly predict patterns in variability in nature from knowledge of their determinants. Getting there will require a renewed focus on large-scale patterns of variability in which hypothesis generation is informed by current knowledge of development, ecology, and population genetics.

References

Abouheif, E. (1997). Developmental genetics and homology: A hierarchical approach. *Trends in Ecology and Evolution* 12:405–408.

Abzhanov, A., and Kaufman, T. C. (1999). Novel regulation of the homeotic gene *Scr* associated with a crustacean leg-to-maxilliped appendage transformation. *Development* 126:1121–1128.

Acosta, F. J.; J. A. Delgado; F. Lopez; and J. M. Serrano (1997). Functional features and ontogenetic changes in reproductive allocation and partitioning strategies of plant modules. *Plant Ecology* 132:71–76.

Adams, D. S.; R. E. Keller; and M. A. R. Koehl (1990). The mechanics of notochord elongation, straightening and stiffening in the embryo of *Xenopus laevis*. *Development* 110:115–130.

Adams, M. B. (1980). Sergei Chetverikov, the Kol'tsov Institute, and the evolutionary synthesis. In *The Evolutionary Synthesis: Perspectives on the Unification of Biology*, ed. E. Mayr and W. B. Provine, 242–278. Cambridge: Harvard University Press.

Adams, M. D. + 194 coauthors (2000). The genome sequence of *Drosophila melanogaster*. *Science* 287:2185–2195.

Adams, N. L., and Shick, J. M. (1996). Mycosporine-like amino acids provide protection against ultraviolet radiation in eggs of the green sea urchin *Strongylocentrotus droebachiensis*. *Photochemistry and Photobiology* 64:149–158.

Affolter, M., and Mann, R. (2001). Development: Legs, eyes, or wings—selectors and signals make the difference. *Science* 292:1080–1081.

Agassiz, L. (1833–1843). *Recherches sur les poissons fossiles*. Neuchatel: Petitpierre, 5 volumes.

Aguinaldo, A. M. A.; J. M. Turbeville; L. S. Linford; M. C. Rivera; J. R. Garey; R. A. Raff; and J. A. Lake (1997). Evidence for a clade of nematodes, arthropods and other moulting animals. *Nature* 387:489–493.

Akam, M. (1998). Hox genes, homeosis and the evolution of segment identity: No need for hopeful monsters. *International Journal of Developmental Biology* 42:445–451.

Akarsu, A. N.; I. Stoilov; E. Yilmaz; B. S. Sayli; and M. Sarfarazi (1996). Genomic structure of HOXD 13 gene: A nine polyalanine duplication causes synpolydactyly in two unrelated families. *Human Molecular Genetics* 5:945–952.

Alberch, P. (1982). Developmental constraints in evolutionary processes. In *Evolution and Development,* ed. J. T. Bonner, 313–332. Berlin: Springer-Verlag.

Alberch, P., and Gale, E. A. (1985). A developmental analysis of an evolutionary trend: Digital reduction in amphibians. *Evolution* 39:8–23.

Alberch, P.; S. J. Gould; G. F. Oster; and D. B. Wake (1979). Size and shape in ontogeny and phylogeny. *Paleobiology* 5:296–317.

Albert, V. A., and Mishler, B. D. (1992). On the rationale and utility of weighting nucleotide sequence data. *Cladistics* 8:73–83.

Albert, V. A.; B. D. Mishler; and M. W. Chase (1992). Character-state weighting for restriction site data in phylogenetic reconstruction, with an example from chloroplast DNA. In *Molecular Systematics of Plants,* ed. P. S. Soltis, D. E. Soltis, and J. J. Doyle, 369–403. New York: Chapman and Hall.

Albert, V. A.; M. W. Chase; and B. D. Mishler (1993). Character-state weighting for cladistic analysis of protein-coding DNA sequences. *Annals Missouri Botanical Garden* 80:752–766.

Alberts, B.; D. Bray; J. Lewis; M. Raff; K. Roberts; and J. Watson (1989). *Molecular Biology of the Cell,* 2nd ed. New York, NY: Garland Publishing.

Albertson, R. C.; J. A. Markert; P. D. Danley; and T. D. Kocher (1999). Phylogeny of a rapidly evolving clade: The cichlid fishes of Lake Malawi, East Africa. *Proceedings of the National Academy of Sciences, USA* 96:5107–5110.

Allen, D. M.; H. van Praag; J. Ray; Z. Weaver; C. J. Winrow; T. A. Carter; R. Braquet; E. Harrington; T. Ried; K. D. Brown; F. H. Gage; and C. Barlow (2001). Ataxia telangiectasia mutated is essential during adult neurogenesis. *Genes and Development* 15:554–566.

Allen, G. E. (1986). T. H. Morgan and the Split Between Embryology and Genetics, 1910–1935. In *A History of Embryology,* ed. T. J. Horder, J. A. Witkowski, and C. C. Wylie, 113–146. Cambridge: Cambridge University Press.

Altig, R., and Johnston, G. F. (1986). Guilds of anuran larvae: Relationships among developmental modes, morphologies, and habitats. *Herpetological Monographs* 3:81–109.

Alvarez, H. P. (2000). Grandmother hypothesis and primate life histories. *American Journal of Physical Anthropology* 113:435–450.

Ambros, V. (2000). Control of developmental timing in *Caenorhabditis elegans. Current Opinion in Genetics & Development* 10:428–433.

Ambrose, B. A.; D. R. Lerner; P. Ciceri; C. M. Padilla; M. F. Yanofsky; and R. J. Schmidt (2000). Molecular and genetic analysis of the *Silky1* gene reveals conservation in floral organ specification between eudicots and monocots. *Molecular Cell* 5:569–579.

Amores, A.; A. Force; Y. Yan; Y. L. Joly; C. Amemiya; A. Fritz; R. K. Ho; J. Langeland; V. Prince; Y. Wang; M. Westerfield; M. Ekker; and J. H. Postlethwait (1998). Zebrafish Hox clusters and vertebrate genome evolution. *Science* 282:1711–1714.

Amprino, R. (1947). La structure du tissu osseux envisagée comme expression de différences dans la vitesse de l'acroissement. *Archives de Biologie* 58:315–330.

Amundson, R. (1994). Two concepts of constraint: adaptationism and the challenge from developmental biology. *Philosophy of Science* 61:556–578.

Amundson, R. (1998). Typology reconsidered: Two doctrines on the history of evolutionary biology. *Biology and Philosophy* 13:153–177.

Amundson, R. (2000). Embryology and evolution 1920–1960: Worlds apart? *History and Philosophy of the Life Sciences* 22:313–330.

Amundson, R. (2001). Adaptation and development: On the lack of common ground. In *Adaptation and Optimality*, ed. S. Orzack and E. Sober, 303–334. Cambridge: Cambridge University Press.

Andrews, J. H. (1995). Fungi and the evolution of growth form. *Canadian Journal of Botany* (Suppl. 1, Section E-H) 73:S1206–S1212.

Antonovics, J., and van Tienderen, P. H. (1991). Ontoecogenophyloconstraints? The chaos of constraint terminology. *Trends in Ecology and Evolution* 6:166–168.

Aparicio, S. (2000). Vertebrate evolution: Recent perspectives from fish. *Trends in Genetics* 16:54–56.

Appel, T. A. (1987). *The Cuvier-Geoffroy Debate: French Biology in the Decades Before Darwin.* Oxford: Oxford University Press.

Aranda-Anzaldo, A. (2000). The Hox-gene research programme and the shortcomings of molecular performationism. *Rivista di Biologia / Biology Forum* 93:57–82.

Archibald, J. D.; A. O. Averianov; and E. G. Ekdale (2001). Late Cretaceous relatives of rabbits, rodents, and other extant eutherian mammals. *Nature* 414:62–65.

Arendt, D., and Nübler-Jung, K. (1997). Dorsal or ventral: Similarities in fate maps and gastrulation patterns in annelids, arthropods and chordates. *Mechanisms of Development* 61:7–21.

Arendt, D., and Nübler-Jung, K. (1999). Rearranging gastrulation in the name of yolk: Evolution of gastrulation in yolk-rich amniote eggs. *Mechanisms of Development* 81:3–22.

Arndt, A.; C. Marquez; P. Lambert; and M. J. Smith (1996). Molecular phylogeny of Eastern Pacific sea cucumbers (Echinodermata: Holothuroidea) based on mitochondrial DNA sequence. *Molecular Phylogenetics & Evolution* 6:425–437.

Arnold, S. J. (1983). Morphology, performance and fitness. *American Zoologist* 23:347–361.

Arnold, S. J. (1992). Constraints on phenotypic evolution. *The American Naturalist* (Suppl) 140:S85–S107.

Arnold, S. J.; P. Alberch; V. Csányi; R. C. Dawkins; S. B. Emerson, B. Fritsch; T. J. Horder; J. Maynard-Smith; M. J. Starck; G. P. Wagner; D. B. Wake (1989). How do complex organisms evolve? In *Complex Organismal Functions: Integration and Evolution in Vertebrates,* ed. D. B. Wake and G. Roth, 403–433. New York: John Wiley & Sons.

Arnone, M. I., and Davidson, E. H. (1997). The hardwiring of development: Organization and function of genomic regulatory systems. *Development* 124:1851–1864.

Arthur, W. (1988). *A Theory of the Evolution of Development.* Chichester, UK: John Wiley & Sons.

Arthur, W. (1997). *The Origin of Animal Body Plans: A Study in Evolutionary Developmental Biology.* Cambridge: Cambridge University Press.

Arthur, W. (2000). The concept of developmental reprogramming and the quest for an inclusive theory of evolutionary mechanisms. *Evolution & Development* 2:49–57.

Arthur, W. (2001). Evolutionary developmental biology: Developmental constraint. In *Encyclopedia of Life Sciences.* London: Nature Publishing Group, Macmillan Publishers.

Atchley, W. R. (1990). Heterochrony and morphological change: a quantitative genetic perspective. *Seminars in Developmental Biology* 1:289–297.

Atchley, W. R., and Hall, B. K. (1991). A model for development and evolution of complex morphological structures and its application to the mammalian mandible. *Biological Reviews* 66:101–157.

Atchley, W. R.; S. Xu; and C. Vogl (1994). Developmental quantitative genetic models of evolutionary change. *Developmental Genetics* 15:92–103.

Averof, M., and Akam, M. (1995a). Hox genes and the diversification of insect and crustacean body plans. *Nature* 376:420–423.

Averof, M., and Akam, M. (1995b). Insect-crustacean relationships: Insights from comparative developmental and molecular studies. *Philosophical Transactions of the Royal Society of London, Series B* 347:293–303.

Averof, M., and Cohen S. M. (1997). Evolutionary origin of insect wings from ancestral gills. *Nature* 385:627–630.

Averof, M., and Patel, N. H. (1997). Crustacean appendage evolution associated with changes in Hox gene expression. *Nature* 388:682–686.

Ayala, F. J. (2000). Neutralism and selectionism: The molecular clock. *Gene* 261:27–33.

Baatz, M., and Wagner, G. P. (1997). Adaptive inertia caused by hidden pleiotropic effects. *Theoretical Population Biology* 51:49–66.

Baguñà, J.; E. Saló; and C. Auladell (1989). Regeneration and pattern formation in planarians. III. Evidence that neoblasts are totipotent stem cells and the source of blastema cells. *Development* 107:77–86.

Balam, G., and Gurri, F. (1994). A physiological adaptation to undernutrition. *Annals of Human Biology* 21:483–489.

Baldwin, B. G., and Sanderson, M. J. (1998). Age and rate of diversification of the Hawaiian silversword alliance (Compositae). *Proceedings of the National Academy of Sciences, USA* 95:9402–9406.

Baldwin, I. T., and Schmelz, E. A. (1996). Immunological "memory" in the induced accumulation of nicotine in wild tobacco. *Ecology* 77:236–246.

Baldwin, J. M. (1896). A new factor in evolution. *American Naturalist* 30:441–451, 536–553.

Baldwin, J. M. (1902). *Development and Evolution.* New York: Macmillan.

Balfour, F. M. (1880). *A Treatise on Comparative Embryology,* vol. 1. London: Macmillan.

Balfour, F. M. (1881). *A Treatise on Comparative Embryology,* vol. 2. London: Macmillan.

Banerjee, D., and Slack, F. (2002). Control of developmental timing by small temporal RNAs: A paradigm for RNA-mediated regulation of gene expression. *BioEssays* 24:119–129.

Bard, J. B. L. (1991). The fifth day of creation. *BioEssays* 12:303–306.

Bartolomaeus, T. (1994). On the ultrastructure of the coelomic lining in the Annelida, Sipuncula and Echiura. *Microfauna Marina* 9:171–220.

Bass, B. L. (2001). RNA Editing. In *Frontiers in Molecular Biology,* ed. B. D. Hames and D. M. Glover, 187. Oxford: Oxford University Press.

Bateson, W. (1894). *Materials for the Study of Variation Treated with Especial Regard to Discontinuity in the Origin of Species* (reprinted in 1992). Baltimore and London: Johns Hopkins University Press.

Bayer, M. M, and Todd, C. D. (1997). Evidence for zooid senescence in the marine bryozoan *Electra pilosa. Invertebrate Biology* 116:331–340.

Beeman, R. W. (1987). A homoeotic gene cluster in the red flour beetle. *Nature* 327:247–249.

Behara, N., and Nanjundiah, V. (1995). An investigation into the role of phenotypic plasticity in evolution. *Journal of Theoretical Biology* 172:225–234.

Beisson, J., and Sonneborn, T. M. (1965). Cytoplasmic inheritance of the organization of the cell cortex in *Paramecium aurelia. Proceedings of the National Academy of Sciences USA* 53:275–282.

Beldade, P.; P. M. Brakefield; and A. D. Long (2002). Contribution of *Distal-less* to quantitative variation in butterfly eyespots. *Nature* 415:315–318.

Bell, A. D. (1991). *Plant Form. An Illustrated Guide to Flowering Plant Morphology.* Oxford: Oxford University Press.

Bellairs, A. d'A., and C. Gans. (1983). A reinterpretation of the amphisbaenian orbitosphenoid. *Nature* 302:243–244.

Bennett, M. D. (1972). Nuclear DNA content and minimum generation time in herbaceous plants. *Proceedings of the Royal Society of London Series B* 181:109–135.

Bennett, M. D., and Leitch, I. J. (2001). Angiosperm DNA C-values database (release 3.1, Sept. 2001). http://www.rbgkew.org.uk/cval/homepage.html.

Bensasson, D.; D. A. Petrov; D.-X. Zhang; D. L. Hartl; and G. M. Hewitt (2001). Genomic gigantism: DNA loss is slow in mountain grasshoppers. *Molecular Biology and Evolution* 18:246–253.

Berger, C.; J. Urban; and G. M. Technau (2001). Stage-specific inductive signals in the *Drosophila* neuroectoderm control the temporal sequence of neuroblast specification. *Development* 128:3243–3251.

Berleth, T., and Sachs, T. (2001). Plant morphogenesis: Long-distance coordination and local patterning. *Current Opinion in Plant Biology* 4:57–62.

Bertalanffy, L. von (1960). Principles and theory of growth. In *Fundamental Aspects of Normal and Malignant Growth,* ed. W. Nowinski, 137–259. Amsterdam: Elsevier.

Bird, A. (1992). The essentials of DNA methylation. *Cell* 70:5–8.

Bird, A. (1995). Gene number, noise reduction, and biological complexity. *Trends in Genetics* 11:94–100.

Bisgrove, B. W., and Raff, R. A. (1989). Evolutionary conservation of the larval serotonergic nervous system in a direct-developing sea urchin. *Development, Growth & Differentiation* 31:363–370.

Blaustein, A. R.; P. D. Hoffman; D. G. Hokit; J. M. Kiesecker; S. C. Walls; and J. B. Hays (1994). UV repair and resistance to solar UV-B in amphibian eggs: A link to population declines? *Proceedings of the National Academy of Sciences, USA* 91:1791–1795.

Bohonak, A. J. (1999). Dispersal, gene flow, and population structure. *Quarterly Review of Biology* 74:21–45.

Bolker, J. A. (2000). Modularity in development and why it matters to evo-devo. *American Zoologist* 40:770–776.

Bolker, J. A., and Raff, R. A. (1996). Developmental genetics and traditional homology. *BioEssays* 18:489–494.

Bonner, J. T. (1965). *Size and Cycle.* Princeton: Princeton University Press.

Bonner, J. T. (1983). How behavior came to affect the evolution of body shape. *Scientia* 118:175–183.

Bonner, J. T. (1988). *The Evolution of Complexity by Means of Natural Selection.* Princeton: Princeton University Press.

Bookstein, F. (1978). *The Measurement of Biological Shape and Shape Change.* New York: Springer-Verlag.

Bookstein, F. (1989). Principal warps: Thin plate splines and the decomposition of deformations. *IEEE Transactions on Pattern Analysis and Machine Intelligence* 11:567–585.

Bookstein, F.; B. Chernoff; R. Elder; J. Humphries; G. Smith; and R. Strauss, eds. (1985). *Morphometrics in Evolutionary Biology.* Philadelphia: The Academy of Natural Sciences.

Boore, J. L., and Brown, W. M. (2000). Mitochondrial genomes of *Galathealinum, Helobdella,* and *Platynereis:* Sequence and gene arrangement comparisons indicate that Pogonophora is not a phylum and Annelida and Arthropoda are not sister taxa. *Molecular Biology and Evolution* 17:87–106.

Bouché, N., and Bouchez, D. (2001). *Arabidopsis* gene knockout: Phenotypes wanted. *Current Opinion in Plant Biology* 4:111–117.

Bowe, L. M.; G. Coat; and C. W. dePamphilis (2000). Phylogeny of seed plants based on all three genomic compartments: Extant gymnosperms are monophyletic and Gnetales' closest relatives are conifers. *Proceedings of the National Academy of Sciences, USA* 97:4092–4097.

Bowerman, B. (1999). Maternal control of polarity and patterning during embryogenesis in the nematode *Caenorhabditis elegans.* In *Cell Lineage and Fate Determination,* ed. S. A. Moody, 97–117. San Diego: Academic Press.

Bowler, P. J. (1983). *The Eclipse of Darwinism.* Baltimore: Johns Hopkins University Press.

Bowler, P. J. (1996). *Life's Splendid Drama.* Chicago: University of Chicago Press.

Boyer, B. C., and Henry, J. Q. (1998). Evolutionary modifications of the spiralian developmental program. *American Zoologist* 38:621–633.

Boyer, B. C.; J. Q. Henry; and M. Q. Martindale (1996). Dual origins of mesoderm in a basal spiralian: Cell lineage analyses in the polyclad turbellarian *Hoploplana inquilina. Developmental Biology* 179:329–338.

Bradshaw, A. D. (1965). Evolutionary significance of phenotypic plasticity in plants. *Advances in Genetics* 13:115–155.

Brakefield, P. M. (1996). Seasonal polyphenism in butterflies and natural selection. *Trends in Ecology and Evolution* 11:275–277.

Brakefield, P. M., and French, V. (1999). Butterfly wings: The evolution and development of colour patterns. *BioEssays* 21:391–401.

Brakefield, P. M., and Mazzotta, V. (1995). Matching field and laboratory environments: Effects of neglecting daily temperature variation on insect reaction norms. *Journal of Evolutionary Biology* 8:559–573.

Brakefield, P. M., and Reitsma, N. (1991). Phenotypic plasticity, seasonal climate and the population biology of *Bicyclus* butterflies. *Ecological Entomology* 16:291–303.

Brakefield P. M.; J. Gates; D. Keys; F. Kesbeke; P. J. Wijngaarden; A. Monteiro; V. French; and S. B. Carroll (1996). Development, plasticity and evolution of butterfly eyespot patterns. *Nature* 384:236–242.

Brakefield, P. M.; F. Kesbeke; and P. B. Koch (1998). The regulation of phenotypic plasticity of eyespots in the butterfly *Bicyclus anynana. American Naturalist* 152:853–860.

Brand, U.; J. C. Fletcher; M. Hobe; E. M. Meyerowitz; and R. Simon (2000). Dependence of stem cell fate in *Arabidopsis* on a feedback loop regulated by CLV3 activity. *Science* 289:617–619.

Brandon, R. (1982). The levels of selection. In *PSA 1982,* vol. 1, ed. P. Ashquit and T. Nickles, 315–324. East Lansing, Mich.: Philosophy of Science Association.

Brandon, R. N. (1999). The units of selection revisited: The modules of selection. *Biology and Philosophy* 14:167–180.

Branicky, R.; C. Bénard; and S. Hekimi (2000). *clk-1,* mitochondria, and physiological rates. *BioEssays* 22:48–56.

Bray, D. (2001). *Cell Movements: From Molecules to Motility,* 2nd ed. New York: Garland Publishing.

Bretschneider, T.; B. Vasiev; and C. J. Weijer (1999). A model for *Dictyostelium* slug movement. *Journal of Theoretical Biology* 199:125–136.

Brink, R. A. (1960). Paramutation and chromosome organization. *Quarterly Review of Biology* 35:120–137.

Brinkmann, R. (1929). Statistisch-biostratigraphische Untersuchungen an Miteljurassischen Ammoniten über Artbegriff und Stammesentwicklung. *Abhandlungen der Gesellschaft der Wissenschaften zu Göttingen, Mathematische-physische Klasse,* NF 13(3):1–330.

Broadbent, B. S.; B. J. Broadbent; and W. Golden (1975). *Bolton Standards of Dentofacial Developmental Growth.* St. Louis: Mosby.

Brockes, J. P. (1994). New approaches to amphibian limb regeneration. *Trends in Genetics* 10:169–173.

Bromham, L., and Degnan, B. M. (1999). Hemichordates and deuterostome evolution: Robust molecular phylogenetic support for a hemichordate + echinoderm clade. *Evolution & Development* 1:166–171.

Bromham, L.; A. Rambaut; R. Fortey; A. Cooper; and D. Penny (1998). Testing the Cambrian explosion hypothesis by using a molecular dating technique. *Proceedings of the National Academy of Sciences USA* 95:12386–12389.

Brøndsted, H. V. (1969). *Planarian Regeneration.* London: Pergamon Press.

Bronner-Fraser, M., and Fraser, S. E. (1988). Cell lineage analysis reveals multipotency of some avian neural crest cells. *Nature* 335:161–164.

Brooke, N. M.; J. Garcia-Fernandez; and P. W. H. Holland (1998). The ParaHox gene cluster is an evolutionary sister of the Hox gene cluster. *Nature* 392:920–922.

Brookfield, J. F. Y. (2000). Genomic sequencing: The complexity conundrum. *Current Biology* 10:R514–R515.

Brooks, D. R. (1990). Parsimony analysis in historical biogeography and coevolution: Methodological and theoretical update. *Systematic Zoology* 39:14–30.

Brooks, D. R., and McLennan, D. (1991). *Phylogeny, Ecology, and Behavior.* Chicago: University of Chicago Press.

Brown, C. J.; L. Carrel; and H. F. Willard (1997). Expression of genes from the human active and inactive X chromosomes. *American Journal of Human Genetics* 60:1333–1343.

Brown, S. W., and Chandra, H. S. (1977). Chromosome imprinting and the differential regulation of homologous chromosomes. In *Cell Biology: A Comprehensive Treatise,* vol. I, ed. L. Goldstein and D. M. Prescott, 109–89. New York: Academic Press.

Brown, W. M.; J. M. George; and A. C. Wilson (1979). The rapid evolution of animal mitochondrial DNA. *Proceedings of the National Academy of Sciences, USA* 76:1967–1971.

Brueggeman, P. (2001). Underwater Field Guide to Ross Island & McMurdo Sound, Antarctica. Version 11. http://scilib.ucsd.edu/sio/nsf/fguide/chordata1.html.

Brunetti, C. R.; J. E. Selegue; A. Monteiro; V. French; P. M. Brakefield; and S. B. Carroll (2001). The generation and diversification of butterfly eyespot color patterns. *Current Biology* 11:1578–1585.

Brunschwig, K.; C. Wittmann; R. Schnabel; T. R. Burglin; H. Tobler; and F. Muller (1999). Anterior organization of the *Caenorhabditis elegans* embryo by the labial-like Hox gene ceh-13. *Development* 126:1537–1546.

Brusca, R. C., and Brusca, G. J. (1990). *Invertebrates.* Sunderland, Mass.: Sinauer Associates.

Brush, A. H. (1996). On the origin of feathers. *Journal of Evolutionary Biology* 9:131–142.

Bry, L.; P. G. Falk; T. Midtvedt; and J. I. Gordon (1996). A model of host-microbial interactions in an open mammalian ecosystem. *Science* 273:1380–1383.

Brylski, P., and Hall, B. K. (1988). Ontogeny of a macroevolutionary phenotype: The external cheek pouches of geomyoid rodents. *Evolution* 42:391–395.

Budd, G. E. (in press). Arthropods as Ecdysozoa: The fossil evidence. In *Proceedings of the XVIII International Zoological Congress,* ed. T. Legakis.

Budd, G. E., and Jensen, S. (2000). A critical reappraisal of the fossil record of the bilaterian phyla. *Biological Reviews* 75:253–295.

Burd, M. (2001). Adaptation and constraint: Overview. In *Encyclopedia of Life Sciences,* http://www.els.com. London: Nature Publishing Group, Macmillan Publishers.

Burggren, W. W. (1999–2000). Developmental physiology, animal models, and the August Krogh Principle. *Zoology* (Jena) 102:148–156.

Burglin, T. R., and Ruvkun, G. (1993). The *Caenorhabditis elegans* homeobox gene cluster. *Current Opinion in Genetics and Development* 3:615–620.

Burian, R. M. (2000). On the internal dynamics of Mendelian genetics. *Comptes rendus de l'Académie des Sciences, Paris. Serie III, Sciences de la vie / Life Sciences* 324:1127–1137.

Burke, A. C. (2000). Hox genes and the global patterning of the somitic mesoderm. *Current Topics in Developmental Biology* 47:155–181.

Burke, A. C.; C. E. Nelson; B. A. Morgan; and C. Tabin (1995). Hox genes and the evolution of vertebrate axial morphology. *Development* 121:333–346.

Burla, H., and Taylor, C. E. (1982). Increase of phenotypic variance in stressful environments. *Journal of Heredity* 73:142.

Burt, A. (1989). Comparative methods using phylogenetically independent contrasts. *Oxford Surveys in Evolutionary Biology* 6:33–53.

Busch, M. A.; K. Bomblies; and D. Weigel (1999). Activation of a floral homeotic gene in *Arabidopsis. Science* 285:585–587.

Bush, G. L., and Smith, J. J. (1998). The genetics and ecology of sympatric speciation: A case study. *Research in Population Ecology* 40:175–187.

Buss, L. (1987). *The Evolution of Individuality.* Princeton: Princeton University Press.

Butler, A. B., and Saidel, W. M. (2000). Defining sameness: Historical, biological, and generative homology. *BioEssays* 22:846–853.

C. elegans sequencing consortium (1998). Genome sequence of the nematode *C. elegans:* A platform for investigating biology. *Science* 282:2012–2018.

Calder, W. A., III. (1984). *Size, Function, and Life History.* Cambridge: Harvard University Press.

Callahan, H. S.; M. Pigliucci; and C. D. Schlichting (1997). Developmental phenotypic plasticity: Where ecology and evolution meet molecular biology. *BioEssays* 19:519–525.

Callery, E. M., and Elinson, R. P. (2000a). Opercular development and ontogenetic re-organization in a direct-developing frog. *Development, Genes & Evolution* 210:377–381.

Callery, E. M., and Elinson, R. P. (2000b). Thyroid hormone-dependent metamorphosis in a direct-developing frog. *Proceedings of the National Academy of Sciences, USA* 97:2615–2620.

Cameron, R. A.; G. Mahairas; J. P. Rast; P. Martinez; T. R. Biondi; S. Swartzell; J. C. Wallace; A. J. Poustka; B. T. Livingston; G. A. Wray; C. A. Ettensohn; H. Lehrach; R. J. Britten; E. H. Davidson; and L. A. Hood (2000). A sea urchin genome project: Sequence scan, virtual map, and additional resources. *Proceedings of the National Academy of Sciences, USA* 97:9514–9518.

Campbell, K.; J. McWhir; W. Ritchie; and I. Wilmut (1996). Sheep cloned by nuclear transfer from a cultured cell line. *Nature* 380:64–66.

Capecchi, M. R. (1997). Hox genes and mammalian development. *Cold Spring Harbor Symposium on Quantitative Biology* 62:273–281.

Caplan, A. I. (1990). Mesenchymal stem cells. *Journal of Orthopaedic Research* 9:641–650.

Carey, J. C.; R. M. Fineman; and F. A. Ziter (1982). The Robin sequence as a consequence of malformation, dysplasia, and neuromuscular syndromes. *Journal of Pediatrics* 101:858–864.

Carey, J. R. (2001). Demographic mechanism for the evolution of long life in social insects. *Experimental Gerontology* 36:713–722.

Carlson, B. M., and Faulkner, J. A. (1996). The regeneration of noninnervated muscle grafts and marcaine-treated muscles in young and old rats. *Journal of Gerontology* 51B: 43–49.

Carpenter, K. (2001). *Eggs, Nests, and Baby Dinosaurs.* Bloomington: Indiana University Press.

Carr, J.; C. Shashikant; W. J. Bailey; and F. Ruddle (1998). Molecular evolution of Hox gene regulation: Cloning and transgenic analysis of the lamprey HoxQ8 gene. *Journal of Experimental Zoology* 280:73–85.

Carroll, S. B.; S. D. Weatherbee; and J. A. Langeland (1995). Homeotic genes and the regulation and evolution of insect wing number. *Nature* 375:58–61.

Carroll, S. B.; J. K. Grenier; and S. D. Weatherbee (2001). *From DNA to Diversity: Molecular Genetics and the Evolution of Animal Design.* Malden, Mass.: Blackwell Science.

Carter, D. R.; M. Wong; and T. E. Orr (1991). Musculoskeletal ontogeny, phylogeny, and functional adaptation. *Journal of Biomechanics* (Suppl) 24: 1:3–16.

Carter J. G. (ed.) (1990). *Skeletal Biomineralization: Patterns, Processes and Evolutionary Trends,* vol. 1. New York: Van Nostrand Reinhold.

Case, T. J. (1978). On the evolution and adaptive significance of postnatal growth rates in the terrestrial vertebrates. *The Quarterly Review of Biology* 53:243–280.

Castanet, J.; H. Francillon-Vieillot; F. J. Meunier; and A. de Ricqlès (1993). Bone and individual aging. In *Bone, Volume 7: Bone Growth,* ed. B. K. Hall, 245–283. London: CRC Press.

Castanet, J.; A. Grandin; A. Abourachid; and A. de Ricqlès (1996). Expression de la dynamique de croissance dans la structure de l'os périostique chez *Anas plathyrhynchos. Comptes Rendus de l'Académie des Sciences, Paris* 319:301–308.

Cavalier-Smith, T. (1985). Introduction: The evolutionary significance of genome size. In *The Evolution of Genome Size,* ed. T. Cavalier-Smith, 1–36. New York: John Wiley & Sons.

Cecchi, C.; A. Mallamaci; and E. Boncinelli (2000). Otx and Emx homeobox genes in brain development. *International Journal of Developmental Biology* 44:663–668.

Chadwick, D. J., and Cardew, G., eds. (1998). *Epigenetics* (Novartis Foundation Symposium 214). Chichester, UK: John Wiley & Sons.

Chapman, M.; D. Michael; and L. Margulis (2000). Centrioles and kinetosomes: Form, function and evolution. *Quarterly Review of Biology* 75:409–429.

Charlesworth, B., and Charlesworth, D. (1999). The genetic basis of inbreeding depression. *Genetic Research* 74:329–340.

Charnov, E. L. (1993). *Life History Invariants: Some Explorations of Symmetry in Evolutionary Ecology.* New York: Oxford University Press.

Chauvet, S.; S. Merabet; D. Bilder; M. P. Scott; J. Pradel; and Y. Graba (2000). Distinct Hox protein sequences determine specificity in different tissues. *Proceedings of the National Academy of Sciences, USA* 97:4064–4069.

Chen, C. S.; M. Mrksich; S. Huang; G. M. Whitesides; and D. E. Ingber (1997). Geometric control of cell life and death. *Science* 276:1425–1428.

Chen, F., and Capecchi, M. R. (1999). Paralogous mouse Hox genes, Hoxa9, Hoxb9, and Hoxd9, function together to control development of the mammary gland in response to pregnancy. *Proceedings of the National Academy of Sciences, USA* 96:541–546.

Chen, J. Y.; P. Oliveri; C. W. Li; G. Q. Zhou; F. Gao; J. W. Hagadorn; K. J. Peterson; and E. H. Davidson (2000). Precambrian animal diversity: Putative phosphatized embryos from the Doushantuo formation of China. *Proceedings of the National Academy of Sciences, USA* 97:4457–4462.

Chen, Y., and Schier, A. F. (2001). The zebrafish Nodal signal Squint functions as a morphogen. *Nature* 411:607–610.

Chernoff, B., and Magwene, P. (1999). Afterward. In *Morphological Integration,* ed. E. Olson and R. Miller, 319–353. Chicago: University of Chicago Press.

Cheshier, S.; S. Morrison; X. Liao; and I. Weissman (1999). *In vivo* proliferation and cell cycle kinetics of long-term self-renewing hematopoietic stem cells. *Proceedings of the National Academy of Sciences, USA* 96:3120–3125.

Cheverud, J. M. (1982). Phenotypic, genetic, and environmental morphological integration in the cranium. *Evolution* 36:499–516.

Cheverud, J. M. (1984). Quantitative genetics and developmental constraints on evolution by selection. *Journal of Theoretical Biology* 110:155–171.

Cheverud, J. M. (1988). The evolution of genetic correlation and developmental constraints. In *Population Genetics and Evolution,* ed. G. de Jong, 94–101. Berlin: Springer-Verlag.

Cheverud, J. M. (1989). A comparative analysis of morphological variation patterns in the papionins. *Evolution* 43:1737–1747.

Cheverud, J. M. (1996). Developmental integration and the evolution of pleiotropy. *American Zoologist* 36:44–50.

Cheverud, J. M., and Dow, M. (1985). An autocorrelation analysis of the effect of lineal fission on genetic variation among social groups. *American Journal of Physical Anthropology* 67:113–121.

Cheverud, J. M., and Richtsmeier, J. (1986). Finite element scaling applied to sexual dimorphism in rhesus macaque *(Macaca mulatta)* facial growth. *Systematic Zoology* 35:109–128.

Cheverud, J. M.; J. Lewis; W. Bachrach; and W. Lew (1983). The measurement of form and variation in form: An application of three-dimensional quantitative morphology by finite-element methods. *American Journal of Physical Anthropology* 62:151–165.

Chipman, A. D.; A. Haas; E. Tchernov; and O. Khaner (2000). Variation in anuran embryogenesis: differences in sequence and timing of early developmental events. *Journal of Experimental Zoology (Molecular and Developmental Evolution)* 288:352–365.

Chuong, C-M., ed. (1998). *Molecular Basis of Epithelial Appendage Morphogenesis*. Austin: R. G. Landes.

Chuong, C-M.; R. B. Widelitz; S. Ting-Berreth; and T. X. Jiang (1996). Early events during avian skin appendage regeneration: Dependence on epithelial-mesenchymal interaction and order of molecular reappearance. *Journal of Investigative Dermatology* 107:639–646.

Cifelli, R. L. (2001). Early mammalian radiations. *Journal of Paleontology* 75:1214–1226.

Clarke, G. M. (1998). The genetic basis of developmental stability. 4. Inter- and intra-individual character variation. *Heredity* 80:562–567.

Claus, C., and Grobben, K. (1917). *Lehrbuch der Zoologie*. Marbug: Elwert'sche verlagsbuchhandlung.

Clearwater, M. J., and Gould, K. S. (1994). Comparative leaf development of juvenile and adult *Pseudopanax crassifolius*. *Canadian Journal of Botany* 72:658–670.

Coen, E. S., and Meyerowitz, E. M. (1991). The war of the whorls: genetic interactions controlling flower development. *Nature* 353:31–37.

Coen, E. S.; J. M. Romero; S. Doyle; R. Elliot; G. Murphy; and R. Carpenter (1990). *floricaula:* A homeotic gene required for flower development in *Antirrhinum majus*. *Cell* 63:1311–1322.

Cohen, J., and Massey, B. (1982). *Living Embryos*. Oxford: Pergamon.

Cohn, M. J., and Tickle, C. (1999). Developmental basis of limblessness and axial patterning in snakes. *Nature* 399:474–479.

Cohn, M. J.; K. Patel; R. Krumlauf; D. G. Wilkinson; J. D. W. Clarke; and C. Tickle (1997). Hox9 genes and vertebrate limb specification. *Nature* 387:97–101.

Colbert, M. W., and Rowe, T. (2001). Ontogenetic sequence analysis: Using parsimony to characterize developmental hierarchies. *Journal of Morphology* 248:218.

Cole, T., III (1996). Historical note: Early anthropological contributions to "geometric morphometrics." *American Journal of Physical Anthropology* 101:291–296.

Cole, T., III; S. Lele; and J. Richtsmeier (2002). A parametric bootstrap approach to the detection of phylogenetic signals in landmark data. In *Morphometrics: Shape and Phylogenetics*, ed. N. MacLeod and P. Forey, 194–219. London: Taylor and Francis.

Coleman, W. (1964). *Georges Cuvier, zoologist; A study in the history of evolution theory*. Cambridge, Mass.: Harvard University Press.

Collazo, A. (2000). Developmental variation, homology, and the pharyngula stage. *Systematic Biology* 49:3–18.

Collazo, A., and Fraser, S. E. (1996). Integrating cellular and molecular approaches into studies of development and evolution: The issue of morphological homology. *Aliso* 14:237–262.

Collazo, A.; M. Bronner-Fraser; and S. E. Fraser (1993). Vital dye labelling of *Xenopus laevis* trunk neural crest reveals multipotency and novel pathways of migration. *Development* 118:363–376.

Collazo, A.; J. A. Bolker; and R. Keller (1994). A phylogenetic perspective on teleost gastrulation. *American Naturalist* 144:133–152.

Collin, R. (2001). The effects of mode of development on phylogeography and population structure of North Atlantic *Crepidula* (Gastropoda: Calyptraeidae). *Molecular Ecology* 10:2249–2262.

Collins, A. G. (1998). Evaluating multiple alternative hypotheses for the origin of Bilateria: An analysis of 18S rRNA molecules. *Proceedings of the National Academy of Sciences, USA* 95:15458–15463.

Condie, B. G., and Capecchi, M. R. (1993). Mice with targeted disruptions in the paralogous genes hoxa-3 and hoxd-3 reveal synergistic interactions. *Nature* 370:304–307.

Conklin, E. G. (1897). The embryology of *Crepidula,* a contribution to the cell lineage and early development of some marine gastropods. *Journal of Morphology* 13:1–266.

Conklin, E. G. (1905). Mosaic development in ascidian eggs. *Journal of Experimental Zoology* 2:145–223.

Conway Morris, S. (1998). *The Crucible of Creation.* Oxford: Oxford University Press.

Conway Morris, S., and Peel, J. S. (1995). Articulated halkieriids from the Lower Cambrian of North Greenland and their role in early protostome evolution. *Philosophical Transactions of the Royal Society of London B* 347:305–358.

Cook, C. E.; M. L. Smith; M. J. Telford; A. Bastianello; and M. Akam (2001). Hox genes and the phylogeny of the arthropods. *Current Biology* 11:759–763.

Cope, E. D. (1889). On inheritance in evolution. *American Naturalist* 23:1058–1071.

Coulier, F.; C. Popovici; R. Villet; and D. Birnbaum (2000). MetaHox gene clusters. *Journal of Experimental Zoology (Molecular and Developmental Evolution)* 288:345–351.

Coy, J. F.; Z. Sedlacek; D. Bachner; H. Delius; and A. Poustka (1999). A complex pattern of evolutionary conservation and alternative polyadenylation within the long 3′-untranslated region of the methyl-CpG-binding protein 2 gene (MECP2) suggests a regulatory role in gene expression. *Human Molecular Genetics* 8:1253–1262.

Coyne, J. A. (1983). Genetic basis of differences in genital morphology among three sibling species of *Drosophila. Evolution* 37:1101–1118.

Crespi, B. J. (1989). Facultative viviparity in a thrips. *Nature* 337:357–358.

Cronin, D. G., and Hunter, W. S. (1980). Craniofacial morphology in twins discordant for cleft lip and/or palate. *Cleft Palate Journal* 17:116–126.

Cubas, P.; C. Vincent; and E. Coen (1999). An epigenetic mutation responsible for natural variation in floral symmetry. *Nature* 401:157–161.

Cusick, F. (1966). On phylogenetic and ontogenetic fusions. In *Trends in Plant Morphogenesis,* ed. E. G. Cutter, 170–183. London: Longmans.

Cuvier, G. (1812). *Recherches sur les Ossemens Fossiles.* Paris: Déterville.

Czerny, T.; G. Halder; U. Kloter; A. Souabni; W. J. Gehring; and M. Busslinger (1999). *twin of eyeless,* a second Pax-6 gene of *Drosophila,* acts upstream of *eyeless* in the control of eye development. *Molecular Cell* 3:297–307.

Dahms, H.-U. (2000). Phylogenetic implications of the crustacean nauplius. *Hydrobiologia* 417:91–99.

Damen, W. G. M.; M. Weller; and D. Tautz (2000). Expression patterns of *hairy, even-skipped,* and *runt* in the spider *Cupiennius salei* imply that these genes were segmentation genes in a basal arthropod. *Proceedings of the National Academy of Sciences, USA* 97:4515–4519.

Darwin, C. (1859). *On the Origin of Species by Means of Natural Selection, or the Preservation of Favoured Races in the Struggle for Life.* London: John Murray. (A facsimile of the first edition, published by Harvard University Press in 1964.)

Darwin, C. (1874). *The Descent of Man and Selection in Relation to Sex,* 2nd ed. London: John Murray.

Davidson, E. H. (1990). How embryos work: A comparative view of diverse modes of cell fate specification. *Development* 108:365–389.

Davidson, E. H. (1991). Spatial mechanisms of gene regulation in metazoan embryos. *Development* 113:1–26.

Davidson, E. H. (2001). *Genomic Regulatory Systems: Development and Evolution.* San Diego: Academic Press.

Davidson, E. H., and Ruvkun, G. (1999). Themes from a NASA workshop on gene regulatory processes in development and evolution. *Journal of Experimental Zoology* 285:104–115.

Davidson, E. H.; K. J. Peterson; and R. A. Cameron (1995). Origin of bilaterian body plans—evolution of developmental regulatory mechanisms. *Science* 270:1319–1325.

Davidson, L. A.; M. A. R. Koehl; R. Keller; and G. F. Oster (1995). How do sea urchins invaginate: Using biomechanics to distinguish between mechanisms of primary invagination. *Development* 121:2005–2018.

Davis, A. P.; D. P. Witte; H. M. Hsieh-Li; S. S. Potter; and M. R. Capecchi (1995). Absence of radius and ulna in mice lacking hoxa-11 and hoxd-11. *Nature* 375:791–795.

Davis, G. K. (1999). The origin and evolution of segmentation. *Trends in Cell Biology* 9: M68–72.

Davis, G. K., and Patel, N. H. (2002). Short, long, and beyond: Molecular and embryological approaches to insect segmentation. *Annual Review of Entomology* 47:669–699.

Dawkins, R. (1976). *The Selfish Gene.* Oxford: Oxford University Press.

Dawkins, R. (1982). *The Extended Phenotype.* Oxford: Oxford University Press.

Dawkins, R. (1986). *The Blind Watchmaker.* London: Longman.

Dawkins, R. (1989). *The Selfish Gene,* New edition. Oxford: Oxford University Press.

De Beer, G. R. (1930). *Embryology and Evolution.* Oxford: Clarendon Press.

de Duve, C. (1996). The birth of complex cells. *Scientific American* 274(4):50–57.

de Gennes, P. G. (1992). Soft matter. *Science* 256:495–497.

DeLeon, V.; M. Zumpano; and J. Richtsmeier (2001). The effect of neurocranial surgery on basicranial morphology in isolated sagittal craniosynostosis. *Cleft Palate-Craniofacial Journal* 38:134–146.

De Robertis, E. M., and Sasai, Y. (1996). A common plan for dorsoventral patterning in Bilateria. *Nature* 380:37–40.

De Robertis E. M.; A. Fainsod; L. K. Gont; and H. Steinbeisser (1994). The evolution of vertebrate gastrulation. *Development* (Suppl): 117–124.

de Rosa, R.; J. K. Grenier; T. Andreevas; C. E. Cook; A. Adoutte; M. Akam; S. B. Carroll; and G. Balavoine (1999). Hox genes in brachiopods and priapulids and protostome evolution. *Nature* 399:772–776.

DeWitt, T. J. (1998). Costs and limits of phenotypic plasticity: Tests with predator-induced morphology and life-history in a freshwater snail. *Journal of Evolutionary Biology* 11:465–480.

DeWitt, T. J.; A. Sih; and D. S. Wilson (1998). Costs and limits of phenotypic plasticity. *Trends in Ecology and Evolution* 13:77–81.

Debat, V., and David, P. (2001). Mapping phenotypes: Canalization, plasticity and developmental stability. *Trends in Ecology and Evolution* 16:555–561.

Denenberg, V. H. (1969). The effects of early experience. In *The Behaviour of Domestic Animals,* 2nd ed., ed. E. S. E. Hafez, 95–130. Baltimore: Williams and Wilkins.

Denenberg, V. H., and Rosenberg, K. M. (1967). Nongenetic transmission of information. *Nature* 216:549–550.

Depéret, C. (1907). *Les Transformations du Monde Animal.* Paris: Flammarion.

Desalle, R. and Grimaldi, D. (1993). Phylogenetic pattern and developmental process in *Drosophila. Systematic Biology* 42:458–475.

Deschamps, J.; E. van den Akker; S. Forlani; W. De Graaff; T. Oosterveen; B. Roelen; and J. Roelfsema (1999). Initiation, establishment and maintenance of Hox gene expression patterns in the mouse. *International Journal of Developmental Biology* 43:635–650.

Desmond, A. (1979). Designing the dinosaur: Richard Owen's response to Robert Edmond Grant. *Isis* 70:224–234.

Desmond, A. (1982). *Archetypes and Ancestors.* Chicago: University of Chicago Press.

Detlaff, T. A., and Detlaff, A. A. (1961). On relative dimensionless characteristics of the development duration in embryology. *Archives of Biology* 72: 1–16.

Dhawan, R. R.; T. J. Schoen; and D. C. Beebe (1997). Isolation and expression of homeobox genes from the embryonic chicken eye. *Molecular Vision* 3:7–14.

Dhouailly, D. (1973). Dermo-epidermal interactions between birds and mammals: Differentiation of cutaneous appendages. *Journal of Embryology and Experimental Morphology* 30:587–603.

Di Gregorio, A.; A. Spagnuolo; F. Ristoratore; M. Pischetola; F. Aniello; M. Branno; L. Cariello; and R. Di Lauro (1995). Cloning of ascidian homeobox genes provides evidence for a primordial chordate cluster. *Gene* 156:253–257.

Dickinson, W. J. (1995). Molecules and morphology: Where's the homology? *Trends in Genetics* 11:119–121.

Dieckmann, U., and Doebeli, M. (1999). On the origin of species by sympatric speciation. *Nature* 400:354–357.

Diggle, P. (1994). The expression of andromonoecy in *Solanum hirtum* (Solanaceae): Phenotypic plasticity and ontogenetic contingency. *American Journal of Botany* 81:1354–1365.

Diggle, P.; K-Y. Liang; and S. Zeger (1994). *Analysis of Longitudinal Data.* Oxford: Oxford University Press.

DiNardo, S.; J. Heemskerk; S. Dougan; and P. H. O'Farrell (1994). The making of a maggot: Patterning the *Drosophila* embryonic epidermis. *Current Opinion in Genetics and Development* 4:529–534.

Dingle, H., and Hegmann, J. (1982). *Evolution and Genetics of Life Histories.* Berlin: Springer-Verlag.

Dixon, K. E. (1994). Evolutionary aspects of primordial germ cell formation. *Ciba Foundation Symposium* 182:92–120.

Dobzhansky, T. (1937). *Genetics and the Origin of Species.* New York: Columbia University Press.

Dobzhansky, T. (1951). *Genetics and the Origin of Species,* 3rd ed. New York: Columbia University Press.

Donoghue, M. J. (1989). Phylogenies and analysis of evolutionary sequences, with examples from seed plants. *Evolution* 43:1137–1156.

Donovan, D. T. (1973). The influence of theoretical ideas on ammonite classification from Hyatt to Trueman. *Paleontological Contributions of the University of Kansas* 62:1–16.

Doolittle, W. F., and Sapienza, C. (1980). Selfish genes, the phenotype paradigm and genome evolution. *Nature* 284:601–603.

Dover, G. (2000). How genomic and developmental dynamics affect evolutionary processes. *BioEssays* 22:1153–1159.

Driscoll, D. J., and Migeon, B. R. (1990). Sex difference in methylation of single copy genes in human meiotic germ cells. *Somatic Cell and Molecular Genetics* 16:267–268.

Duboule, D. (1994). Temporal colinearity and the phylotypic progression: A basis for the stability of a vertebrate Bauplan and the evolution of morphologies through heterochrony. *Development* (Supp): 1994:135–142.

Duboule, D. (1998). Vertebrate hox genes: Clustering and/or colinearity? *Current Opinions in Genetics and Development* 8:514–518.

Duboule, D., and Morata, G. (1994). Colinearity and functional hierarchy among genes of the homeotic complexes. *Trends in Genetics* 10:358–364.

Duboule, D., and Wilkins, A. S. (1998). The evolution of "bricolage." *Trends in Genetics* 14:54–59.

Dubrulle, J.; M. J. McGrew; and O. Pourquie (2001). FGF signaling controls somite boundary position and regulates segmentation clock control of spatiotemporal Hox gene activation. *Cell* 106:219–232.

Duda, T. F., and Palumbi, S. R. (1999). Developmental shifts and species selection in gastropods. *Proceedings of the National Academy of Sciences, USA* 96:10272–10277.

Dunnett, D.; A. Goodbody; and M. Stanisstreet (1991). Computer modelling of neural tube defects. *Acta Biotheoretica* 39:63–79.

Dyer, B., and Obar, R. (1994). *Tracing the History of Eukaryotic Cells.* New York: Columbia University Press.

Eberhard, W. G. (2001). Multiple origins of a major novelty: Moveable abdominal lobes in male sepsid flies (Diptera: *Sepsidae*) and the question of developmental constraints. *Evolution & Development* 3:206–222.

Echeverri, K.; J. D. W. Clarke; and E. Tanaka (2001). *In vivo* imaging indicates muscle fiber dedifferentiation is a major contributor to the regenerating tail blastema. *Developmental Biology* 236:151–164.

Eddy, E. M. (1975). Germ plasm and the differentiation of the germ cell line. *International Review of Cytology* 43:229–281.

Eguchi, G., and Okada, T. (1973). Differentiation of lens tissue from the progeny of chick retinal pigment cells cultured *in vitro:* A demonstration of a switch of cell types in clonal cell culture. *Proceedings of the National Academy of Sciences, USA* 70:1495–1499.

Eicher, E. M., and Washburn, L. (1989). Normal testis determination in the mouse depends on genetic interaction of a locus on chromosome 17 and the Y chromosome. *Genetics* 123:173–179.

Eldredge, N., and Gould, S. J. (1972). Punctuated equilibria: An alternative to phyletic gradualism. In *Models in Paleontology,* ed. T. J. M. Schopf, 82–115. San Francisco: Freeman.

Elinson, R. P. (1987). Change in developmental patterns: Embryos of amphibians with large eggs. In *Development as an Evolutionary Process,* ed. R. A. Raff and E. C. Raff, 1–21. New York: Alan R. Liss.

Elinson, R. P. (1990). Direct development in frogs: Wiping the recapitulationist slate clean. *Seminars in Developmental Biology* 1:263–270.

Eliot, T. S. (1930). Ash-Wednesday. In *The Waste Land and Other Poems* (1990), 53. London: Faber and Faber Limited.

Elul, T., and Keller, R. (2000). Monopolar protrusive activity: A new morphogenetic cell behavior in the neural plate dependent on vertical interactions with the mesoderm in *Xenopus. Developmental Biology* 224:3–19.

Emerson, A. E. (1958). The evolution of behavior in social insects. In *Behavior and Evolution,* ed. A. E. Roe and G. G. Simpson, 311–335. New Haven: Yale University Press.

Emlen, D. J. (2001). Costs and the diversification of exaggerated animal structures. *Science* 291:1534–1536.

Emlet, R. B. (1989). Apical skeletons of sea urchins (Echinodermata, Echinoidea): Two methods for inferring mode of larval development. *Paleobiology* 15:223–256.

Endler, J. A. (1986). *Natural Selection in the Wild.* Princeton: Princeton University Press.

Endler, J. A. (1992). Natural selection: Current usages. In *Keywords in Evolutionary Biology,* ed. E. F. Keller and L. Lloyd, 220–224. Cambridge, Mass.: Harvard University Press.

Endo, K.; T. Masaki; and K. Kumagai (1988). Neuroendocrine regulation of the development of seasonal morphs in the Asian comma butterfly, *Polygonia c-aureum* L.: Difference in activity of summer-morph-producing hormone from brain extracts of the long-day and short-day pupae. *Zoological Science* 5:145–152.

Endress, P. K.; P. Baas; and M. Gregory (2000). Systematic plant morphology and anatomy: Fifty years of progress. *Taxon* 49:401–434.

Erickson, G.; K. Curry-Rogers; and S. A. Yerby (2001). Dinosaurian growth patterns and rapid avian growth rates. *Nature* 412:429–433.

Erwin, D. H. (1993). Early introduction of major morphological innovations. *Acta Palaeontologica Polonica* 38:281–294.

Erwin, D. H. (1999). The origin of body plans. *American Zoologist* 39:617–629.

Erwin, D. H. (2000). Macroevolution is more than repeated rounds of microevolution. *Evolution & Development* 2:78–84.

Eshel, I., and Matessi, C. (1998). Canalization, genetic assimilation and preadaptation: A quantitative genetic model. *Genetics* 149:2119–2133.

Evans, J. D., and Wheeler, D. E. (2001). Gene expression and the evolution of insect polyphenisms. *BioEssays* 23:62–68.

Evans, S. M.; W.Yan; M. P. Murillo; J. Ponce; and N. Papalopulu (1995). Tinman, a *Drosophila* Homeobox gene required for heart and visceral mesoderm specification, may be represented by a family of genes in vertebrates—*Xnkx-2.3,* a 2nd vertebrate homolog of *Tinman. Development* 121:3889–3899.

Falciani, F.; B. Hausdorf; R. Schroder; M. Akam; D. Tautz; R. Denell; and S. Brown (1996). Class 3 Hox genes in insects and the origin of zen. *Proceedings of the National Academy of Sciences, USA* 93:8479–8484.

Falconer, D. S., and Mackay, T. F. C. (1996). *Introduction to Quantitative Genetics.* Harlow: Longman.

Falsetti, A. B., and Cole, T. M. (1992). Relative growth of the postcranial skeleton in callitrichines. *Journal of Human Evolution* 23:79–92.

Fan, H.-Y.; Y. Hu; M. Tudor; and H. Ma (1997). Specific interactions between the K domains of AG and AGLs, members of the MADS domain family of DNA binding proteins. *Plant Journal* 12:999–1010.

Fang, H., and Elinson, R. P. (1996). Patterns of distal-less gene expression and inductive interactions in the head of the direct developing frog *Eleutherodactylus coqui. Developmental Biology* 179:160–172.

Fang, H., and Elinson, R. P. (1999). Evolutionary alteration in anterior patterning: *Otx2* expression in the direct developing frog *Eleutherodactylus coqui. Developmental Biology* 205:233–239.

Fang H.; Y. Marikawa; and R. P. Elinson (2000). Ectopic expression of *Xenopus* noggin RNA induces complete secondary body axes in embryos of the direct developing frog *Eleutherodactylus coqui. Development, Genes & Evolution* 210:21–27.

Favier, B., and Dolle, P. (1997). Developmental functions of mammalian Hox genes. *Molecular Human Reproduction* 3:115–131.

Feder, J. L.; J. B. Roethele; B. Wlazlo; and S. H. Berlocher (1997). Selective maintenance of allozyme differences among sympatric host races of the apple maggot fly. *Proceedings of the National Academy of Sciences, USA* 94:11417–11421.

Felix, M. A.; P. De Ley; R. J. Sommer; L. Frisse; S. A. Nadler; W. K. Thomas; J. Vanfleteren; and P. W. Sternberg (2000). Evolution of vulva development in the Cephalobina (Nemotoda). *Developmental Biology* 221:68–86.

Felsenstein, J. (1978). Cases in which parsimony and compatibility will be positively misleading. *Systematic Zoology* 27:401–410.

Felsenstein, J. (1985). Phylogenies and the comparative method. *American Naturalist* 125:1–15.

Ferguson, E. L. (1996). Conservation of dorsal-ventral patterning in arthropods and chordates. *Current Opinion in Genetics and Development* 6:424–431.

Ferkowicz, M. J., and Raff, R. A. (2001). Wnt gene expression in sea urchin development: Heterochronies associated with the evolution of developmental mode. *Evolution & Development* 3:24–33.

Ferreira, H. B.; Y. Zhang; C. Zhao; and S. W. Emmons (1999). Patterning of *Caenorhabditis elegans* posterior structures by the Abdominal-B homolog, egl-5. *Developmental Biology* 207:215–228.

Ferrier, D. E. K., and Holland, P. W. H. (2001). Ancient origin of the Hox gene cluster. *Nature Reviews, Genetics* 2:33–38.

Ferrier, D. E. K.; C. Minguillon; P. W. H. Holland; and J. Garcia-Fernandez (2000). The amphioxus Hox cluster: Deuterostome posterior flexibility and Hox14. *Evolution & Development* 2:284–293.

Farris, J. S. (1970). Methods for computing Wagner trees. *Systematic Zoology* 19:83–92.

Farris, J. S. (1983). The logical basis of phylogenetic analysis. In *Advances in Cladistics,* vol 2., ed. N. Platnick and V. Funk, 7–36. New York: Columbia University Press.

Fink, W. L. (1982). The conceptual relationship between ontogeny and phylogeny. *Paleobiology* 8:254–264.

Finnerty, J. R., and Martindale, M. (1998). The evolution of the Hox cluster: Insights from outgroups. *Current Opinion in Genetics and Development* 8:681–687.

Finnerty, J. R., and Martindale, M. Q. (1999). Ancient origins of axial patterning genes: Hox genes and ParaHox genes in the Cnidaria. *Evolution & Development* 1:16–23.

Fisher, D. C. (1986). Progress in organismal design. In *Patterns and Processes in the History of Life,* ed. D. M. Raup and D. Jablonski, 99–117. Berlin: Springer.

Fisher, R. A. (1918). The correlation between relatives on the supposition of Mendelian inheritance. *Journal of the Royal Society of Edinburgh* 52:399–432.

Fisher, R. A. (1937). The relation between variability and abundance shown by the measurements of the eggs of British nesting birds. *Proceedings of the Royal Society of London (B)* 122:1–26.

Fisher, R. A., and Ford, E. B. (1928). The variability of species in the Lepidoptera, with reference to abundance and sex. *Transactions of the Entomological Society of London* 76:367–384.

Fitch, W. M. (2000). Homology, a personal view on some of the problems. *Trends in Genetics* 16:227–231.

Fitzhugh, K. (1997). The abduction of cladistics. *Cladistics* 13:170–171.

Foote, M. (1993). Discordance and concordance between morphological and taxonomic diversity. *Paleobiology* 19:185–204.

Foote, M. (1997). The evolution of morphological diversity. *Annual Review of Ecology and Systematics* 28:129–152.

Foote, M.; J. P. Hunter; C. M. Janis; and J. Sepkoski (1999). Evolutionary and preservational constraints on origins of biologic groups: Divergence times of eutherian mammals. *Science* 283:1310–1314.

Frankel, J. (1989). *Pattern Formation: Ciliate Studies and Models.* New York: Oxford University Press.

Frankel, J. (1992). The patterning of ciliates. *Journal of Protozoology* 38:519–525.

Fraser, S. E., and Bronner-Fraser, M. (1991). Migrating neural crest cells in the trunk of the avian embryo are multipotent. *Development* 112:913–920.

Freeman, G. (1979). The multiple roles which cell division can play in the localization of developmental potential. In *Determinants of Spatial Organization,* ed. S. Subtleny and I. Konigsberg, 53–76. New York: Academic Press.

Freeman, G., and Lundelius, J. W. (1992). Evolutionary implications of the mode of D quadrant specification in coelomates with spiral cleavage. *Journal of Evolutionary Biology* 5:205–247.

French, V. (1986). Interaction between the leg and surrounding thorax in the beetle. *Journal of Embryology and Experimental Morphology* 91:227–250.

French, V. (2001). Insect segmentation: Genes, stripes and segments in "Hoppers." *Current Biology* 11:R910–R913.

Frohlich, M. W. (2001). A detailed scenario and possible tests of the Mostly Male theory of flower evolutionary origins. In *Beyond Heterochrony: The Evolution of Development,* ed. M. L. Zelditch, 59–104. New York: John Wiley & Sons.

Frohlich, M. W., and Parker, D. S. (2000). The Mostly Male theory of flower evolutionary origins. *Systematic Botany* 25:155–170.

Funk, V. A., and Brooks, D. R. (1990). *Phylogenetic Systematics as the Basis of Comparative Biology.* Washington: Smithsonian Institution Press.

Fusco, G. (2001). How many processes are responsible for phenotypic evolution? *Evolution & Development* 3:279–286.

Futuyma, D. J. (1979). *Evolutionary Biology.* Sunderland, Mass.: Sinauer.

Futuyma, D. J. (1988). Sturm und Drang and the evolutionary synthesis. *Evolution* 42:217–226.

Futuyma, D. J. (1998). *Evolutionary Biology,* 3rd ed. Sunderland, Mass.: Sinauer.

Gaeth, A. P.; R. V. Short; and M. B. Renfree (1999). The developing renal, reproductive, and respiratory systems of the African elephant suggest an aquatic ancestry. *Proceedings of the National Academy of Sciences, USA* 96:5555–5558.

Galant, R., and Carroll, S. B. (2002). Evolution of a transcriptional repression domain in an insect Hox protein. *Nature* 415:910–913.
Galis, F. (1999). Why do almost all mammals have seven cervical vertebrae? Developmental constraints, *Hox* genes, and cancer. *Journal of Experimental Zoology (Molecular and Developmental Evolution)* 285:19–26.
Galis, F., and Drucker, E. G. (1996). Pharyngeal biting mechanics in centrarchid and cichlid fishes: Insights into a key evolutionary innovation. *Journal of Evolutionary Biology* 9:641–670.
Galis, F., and Metz, J. A. J. (2001). Testing the vulnerability of the phylotypic stage: On modularity and evolutionary conservatism. *Journal of Experimental Zoology (Molecular and Developmental Evolution)* 291:195–204.
Galis, F.; J. J. M. van Alphen; and J. A. J. Metz (2001). Why five fingers? Evolutionary constraints on digit numbers. *Trends in Ecology and Evolution* 16:637–646.
Gans, C., and Northcutt, R. G. (1983). Neural crest and the origin of vertebrates: A new head. *Science* 220:268–274.
Gans, C., and Northcutt, R. G. (1985). Neural crest: The implications for comparative anatomy. In *Functional Morphology of Vertebrates,* ed. H.-R. Duncker and G. Fleischer, Fortschritte der Zoologie 30:507–514. Stuttgart: G. Fischer-Verlag.
Garcia-Bellido, A. (1975). Genetic control of wing disk development in Drosophila. In *Ciba Foundation Symposium 29: Cell Patterning,* ed. J. Rivers and R. Porter, 161–182. Amsterdam: Elsevier.
García-Cardeña, G.; J. Comander; K. R. Anderson; B. R. Blackman; and M. A. Gimbrone, Jr. (2001). Biomechanical activation of vascular endothelium as a determinant of its functional phenotype. *Proceedings of the National Academy of Sciences, USA* 98:4478–4485.
Gardner, R. L. (1968). Mouse chimeras obtained by the injection of cells into the blastocyst. *Nature* 220:596–597.
Garey, J. R.; T. J. Near; M. R. Nonnemacher; and S. A. Nadler (1996). Molecular evidence for Acanthocephala as a subtaxon of Rotifera. *Journal of Molecular Evolution* 43:287–292.
Garstang, W. (1922). The theory of recapitulation: A critical restatement of the biogenetic law. *Journal of the Linnean Society, Zoology* 35:81–101.
Gartner, B. L. (1999). *Physiological Ecology Series: Plant Stems: Physiology and Functional Morphology.* San Diego: Academic Press.
Gasser, M.; M. Kaiser; D. Berrigan; and S. C. Stearns (2000). Life-history correlates of evolution under high and low adult mortality. *Evolution* 54:1260–1272.
Gauchat, D.; F. Mazet; C. Berney; M. Schummer; S. Kreger; J. Pawlowski; and B. Galliot (2000). Evolution of Antp-class genes and differential expression of *Hydra* Hox/paraHox genes in anterior patterning. *Proceedings of the National Academy of Sciences, USA* 97:4493–4498.
Gaudry, A. (1866). Considérations générales sur les animaux fossiles de Pikermi. *Librairie de la Société Géologique de France, Paris:* F. Savy.
Gaunt, S. J. (1994). Conservation in the *Hox* code during morphological evolution. *International Journal of Developmental Biology* 38:549–552.
Gaunt, S. J., and Strachan, L. (1994). Forward spreading in the establishment of a vertebrate Hox expression boundary: The expression domain separates into an-

terior and posterior zones, and the spread occurs across implanted glass barriers. *Developmental Dynamics* 199:229–240.

Gauthier, J. (1986). Saurischian monophyly and the origin of birds. *Memoirs of the California Academy of Sciences* 8:1–55.

Gee, H. (1996). *Before the Backbone: Views on the Origin of the Vertebrates.* New York: Chapman and Hall.

Gegenbaur, C. (1870). *Grundzüge der vergleichenden Anatomie,* 2nd ed. Leipzig.

Gehring, W. J., and Ikeo, K. (1999). Pax 6: Mastering eye morphogenesis and eye evolution. *Trends in Genetics* 15:371–377.

Gerhart, J., and Kirschner, M. (1997). *Cells, Embryos, and Evolution: Towards a Cellular and Developmental Understanding of Phenotypic Variation and Evolutionary Adaptability.* Malden, Mass: Blackwell Science.

Gibert, P.; B. Moreteau; and J. R. David (2000). Developmental constraints on an adaptive plasticity: Reaction norms of pigmentation in adult segments of *Drosophila melanogaster. Evolution & Development* 2:249–260.

Gilbert, S. F. (2000a). *Developmental Biology,* 6th ed. Sunderland, Mass.: Sinauer.

Gilbert, S. F. (2000b). Genes classical and genes developmental: The different uses of genes in evolutionary syntheses. In *The Concept of the Gene in Development and Evolution,* ed. P. Beurton, R. Falk, and H-J. Rheinberger, 178–192. New York: Cambridge University Press.

Gilbert, S. F. (2001). Ecological developmental biology: Developmental biology meets the real world. *Developmental Biology* 233:1–12.

Gilbert, S. F., and Bolker, J. A. (2001). Homologies of process and modular elements of embryonic construction. *Journal of Experimental Zoology (Molecular and Developmental Evolution)* 291:1–12.

Gilbert, S. F., and Raunio, A. M. (1997). *Embryology: Constructing the Organism.* Sunderland, Mass.: Sinauer.

Gilbert, S. F.; J. M. Opitz; and R. A. Raff (1996). Resynthesizing evolutionary and developmental biology. *Developmental Biology* 173:357–372.

Gilroy, S., and Trewavas, A. (2001). Signal processing and transduction in plant cells: The end of the beginning? *Nature Reviews, Molecular Cell Biology* 2:307–314.

Gingerich, P. (1979). The stratophenetic approach to phylogenetic reconstructuon in paleontology. In *Phylogenetic Analysis and Paleontology,* ed. J. Cracraft and N. Eldredge, 41–77. New York: Columbia University Press.

Gionti, M.; F. Ristoratore; A. Di Gregorio; F. Aniello; M. Branno; and R. Di Lauro (1998). Cihox5, a new *Ciona intestinalis* Hox-related gene, is involved in regionalization of the spinal cord. *Development, Genes & Evolution* 207:515–523.

Giribet, G. (2001). Exploring the behavior of POY, a program for direct optimization of molecular data. *Cladistics* 17:S60–70.

Givnish, T. J., and Sytsma, K. J., eds. (1997). *Molecular Evolution and Adaptive Radiation.* Cambridge: Cambridge University Press.

Glazier, D. S. (1990). Reproductive efficiency and the timing of gestation and lactation in rodents. *The American Naturalist* 135:269–277.

Gleissberg, S., and Kandereit, J. W. (1999). Evolution of leaf morphogenesis: Evi-

dence from developmental and phylogenetic data in Papaveraceae. *International Journal of Plant Science* 160:787–794.
Godfrey, L., and Sutherland, M. R. (1995). What's growth got to do with it? Process and product in the evolution of ontogeny. *Journal of Human Evolution* 29:405–431.
Godfrey, L., and Sutherland, M. R. (1996). Paradox of peramorphic paedomorphosis: Heterochrony and human evolution. *American Journal of Physical Anthropology* 99:17–42.
Godfrey-Smith, P. (1992). Additivity and the units of selection. In *PSA 1992,* vol. 1, ed. D. Hull, M. Forbes, and K. Okruhlik, 315–328. East Lansing, Mich.: Philosophy of Science Association.
Godwin, A. R., and Capecchi, M. R. (1999). Hair defects in Hoxc13 mutant mice. *Journal of Investigative Dermatology Symposium Proceedings* 4:244–247.
Goethe, J. W. v. (1790). *Versuch die Metamorphose der Pflanzen zu erklären.* Gotha: C. W. Ettinger. Translated in Arber, A. (1946). Goethe's botany. *Chronica Botanica* 10:63–126.
Goffeau, A.; B. G. Barell; H. Bussey; R. W. Davis; B. Dujon; H. Feldmann; J. Galibert; C. Hoheisel; C. Jacq; M. Johnston; E. J. Louis; H. W. Mewes; Y. Murakami; P. Philippsen; H. Tettelin; and S. G. Oliver (1996). Life with 6000 genes. *Science* 274:546–567.
Goin, O. B.; C. J. Goin; and K. Bachmann (1968). DNA and amphibian life history. *Copeia* 1968:532–540.
Goldschmidt, R. (1940). *The Material Basis of Evolution.* New Haven and London: Yale University Press.
Goldsmith, M. H. M. (1977). The polar transport of auxin. *Annual Review of Plant Physiology* 28:439–478.
Goodall, C. (1991). Procrustes methods in the statistical analysis of shape. *Journal of the Royal Statistical Society,* Ser. B 53:285–339.
Goodall, C., and Green, P. (1986). Quantitative analysis of surface growth. *Botanical Gazette* 147:1–15.
Goodwin, B. C.; S. A. Kauffman; and J. D. Murray (1993). Is morphogenesis an intrinsically robust process? *Journal of Theoretical Biology* 163:135–144.
Gordon, D. M. (1992). Phenotypic plasticity. In *Keywords in Evolutionary Biology,* ed. E. F. Keller and E. A. Lloyd, 255–262. Cambridge: Harvard University Press.
Gordon, J. I.; L. V. Hooper; M. S. McNevin; M. Wong; and L. Bry (1997). Epithelial cell growth and differentiation. 3. Promoting diversity in the intestine: Conversations between the microflora, epithelium, and diffuse GALT. *American Journal of Physiology—Gastrointestinal and Liver Physiology* 36:G565–570.
Goss, R. (1956). Regenerative inhibition following limb amputation and immediate insertion into the body cavity. *Anatomical Record* 126:15–27.
Gotthard, K., and Nylin, S. (1995). Adaptive plasticity and plasticity as an adaptation: A selective review of plasticity in animal morphology and life history. *Oikos* 74:3–17.
Gottlieb, G. (1992). *Individual Development and Evolution: The Genesis of Novel Behavior.* Oxford: Oxford University Press.
Gottlieb, G. (2002). *Individual Development and Evolution: The Genesis of Novel*

Behavior. Mahwah, NJ: Erlbaum. (Reprint of 1992 edition, originally published by Oxford University Press.)

Gottlieb, L. D. (1984). Genetics and morphological evolution in plants. *American Naturalist* 123:681–709.

Gould, S. J. (1977). *Ontogeny and Phylogeny.* Cambridge, Mass.: Harvard University Press.

Gould, S. J. (1989a). *Wonderful Life: The Burgess Shale and the Nature of History.* New York: W. W. Norton.

Gould, S. J. (1989b). A developmental constraint in *Cerion,* with comments on the definition and interpretation of constraint in evolution. *Evolution* 43:516–539.

Gould, S. J. (2000). Of coiled oysters and big brains: How to rescue the terminology of heterochrony, now gone astray. *Evolution & Development* 62:1–7.

Gould, S. J., and Lewontin, R. C. (1979). The spandrels of San Marco and the Panglossian paradigm: A critique of the adaptationist programme. *Proceedings of the Royal Society of London B* 205:581–598.

Graba, Y.; D. Aragnol; and J. Pradel (1997). *Drosophila* Hox complex downstream targets and the function of homeotic genes. *BioEssays* 19:379–388.

Graham, L. E. (1993). *Origin of the Land Plants.* New York: John Wiley & Sons.

Graham, L. E., and Wilcox, L. W. (2000). *Algae.* Upper Saddle River, NJ: Prentice Hall.

Grapin-Botton, A., and Melton, D. A. (2000). Endoderm development from patterning to organogensis. *Trends in Genetics* 16:124–130.

Grauer, D., and Li, W.-H. (2000). *Fundamentals of Molecular Evolution,* 2nd ed. Sunderland, Mass.: Sinauer Associates.

Graveley, B. R. (2001). Alternative splicing: Increasing diversity in the proteomic world. *Trends in Genetics* 17:100–107.

Gray, M. (1992). The endosymbiont hypothesis revisited. *International Review of Cytology* 141:233–257.

Gray, M.; G. Burger; and F. Lang (1999). Mitochondrial evolution. *Science* 283:1476–1481.

Gray, R. D. (1992). Death of the gene: Developmental systems strike back. In *Trees of Life: Essays in the Philosophy of Biology,* ed. P. E. Griffiths, 165–209. Dordrecht, The Netherlands: Kluwer.

Gray, R. D. (2001). Selfish genes or developmental systems? In *Thinking About Evolution: Historical, Philosophical, and Political Perspectives,* ed. R. S. Singh, C. B. Krimbas, D. B. Paul, and J. Beatty, 184–207. Cambridge: Cambridge University Press.

Grbić, M.; L. M. Nagy; and M. R. Strand (1998). Development of polyembryonic insects: A major departure from typical insect embryogenesis. *Development, Genes & Evolution* 208:69–81.

Greene, H. W. (1983). Dietary correlates of the origin and radiation of snakes. *American Zoologist* 23:431–441.

Greenspan, R. J. (2001). The flexible genome. *Nature Reviews, Genetics* 2: 383–387.

Gregory, T. R. (2001a). Animal genome size database. http://www.genomesize.com.

Gregory, T. R. (2001b). Coincidence, coevolution, or causation? DNA content, cell size, and the C-value enigma. *Biological Reviews Cambridge* 76:65–101.

Grenier, J. K.; T. L. Garber; R. Warren; P. M. Whitington; and S. Carroll (1997). Evolution of the entire arthropod Hox gene set predated the origin and radiation of the onychophoran/arthropod clade. *Current Biology* 7:547–553.

Griffiths, F. (1928). The significance of Pneumonococcal types. *Journal of Hygiene* 27:113–159.

Griffiths, P. E., and Gray, R. D. (1994). Developmental systems and evolutionary explanation. *Journal of Philosophy* 91:277–304.

Griffiths, P. E., and Gray, R. D. (2001). Darwinism and developmental systems. In *Cycles of Contingency: Developmental Systems and Evolution,* ed. S. Oyama, P. E. Griffiths, and R. D. Gray, 195–218. Cambridge: MIT Press.

Grimes, G. W. (1982). Pattern determination in Hypotrich ciliates. *American Zoologist* 22:35–46.

Gummer, D. L., and Brigham, R. M. (1995). Does fluctuating asymmetry reflect the importance of traits in little brown bats *(Myotis lucifugus). Canadian Journal of Zoology* 73:990–992.

Guralnick, R. P., and Lindberg, D. R. (2001). Reconnecting cell and animal lineages: What do cell lineages tell us about the evolution and development of Spiralia? *Evolution* 55:1501–1519.

Gurdon, J. B. (1962). The developmental capacity of nuclei taken from intestinal epithelial cells of feeding tadpoles. *Journal of Embryology and Experimental Morphology* 10:622–640.

Gurdon, J. B., and Bourillot, P.-Y. (2001). Morphogen gradient interpretation. *Nature* 413:797–803.

Gurdon, J. B.; P. Harger; A. Mitchell; and P. Lemaire (1994). Activin signalling and response to a morphogen gradient. *Nature* 371:487–492.

Guss, K. A.; C. E. Nelson; A. Hudson; M. E. Kraus; and S. B. Carroll (2001). Control of a genetic regulatory network by a selector gene. *Science* 292:1164–1167.

Gussoni, E.; Y. Soneoka; C. D. Strickland; E. A. Buzney; M. K. Khan; A. F. Flint; L. M. Kunkel; and R. C. Mulligan (1999). Dystrophin expression in the mdx mouse restored by stem cell transplantation. *Nature* 401:390–394.

Gyllensten, U.; D. Wharton; A. Josefsson; and A. C. Wilson (1991). Paternal inheritance of mitochondrial DNA in mice. *Nature* 352:255–257.

Haag, E. S., and True, J. R. (2001). Perspective: From mutants to mechanisms? Assessing the candidate gene paradigm in evolutionary biology. *Evolution* 55:1077–1084.

Haag, E. S.; B. J. Sly; M. E. Andrews; and R. A. Raff (1999). Apextrin, a novel extracellular protein associated with larval ectoderm evolution in *Heliocidaris erythrogramma. Developmental Biology* 211:77–87.

Haeckel, E. (1866). *Generelle Morphologie der Organismen: Allgemeine Grundzuege der organischen Formen-Wissenschaft, mechanisch begruendet durch die von Charles Darwin reformirte Descendenz-Theorie,* 2 volumes. Berlin: Georg Reimer.

Haeckel, E. (1875). Die Gastrula und die Eifurchung der Thiere. *Jenaische Zeitschrift fur Naturwissenschaft* 9:402–508.

Hagberg, C.; O. Larson; and J. Milerad (1998). Incidence of cleft lip and palate and risks of additional malformations. *Cleft Palate-Craniofacial Journal* 35:40–45.

Hagemann, W. (1999). Towards an organismic concept of land plants: the marginal blastozone and the development of the vegetative body of selected frondose gametophytes of liverworts and ferns. *Plant Systematics and Evolution* 206:81–133.

Halayko, A. J., and Solway, J. (2001). Molecular mechanisms of phenotypic plasticity in smooth muscle cells. *Journal of Applied Physiology* 90:358–368.

Haldane, J. B. S. (1932). *The Causes of Evolution*. London: Longmans Green.

Halder, G.; P. Callaerts; and W. J. Gehring (1995). New perspectives on eye evolution. *Current Opinion in Genetics and Development* 5:602–609.

Hall, B. K. (1984). Developmental mechanisms underlying the formation of atavisms. *Biological Reviews* 59:89–124.

Hall, B. K. (1992a). Waddington's legacy in development and evolution. *American Zoologist* 32:113–122.

Hall, B. K. (1992b). *Evolutionary Developmental Biology*. London: Chapman and Hall.

Hall, B. K. (ed.) (1994). *Homology: The Hierarchical Basis of Comparative Biology*. New York: Academic Press.

Hall, B. K. (1995). Atavisms and atavistic mutations. *Nature Genetics* 10:126–127.

Hall, B. K. (1996a). Baupläne, phylotypic stages, and constraint: Why are there so few types of animals? *Evolutionary Biology* 29:215–261.

Hall, B. K. (1996b). Evolutionary developmental biology. *McGraw-Hill Yearbook of Science and Technology*, 110–112. New York: McGraw-Hill.

Hall, B. K. (1998). *Evolutionary Developmental Biology*, 2nd ed. New York: Chapman & Hall.

Hall, B. K. (1999a). *Evolutionary Developmental Biology*, 2nd ed. Dordrecht, The Netherlands: Kluwer.

Hall, B. K. (1999b). *The Neural Crest in Development and Evolution*. New York: Academic Press.

Hall, B. K. (2000a). Balfour, Garstang and de Beer: The first century of evolutionary embryology. *American Zoologist* 40:718–728.

Hall, B. K. (2000b). Evo-devo or Devo-evo: Does it matter? *Evolution & Development* 2:177–178.

Hall, B. K. (2002). Descent with modification: The unity underlying homology and homoplasy as seen through an analysis of development and evolution. *Biological Reviews of the Cambridge Philosophical Society* 77 (in press).

Hall, B. K., and Wake, M. H. (1999). Larval development, evolution, and ecology. In *The Origin and Evolution of Larval Forms*, ed. B. K. Hall and M. H. Wake, 1–21. San Diego: Academic Press.

Hall, J. G. (1990). Genomic imprinting: Review and relevance to human diseases. *American Journal of Human Genetics* 46:857–873.

Hall, J. L., and Luck, D. J. (1995). Basal body-associated DNA: *In situ* studies in *Chlamydomonas reinhardtii*. *Proceedings of the National Academy of Sciences, USA* 92:5129–5133.

Hallgrímsson, B., and Maiorana, V. C. (2000). Variability and size in mammals and birds. *Biological Journal of the Linnean Society* 70:571–595.

Hamburger, V. (1980). Embryology and the Modern Synthesis in evolutionary theory. In *The Evolutionary Synthesis: Perspectives on the Unification of Biology*, ed. E. Mayr and W. Provine, 97–112. New York: Cambridge University Press.

Hammond, K. A.; J. Szewczak; and E. Krol (2001). Effects of altitude and temperature on organ phenotypic plasticity along an altitudinal gradient. *Journal of Experimental Biology* 204:1991–2000.
Hampé, A. (1960). La competition entre les elements osseux du zeugopode de Poulet. *Journal of Embryology and Experimental Morphology* 8:241–245.
Hancock, J. M. (1996). Simple sequences and the expanding genome. *BioEssays* 18:421–425.
Hanken, J. (1999). Larvae in amphibian development and evolution. In *The Origin and Evolution of Larval Forms,* ed. B. K. Hall and M. H. Wake, 61–108. San Diego: Academic Press.
Hanken, J.; M. W. Klymkowsky; K. E. Alley; and D. H. Jennings (1997). Jaw muscle development as evidence for embryonic repatterning in direct-developing frogs. *Proceedings of the Royal Society of London B* 264:1349–1354.
Hanken, J.; M. W. Klymkowsky; C. H. Summers; D. W. Seufert; and N. Ingebrigtsen (1992). Cranial ontogeny in the direct-developing frog *Eleutherodactylus coqui* (Anura: Leptodactylidae) analyzed using whole-mount immunohistochemistry. *Journal of Morphology* 211:95–118.
Hardisty, M. W., and Potter, J. C., eds. (1971). *The Biology of Lampreys.* London: Academic Press.
Hardy, A. C. (1965). *The Living Stream.* London: Collins.
Harold, F. (1990). To shape a cell: An inquiry into the causes of morphogenesis of microorganisms. *Microbiological Reviews* 54:381–431.
Harris, J. R. (1998). Placental Endogenous Retrovirus (ERV): Structural, functional, and evolutionary significance. *BioEssays* 20:307–316.
Harris, M. P.; J. F. Fallon; and R. O. Prum (2002). *Shh-Bmp2* signaling module and the evolutionary origin and diversification of feathers. *Journal of Experimental Zoology (Molecular Development and Evolution)* 294:160–176.
Harrison, F. W., eds. (1991–1999). *Microscopic Anatomy of Invertebrates,* vols. 1–15. New York: Wiley-Liss.
Harrison, R. G. (1918). Experiments on the development of the forelimb of *Amblystoma,* a self-differentiating, equipotential system. *Journal of Experimental Zoology* 25:413–462.
Harrison, R. G. (1937). Embryology and its relations. *Science* 85:369–374.
Hart, M. W. (2000). Phylogenetic analyses of mode of larval development. *Seminars in Cell and Developmental Biology* 11:411–418.
Hartman, J. L.; B. Garvik; and L. Hartwell (2001). Principles for the buffering of genetic variation. *Science* 291:1001–1004.
Hartwell, L. H.; J. J. Hopfield; S. Leibler; and A. W. Murray (1999). From molecular to modular cell biology. *Nature* 402:C47–C52.
Harvey, P. H., and Pagel, M. D. (1991). *The Comparative Method in Evolutionary Biology.* Oxford: Oxford University Press.
Harvey, W. (1651). *Exercitationes de generatione animalium.* Amsterdam: Ravesteynium. Reprinted in *The Works of William Harvey,* trans. R. Willis. London: Sydenham Society, 1847.
Haszprunar, G. (1988). On the origin and evolution of major gastropod groups, with special reference to the Streptoneura. *Journal of Molluscan Studies* 54:367–442.

Haszprunar, G.; L. V. Salvini-Plawen; and R. M. Rieger (1995). Larval planktotrophy: A primitive trait in the Bilateria? *Acta Zoologica* (Copenhagen) 76:141–154.

Hatchwell, B. J., and Komdeur, J. (2000). Ecological constraints, life history traits and the evolution of cooperative breeding. *Animal Behavior* 59:1079–1086.

Haun, C.; J. Alexander; D. Y. Stainier; and P. G. Okkema (1998). Rescue of *Caenorhabditis elegans* pharyngeal development by a vertebrate heart specification gene. *Proceedings of the National Academy of Sciences, USA* 95:5072–5075.

Hebb, D. O. (1946). On the nature of fear. *Psychological Review* 53:250–275.

Hedges, S. B.; P. H. Parker; C. G. Sibley; and S. Kumar (1996). Continental breakup and the ordinal diversification of birds and mammals. *Nature* 381:226–229.

Hennig, W. (1966). *Phylogenetic Systematics.* Urbana: University of Illinois Press.

Henry, J .J., and Raff, R. A. (1994). Progressive determination of cell fates along the dorsoventral axis in the sea urchin *Heliocidaris erythrogramma. Roux's Archives of Developmental Biology* 204:62–69.

Henry, J. Q., and Martindale, M. Q. (1994). Establishment of the dorsoventral axis in nemertean embryos: Evolutionary considerations of spiralian development. *Developmental Genetics* 15:64–78

Henry, J. Q.; M. Q. Martindale; and B. C. Boyer (2000). The unique developmental program of the acoel flatworm, *Neochildia fusca. Developmental Biology* 220:285–295.

Hernandez, L. P. (2000). Intraspecific scaling of feeding mechanics in an ontogenetic series of zebrafish, *Danio rerio. Journal of Experimental Biology* 203:3033–3043.

Herring, S. W. (1994). Development of functional interactions between skeletal and muscular systems. In *Bone,* ed. B. K. Hall, 165–191. Boca Raton: CRC Press.

Hertwig, O. (1894). *Präformation oder Epigenese?* Jena: Gustav Fischer.

Hickman, C. S. (1999). Adaptive function of gastropod larval shell features. *Invertebrate Biology* 118:346–356.

Higashi M.; G. Takimoto; and N. Yamamura (1999). Sympatric speciation by sexual selection. *Nature* 402:523–526.

Higgins, J., and Azad, A. F. (1995). Use of polymerase chain reaction to detect bacteria in arthropods: A review. *Journal of Medical Entomology* 32:13–22.

Hilgendorf F. (1867). Uber *Planorbis multiformis* im Steinheimer Süsswasserkalk. *Monatsberichte koeniglinke der Preussische Akademie der Wissenschaften Zu Berlin* 1866:474–504.

Hillis, D. M. (1994). Homology in molecular biology. In *Homology: The Hierarchical Basis of Comparative Biology,* ed. B. K. Hall, 339–368. New York: Academic Press.

Hinegardner, R., and Engleberg, J. (1983). Biological complexity. *Journal of Theoretical Biology* 104:7–20.

Hjelm, K. K. (1986). Is non-genic inheritance involved in carcinogenesis? A cytotactic model of transformation. *Journal of Theoretical Biology* 119:89–101.

Hobmayer, B.; F. Rentzsch; K. Kuhn; C. M. Happel; C. C. von Laue; P. Snyder; U. Rothbacher; and T. W. Holstein (2000). WNT signalling molecules act in axis formation in the diploblastic metazoan Hydra. *Nature* 407:186–189.

Hodin, J. (2000). Plasticity and constraints in development and evolution. *Journal of Experimental Zoology (Molecular and Developmental Evolution)* 288:1–20.
Hoefs, M., and Nowlan, U. (1997). Comparison of horn growth in captive and free-ranging Dall's rams. *Journal of Wildlife Management* 61:1154–1160.
Hoffmann, A. A., and J. Merilä (1999). Heritable variation and evolution under favourable and unfavourable conditions. *Trends in Ecology and Evolution* 14:96–101.
Holland, L. Z. (2000). Body-plan evolution in the Bilateria: Early antero-posterior patterning and the deuterostome-protostome dichotomy. *Current Opinion in Genetics and Development* 10:434–442.
Holland, L. Z., and Holland, N. D. (1999). Chordate origins of the vertebrate central nervous system. *Current Opinion in Neurobiology* 9:596–602.
Holland, N. D., and Chen, J. (2001). Origin and early evolution of the vertebrates: New insights from advances in molecular biology, anatomy, and palaeontology. *BioEssays* 23:142–151.
Holland, P. W. H. (2000). The future of evolutionary developmental biology. *Nature* 402 (Suppl): C41–C44.
Holland, P. W. H. (2001). Beyond the Hox: How widespread is homeobox gene clustering? *Journal of Anatomy* 199:13–23.
Holley, S. A.; P. D. Jackson; Y. Sasai; B. Lu; E. M. De Robertis; F. M. Hoffmann; and E. L. Ferguson (1995). A conserved system for dorsal-ventral patterning in insects and vertebrates involving sog and chordin. *Nature* 376:249–253.
Holliday, R. (1987). The inheritance of epigenetic defects. *Science* 238:163–170.
Holliday, R. (1994). Epigenetics: An overview. *Developmental Genetics* 15:453–457.
Holloway, G. J.; P. M. Brakefield; and S. Kofman (1993). The genetics of wing pattern elements in the polyphenic butterfly, *Bicyclus anynana. Heredity* 70:179–186.
Holloway, G. J., and Brakefield, P. M. (1995). Artificial selection of reaction norms of wing pattern elements in *Bicyclus anynana. Heredity* 74:91–99.
Honkanen, T., and Haukioja, E. (1998). Intra-plant regulation of growth and plant-herbivore interactions. *Ecoscience* 5:470–479.
Honma, T., and Goto, K. (2001). Complexes of MADS-box proteins are sufficient to convert leaves into floral organs. *Nature* 409:525–529.
Hooper, L. V.; L. Bry; and P. G. Falk (1998). Host-microbial symbiosis in the mammalian intestine: Exploring an internal ecosystem. *BioEssays* 20:336–343.
Hooper, L. V.; M. H. Wong; A. Thelin; L. Hannson; P. G. Falk; and J. I. Gordon (2001). Molecular analysis of commensal host-microbial relationships in the intestine. *Science* 291:881–884.
Horan, G. S. B.; R. Ramirez-Solis; M. S. Featherstone; D. J. Wolgemuth; A. Bradley; and R. R. Behringer (1995). Compound mutants for the paralogous hoxa-4, hoxb-4, and hoxd-4 genes show more complete homeotic transformations and a dose-dependent increase in the number of vertebrae transformed. *Genes & Development* 9:1667–1677.
Horder, T. J. (1983). Embryological bases of evolution. In *Development and Evolution,* ed. G. C. Goodwin, N. Holder, and C. C. Wylie, 315–352. Cambridge: Cambridge University Press.

Horner, J. R.; A. J. de Ricqlès; and K. Padian (1999). Variation in skeletochronological indicators of the hadrosaurid dinosaur *Hypacrosaurus:* Implications for age assessment of dinosaurs. *Paleobiology* 25:295–304.
Horner, J. R.; A. J. de Ricqlès; and K. Padian (2000). The bone histology of the hadrosaurid dinosaur *Maiasaura peeblesorum:* Growth dynamics and physiology based on an ontogenetic series of skeletal elements. *Journal of Vertebrate Paleontology* 20:109–123.
Horner, J. R.; A. J. de Ricqlès; and K. Padian (2001). Comparative osteohistology of some embryonic and perinatal archosaurs: Phylogenetic and behavioral implications for dinosaurs. *Paleobiology* 27:39–58.
Hoskin, M. G. (1997). Effects of contrasting modes of larval development on the genetic structures of populations of three species of prosobranch gastropods. *Marine Biology* (Berlin) 127:647–656.
Houston, D. W., and King, M. L. (2000). Germ plasm and molecular determinants of germ cell fate. *Current Topics in Developmental Biology* 50:155–181.
Huelsenbeck, J. P., and Bull, J. J. (1996). A likelihood ratio test to detect conflicting phylogenetic signal. *Systematic Biology* 45:92–98.
Huelsenbeck, J. P; B. Rannala; and Z. Yang (1997). Statistical tests of host-parasite cospeciation. *Evolution* 51:410–419.
Huey, R. B., and Bennett, A. F. (1987). Phylogenetic studies of co-adaptation: Preferred temperature versus optimal performance temperatures of lizards. *Evolution* 41:1098–115.
Hufford, L. (2001). Ontogenetic sequences: Homology, evolution and the patterning of clade diversity. In *Beyond Heterochrony: The Evolution of Development,* ed. M. L. Zelditch, 27–57. New York: John Wiley & Sons.
Hughes, N. C.; R. E. Chapman; and J. M. Adrain (1999). The stability of thoracic segmentation in trilobites: A case study in developmental and ecological constraints. *Evolution & Development* 1:24–35.
Hull, D. L. (1980). Individuality and selection. *Annual Review of Ecology and Systematics* 12:311–332.
Hull, D. L. (1988). *Science as a Process: An Evolutionary Account of the Social and Conceptual Development of Science.* Chicago: University of Chicago Press.
Hunter, J. P. (1998). Key innovations and the ecology of macroevolution. *Trends in Ecology and Evolution* 13:31–36.
Hunter, J. P., and Jernvall, J. (1995). The hypocone as a key innovation in mammalian evolution. *Proceedings of the National Academy of Sciences, USA* 92:10718–10722.
Hurst, L. D., and Peck, J. R. (1996). Recent advances in understanding of the evolution and maintenance of sex. *Trends in Ecology and Evolution* 11:46–52.
Hutchinson, G. E., and MacArthur, R. H. (1959). A theoretical model of size distributions among species of mammals. *American Naturalist* 93:117–125.
Huysseune, A. (1995). Phenotypic plasticity in the lower pharyngeal jaw dentition of *Astatoreochromis alluaudi* (Teleostei, Cichlidae). *Archives of Oral Biology* 40:1005–1014.
Huxley, J. S. (1932). *Problems of Relative Growth.* London: MacVeagh.
Hyatt, A. (1866). On the parallelism between the different stages of life in the indi-

vidual and those in the entire group of the molluscous order Tetrabranchiata. *Memoirs of the Boston Society of Natural History* 1:193–209.

Hyatt, A. (1894). Phylogeny of an acquired characteristic. *Proceedings of the American Philosophical Society* 32:349–647.

Illingworth, C. (1974). Trapped fingers and amputated finger tips in children. *Journal of Pediatric Surgery* 9:853–858.

Ingham, P. W. (1991). Segment polarity genes and cell patterning within the *Drosophila* body segment. *Current Opinion in Genetics and Development* 1:261–267.

Irish, V. F., and Jenik, P. D. (2001). Cell lineage, cell signaling and the control of plant morphogenesis. *Current Opinion in Genetics and Development* 11:424–430.

Ivanova-Kazac, O. M. (1977). *Asexual Reproduction of Animals* (in Russian). Leningrad: Izdat'elstvo Leningradskovo Universit'eta.

Jablonka, E., and Lamb, M. (1994). *Epigenetic Inheritance and Evolution: The Lamarckian Dimension.* Oxford: Oxford University Press.

Jablonski, D. (1986). Larval ecology and macroevolution in marine invertebrates. *Bulletin of Marine Science* 39:565–587.

Jablonski, D. (2000). Micro- and macroevolution: Scale and hierarchy in evolutionary biology and paleobiology. *Paleobiology* 26: 15–52.

Jacob, F. (1974). *The Logic of Living Systems. A History of Heredity,* trans. B. Spillmann. London: Allen Lane.

Jacob, F. (1977). Evolution and tinkering. *Science* 196:1161–1166.

Jägersten, G. (1972). *Evolution of the Metazoan Life Cycle.* London: Academic Press.

Jahn, I., ed. (2000). *Geschichte der Biologie: Theorie, Methoden, Institutionen, Kurzbiographien.* Heidelberg: Sektrum Verlag.

Jeffery, J. E.; O. R. P. Bininda-Emonds; M. I. Coates; and M. K. Richardson (2001). Heterochrony in amniote evolution. *Journal of Morphology* 248:246.

Jeffery, W. R. (1994). A model for ascidian development and developmental modifications during evolution. *Journal of the Marine Biological Association of the United Kingdom* 74:35–48.

Jeffery, W. R., and Swalla, B. J. (1992). Evolution of alternative modes of development in ascidians. *BioEssays* 14:219–226.

Jeffery, W. R.; B. J. Swalla; N. Ewing; and T. Kusubake (1999). Evolution of the ascidian anural larva: Evidence from embryos and molecules. *Molecular Biology and Evolution* 5:646–654.

Jennings, D. H., and Hanken, J. (1998). Mechanistic basis of life history evolution in anuran amphibians: Thyroid gland development in the direct-developing frog, *Eleutherodactylus coqui. General & Comparative Endocrinology* 111:225–232.

Jeon, K. W. (1983). *Intracellular Symbiosis.* New York: Academic Press.

Jeon, M.; H. F. Gardner; E. A. Miller; J. Deshler; and A. E. Rougvie (1999). Similarity of the *C. elegans* developmental timing protein LIN-42 to circadian rhythym proteins. *Science* 286:1141–1146.

Jernvall, J., and Jung, H. S. (2000). Genotype, phenotype, and developmental biology of molar tooth characters. *American Journal of Physical Anthropology* (Suppl) 31:171–190.

Jernvall, J.; J. P. Hunter; and M. Fortelius (1996). Molar tooth diversity, disparity andecology in Cenozoic ungulate radiations. *Science* 274:1489–1492.

Jernvall, J.; S. V. Keranen; and I.Thesleff (2000). Evolutionary modification of development in mammalian teeth: Quantifying gene expression patterns and topography. *Proceedings of the National Academy of Sciences, USA* 97:14444–14448.

Jiang, J., and Struhl, G. (1998). Regulation of the Hedgehog and Wingless signalling pathways by the F-box/WD40-repeat protein Slimb. *Nature* 391:493–496.

Jockusch, E. L. (1997). An evolutionary correlate of genome size change in plethodontid salamanders. *Proceedings of the Royal Society of London Series B* 264:597–604.

Jockusch, E. L.; C. Nulsen; S. J. Newfeld; and L. M. Nagy (2000). Leg development in flies versus grasshoppers: Differences in *dpp* expression do not lead to differences in the expression of downstream components of the leg patterning pathway. *Development* 127:1617–1626.

Jofuku, K. D.; B. G. W. den Boer; M. Van Montagu; and J. K. Okamuro (1994). Control of *Arabidopsis* flower and seed development by the homeotic gene *APETALA2*. *Plant Cell* 6:1211–1225.

Johannsen, W. (1909). *Elemente der Exakten Erblichkeitslehre.* Jena: Gustav Fischer.

Johnson, A.; R. F. Bachvarova; M. Drum; and T. Masi (2001). Expression of axolotl *dazl* RNA, a marker of germ plasm: Widespread maternal RNA and onset of expression in germ cells approaching the gonad. *Developmental Biology* 234:402–415.

Johnson, J. B., and Belk, M. C. (2001). Predation environment predicts divergent life-history phenotypes among populations of the live-bearing fish *Brachyrhaphis rhabdophora. Oecologia* 126:142–149.

Johnson, M. H., and Day, M. L. (2000). Egg timers: How is developmental time measured in the early vertebrate embryo? *BioEssays* 22:57–63.

Johnson, N. A., and Porter, A. H. (2000). Rapid speciation via parallel, directional selection on regulatory genetic pathways. *Journal of Theoretical Biology* 205:527–542.

Johnson, N. A., and Porter, A. H. (2001). Toward a new synthesis: Population genetics and evolutionary developmental biology. *Genetica* 112:45–58.

Johnston, T. D. (1987). The persistence of dichotomies in the study of behavioral development. *Developmental Review* 7:149–182.

Johnston, T. D., and Gottlieb, G. (1990). Neophenogenesis: A developmental theory of phenotypic evolution. *Journal of Theoretical Biology* 147:471–495.

Jones, C. M.; N. Armes; and J. C. Smith (1996). Signalling by TGF-beta family members: Short-range effects of Xnr-2 and BMP-4 contrast with the long-range effects of activin. *Current Biology* 6:1468–1475.

Jones, J. E., and Corwin, J. T. (1996). Regeneration of sensory cells after laser-ablation in the lateral-line system: Hair cell lineage and macrophage behavior revealed by time-lapse video microscopy. *Journal of Neuroscience* 16:649–662.

Jones, P. A., and Laird, P. W. (1999). Cancer epigenetics comes of age. *Nature Genetics* 21:163–167.

Judson, P. O., and Normark, B. B. (1996). Ancient asexuals. *Trends in Ecology and Evolution* 11:41–46.

Jung, H. S.; P. H. Francis-West; R. B. Widelitz; T. X. Jiang; S. Ting-Berreth; C. Tickle; L. Wolpert; and C. M. Chuong (1998). Local inhibitory action of BMPs and their relationships with activators in feather formation: Implications for periodic patterning. *Developmental Biology* 196:11–23.

Kagan, M.; N. Novoplansky; and T. Sachs (1992). Variable cell lineages during the development of stomatal patterns. *Annals of Botany* 69:303–312.

Kaplan, D. R. (2001). Fundamental concepts of leaf morphology and morphogenesis: A contribution to the interpretation of molecular genetic mutants. *International Journal of Plant Science* 162:467–474.

Kauffman, S. A. (1993). *The Origins of Order: Self-Organization and Selection in Evolution.* New York: Oxford University Press.

Keegan, L. P.; A. Gallo; and M. A. O'Connell (2000). Development: Survival is impossible without an editor. *Science* 290:1707–1709.

Keller, L., and Ross, K. G. (1993). Phenotypic plasticity and "cultural transmission" of alternative social organisations in the fire ant *Solenopsis invicta. Behavioural Ecology and Sociobiology* 33:121–129.

Kempermann, G.; E. P. Brandon; and F. H. Gage (1998). Environmental stimulation of 129/SvJ mice causes increased cell proliferation and neurogenesis in the adult dentate gyrus. *Current Biology* 8:939–942.

Kerr, A. M., and Kim, J. (1999). Bi-penta-bidecaradial symmetry: A review of evolutionary and developmental trends in Holothuroidea (Echinodermata). *Journal of Experimental Zoology* 285:190–203.

Kessel, M., and Gruss, P. (1991). Homeotic transformations of murine vertebrae and concomitant alteration of Hox codes induced by retinoic acid. *Cell* 76:89–104.

Kettle, C.; J. Johnstone; T. Jowett; H. M. Arthur; and W. Arthur (2002). The pattern of segment formation, as revealed by *engrailed* expression, in a centipede with a variable number of segments. *Evolution and Development* 4 (in press).

Keys, D. N.; D. L. Lewis; J. E. Selegue; B. J. Pearson; L. V. Goodrich, R. L. Johnson; J. Gates; M. P. Scott; S. B. Carroll. (1999). Recruitment of a *hedgehog* regulatory circuit in butterfly eyespot evolution. *Science* 283:532–534.

Kiesecker, J. M.; A. R. Blaustein; and L. K. Belden (2001). Complex causes of amphibian population declines. *Nature* 410:681–684.

Kim, C. B.; C. Amemiya; W. Bailey; K. Kawasaki; J. Mezey; W. Miller; S. Minoshima; N. Shimizu; G. Wagner; and F. Ruddle (2000). Hox cluster genomics in the horn shark, *Heterodontus francisci. Proceedings of the National Academy of Sciences, USA* 97:1655–1660.

Kim, M.; W. Caio; S. Kessler; and N. Sinha (2001). Developmental changes due to long-distance movement of a homeobox fusion transcript in tomato. *Science* 293:287–289.

Kim, Y., and Nirenberg, M. (1989). *Drosophila* NK-homeobox genes. *Proceedings of the National Academy of Sciences, USA* 86:7716–7720.

Kimmel, C. B. (1996). Was Urbilateria segmented? *Trends in Genetics* 12:329–331.

Kimura, M. (1983). *The Neutral Theory of Molecular Evolution.* Cambridge: Cambridge University Press.

Kingsolver, J. G. (1995). Fitness consequences of seasonal polyphenism in western white butterflies. *Evolution* 49:942–954.

Kirschner, M., and Gerhart, J. (1998). Evolvability. *Proceedings of the National Academy of Sciences, USA* 95:8420–8427.

Kitazawa, C., and Amemiya, S. (2001). Regulating potential in development of a direct-developing echinoid. *Peronella japonica. Development, Growth & Differentiation* 43:73–82.

Klingenberg, C. P. (1998). Heterochrony and allometry: the analysis of evolutionary change in ontogeny. *Biological Reviews* 73:79–123.

Kloc, M.; S. Bilinski; A. Chan; L. H. Allen; N. R. Zearfoss; and L. Etkin (2001). RNA localization and germ cell determination in *Xenopus. International Review of Cytology* 203:63–90.

Kloc, M.; M. Dougherty; S. Bilinski; A. Chan; E. Brey; M. L. King; C. Parick; and L. Etkin (2002). Three-dimensional ultrastructural analysis of RNA distribution within germinal granules in *Xenopus. Developmental Biology* 241:79–93.

Klueg, K. M.; M. A. Harkey; and R. A. Raff (1997). Mechanisms of evolutionary changes in timing, spatial expression, and mRNA processing in the msp130 gene in a direct-developing sea urchin, *Heliocidaris erythrogramma. Developmental Biology* 182:121–133.

Klug, W. S., and Cummings, M. R. (2000). *Concepts of Genetics,* 6th ed. Upper Saddle River, NJ: Prentice Hall.

Kluge, A., and Farris, J. S. (1969). Quantitative phyletics and the evolution of anurans. *Systematic Zoology* 18:1–32.

Knaut, H.; F. Pelegri; B. Bohmann; H. Schwarz; and C. Nusslein-Volhard (2000). Zebrafish *vasa* RNA but not its protein is a component of the germ plasm and segregates asymmetrically before germ line specification. *Journal of Cell Biology* 149:875–888.

Koch, P. B. (1987). Die Steuerung der saisondimorphen Flügelfärbung von *Araschnia levana* L. (Nymphalidae, Lepidoptera) durch Ecdysteroide. *Mitteilungen der Deutschen Gesellschaft für Allgemeine und Angewandte Entomologie* 5:195–197.

Koch, P. B. (1992). Seasonal polyphenism in butterflies: A hormonally controlled phenomenon of pattern formation. *Zoologische Jahrbücher(Physiologie)* 96:227–240.

Koch, P. B.; P. M. Brakefield; and F. Kesbeke (1996). Ecdysteroids control eyespot size and wing color pattern in the polyphenic butterfly *Bicyclus anynana* (Lepidoptera: Satyridae). *Journal of Insect Physiology* 42:223–230.

Kocher, T. D.; J. A. Conroy; K. R. McKaye; and J. R. Stauffer (1993). Similar morphologies of cichlid fish in Lakes Tanganyika and Malawi are due to convergence. *Molecular Phylogenetics and Evolution* 2:158–165.

Koehl, M. A. R. (1996). When does morphology matter? In *Annual Review of Ecology and Systematics,* ed. D. G. Fautin, 501–542. Palo Alto: Annual Reviews Press.

Kojima, S. (1998). Paraphyletic status of Polychaeta suggested by phylogenetic analyses based on the amino acid sequences of elongation factor-1. *Molecular and Phylogenetic Evolution* 9:255–261.

Kollar, E. J., and Baird, G. R. (1969). The influence of the dental papilla on the development of tooth shape in embryonic mouse tooth germ. *Journal of Embryology and Experimental Morphology* 21:131–148.

Kollar, E. J. and Fisher, C. (1980). Tooth induction in chick epithelium: Expression of quiescent genes for enamel synthesis. *Science* 207:993–995.

Kondo, S.; Y. Kuwahara; M. Kondo; K. Naruse; H. Mitani; Y. Wakamatsu; K. Ozato; S. Asakawa; N. Shimizu; and A. Shima (2001). The Medaka rs-3 locus required for scale development encodes ectodysplasin-A receptor. *Current Biology* 711:1202–1206.

Kondo, T.; P. Dolle; J. Zakany; and D. Duboule (1996). Function of posterior HoxD genes in the morphogenesis of the anal sphincter. *Development* 122:2651–2659.

Kooi, R. E., and Brakefield, P. M. (1999). The critical period for wing pattern induction in the polyphenic tropical butterfly *Bicyclus anynana* (Satyrinae). *Journal of Insect Physiology* 45:201–212.

Kooi, R. E.; P. M. Brakefield; and W. E. M.-Th. Rossie (1996). Effects of food plant on phenotypic plasticity in the tropical butterfly *Bicyclus anynana. Entomologia Experientalis et Applicata* 80:149–151.

Koonin, E. V.; A. V. Mushegian; and P. Bork (1996). Non-orthologous gene displacement. *Trends in Genetics* 12:334–336.

Kopp, A.; I. Duncan; and S. B. Carroll (2000). Genetic control and evolution of sexually dimorphic characters in *Drosophila. Nature* 408:553–559.

Koprunner, M.; C. Thisse; B. Thisse; and E. Raz (2001). A zebrafish nanos-related gene is essential for the development of primordial germ cells. *Genes & Development* 15:2877–2885.

Kornfield, I., and Smith, P. F. (2000). African cichlid fishes: Model systems for evolutionary biology. *Annual Review of Ecology and Systematics* 31:163–196.

Kornfeld, K. (1997). Vulval development in *Caenorhabditis elegans. Trends in Genetics* 13:55–61.

Kostic, D., and Capecchi, M. R. (1994). Targeted disruption of the murine hoxa-4 and Hoxa-6 genes result in homeotic transformations of the vertebral column. *Mechanisms of Development* 46:231–247.

Kourakis, M. J., and Martindale, M. Q. (2000). Combined-method phylogenetic analysis of Hox and ParaHox genes of the Metazoa. *Journal of Experimental Zoology (Molecular and Developmental Evolution)* 288:175–191.

Kowalevsky, A. O. (1871). Weitere Studien über die Entwicklung der einfachen Ascidien. *Archiv für mikroskopische Anatomie* 6:101–130.

Kowalski, C. J. (1972). A commentary on the use of multivariate statistical methods in anthropometric research. *American Journal of Physical Anthropology* 36:119–132.

Kramer, E. M., and Irish, V. F. (1999). Evolution of genetic mechanisms controlling petal development. *Nature* 399:144–148.

Krizek, B. A., and Meyerowitz, E. M. (1996). The *Arabidopsis* homeotic genes

APETALA3 and *PISTILLATA* are sufficient to provide the B class organ identity function. *Development* 122:11–22.

Krumlauf, R. (1994). Hox genes in vertebrate development. *Cell* 78:191–201.

Krupnick, G. A.; K. M. Brown; and A. G. Stephenson (1999). The influence of fruit on the regulation of internal ethylene concentrations and sex expression in *Cucurbita texana. International Journal of Plant Science* 160:321–330.

Kuby, J. (1998). *Immunology.* New York: W. H. Freeman and Company.

Kumar, A.; C. Velloso; Y. Imokawa; and J. Brockes (2000). Plasticity of retrovirus-labelled myotubes in the newt limb regeneration blastema. *Developmental Biology* 218: 125–136.

Kumar, S., and Hedges, S. B. (1998). A molecular timescale for vertebrate evolution. *Nature* 392:917–920.

Kunz, W.; H. H. Treple; and K. Bier (1970). On the function of the germ line chromosomes in the oogenesis of *Wachtiella persiaria* (Lecidomyidae). *Chromosoma* 30:180–192.

Kuo, Z.-Y. (1976). *The Dynamics of Behavior Development: An Epigenetic View,* enlarged ed. New York: Plenum.

Kurland, C. G., and Andersson, S. G. E. (2000). Origin and evolution of the mitochondrial proteome. *Microbiology and Molecular Biology Reviews* 64:786–820.

Lacy, R. C. (1982). Niche breadth and abundance as determinants of genetic variation in populations of mycophagus drosophilid flies (Diptera: Drosophilidae). *Evolution* 36:1265–1275.

Ladoukahis, E. D., and Zouros, E. (2001). Direct evidence for homologous recombination in mussel *(Mytilus galloprovincialis)* mitochondrial DNA. *Molecular Biology and Evolution* 18:1168–1175.

Lajus, D. (2001). Variation patterns of bilateral characters: Variation among characters and among populations in the White Sea Herring *Clupea pallasi marisalbi* (Berg) (Clupeidae, Teleostei). *Biological Journal of the Linnean Society* 74:237–253.

Lamarck, J.-B. de (1802–1806). Mémoires sur les fossiles des environs de Paris, etc. *Annales du Muséum d'Histoire Naturelle,* vols. 1–8.

Lande, R. (1977). On comparing coefficients of variation. *Systematic Zoology* 26:214–217.

Lande, R. (1979). Quantitative genetic analysis of multivariate evolution: Applied to brain:body size allometry. *Evolution* 33:203–215.

Lande, R. (1981). Models of speciation by sexual selection on polygenic traits. *Proceedings of the National Academy of Sciences, USA* 78:3721–3725.

Landman, O. E. (1991). The inheritance of acquired characteristics. *Annual Review of Genetics* 25:1–20.

Lankester, E. R. (1870). On the use of the term homology in modern zoology, and the distinction between homogenetic and homoplastic agreements. *Annals and Magazine of Natural History,* Series 4, 6:34–43.

Lankester, E. R. (1877). Notes on the embryology and classification of the animal kingdom: Comprising a revision of speculations relative to the origin and significance of the germ layers. *Quarterly Journal of Microscopic Science* 17:399–454.

Larson, A.; E. M. Prager; and A. C. Wilson (1984). Chromosomal evolution, speciation and morphological change in vertebrates: The role of social behavior. *Chromosomes Today* 8:215–228.

Lassar, A.; S. Skapek; and B. Novitch (1994). Regulatory mechanisms that coordinate skeletal muscle differentiation and cell cycle withdrawal. *Current Opinion in Cell Biology* 6:788–794.

Latchman, D. (1995). *Gene Regulation: A Eukaryotic Perspective.* London: Chapman & Hall.

Latchman, D. (1998). *Eukaryotic Transciption Factors.* London: Academic Press.

Latham, K. E.; J. McGrath; and D. Solter (1995). Mechanistic and developmental aspects of genetic imprinting in mammals. *International Review of Cytology* 160:53–98.

Laubichler, M. D. (2000). Homology in development and the development of the homology concept. *American Zoologist.* 40:777–788.

Lauder, G. V. (1994). Homology, form, and function. In *Homology: The Hierarchical Basis of Comparative Biology,* ed. B. K. Hall, 151–196. San Diego: Academic Press.

Lauter, N., and Doebley, J. (2002). Genetic variation for phenotypically invariant traits detected in teosinte: Implications for the evolution of novel forms. *Genetics* 160:333–342.

Leemans, R.; T. Loop; B. Egger; H. He; L. Kammermeier; B. Hartmann; U. Certa; H. Reichert; and F. Hirth (2001). Identification of candidate downstream genes for the homeodomain transcription factor Labial in *Drosophila* through oligonucleotide-array transcript imaging. *Genome Biology* 2(5):RESEARCH0015.1–0015.9.

Lele, S., and McCulloch, C. (2002). Invariance, identifiability, and morphometrics. *Journal of American Statistical Association* 97:1–11.

Lele, S., and Richtsmeier, J. (2001). *An Invariant Approach to the Statistical Analysis of Shapes.* Boca Raton, FL: Chapman & Hall/CRC Press.

Lemaire, L., and Kessel, M. (1997). Gastrulation and homeobox genes in chick embryos. *Mechanisms of Development* 67:3–16.

Lenhard, M.; A. Bohnert; G. Jurgens; and T. Laux (2001). Termination of stem cell maintenance in *Arabidopsis* floral meristems by interactions between *WUSCHEL* and *AGAMOUS. Cell* 105:805–814.

Leonovicová, V., and Novák, V. J. A., ed. (1987). *Behaviour as One of the Main Factors of Evolution.* Prague: Czechoslovak Academy of Sciences.

Lerner, I. M. (1954). *Genetic Homeostasis.* New York: Wiley & Sons.

Leroi, A. M. (2000). The scale independence of evolution. *Evolution & Development* 2:67–77.

Leroi, A. M.; M. R. Rose; and G. V. Lauder (1994). What does the comparative method reveal about adaptation? *The American Naturalist* 143:381–402.

Lestrel, P. (1982). A Fourier analytic procedure to describe complex morphological shape. In *Factors and Mechanisms Influencing Bone Growth,* ed. A. D Dixon and B. G. Sarnat, 393–409. New York: Alan R. Liss.

Lestrel, P. (1989). Method for analyzing complex two-dimensional forms: Elliptical Fourier functions. *American Journal of Human Biology* 1:149–164.

Levin, L. A., and Bridges, T. S. (1995). Pattern and diversity in reproduction and development. In *Ecology of Marine Invertebrate Larvae,* ed. L. R. McEdward, 1–48. Boca Raton, FL: CRC Press.

Levine, A.; G. L. Cantoni; and A. Razin (1991). Inhibition of promoter activity by methylation: A possible involvement of a protein mediator. *Proceedings of the National Academy of Sciences, USA* 88:6515–6518.

Levine, A.; G. L. Cantoni; and A. Razin (1992). Methylation in the preinitiation domain affects gene transcription by an indirect mechanism. *Proceedings of the National Academy of Sciences, USA* 89:10119–10123.

Levine, M.; G. M. Rubin; and R. Tjian (1984). Human DNA sequences homologous to a protein coding region conserved between homeotic genes of *Drosophila. Cell* 38:667–673.

Levine, M. (2002). How insects lose their limbs. *Nature* 415:848–849.

Levine, S. (1956). A further test of infantile handling and adult avoidance learning. *Journal of Personality* 25:309–333.

Levins, R. (1968). *Evolution in Changing Environments.* Princeton: Princeton University Press.

Levinton, J. S. (1986). Developmental constraints and evolutionary saltations: A discussion and critique. In *Genetics, Development and Evolution,* ed. J. P. Gustafson, G. L. Stebbins, and F. J. Ayala, 253–288. New York: Plenum.

Lewis, E. B. (1978). A gene complex controlling segmentation in *Drosophila. Nature* 276:565–570.

Lewis, J. G. E. (1981). *The Biology of Centipedes.* Cambridge: Cambridge University Press.

Lewis, J.; W. Lew; and J. Zimmerman (1980). A nonhomogeneous anthropometric scaling method based on finite element principles. *Journal of Biomechanics* 13:815–824.

Lewontin, R. C. (1970). The units of selection. *Annual Review of Ecology and Systematics* 1:1–18.

Lewontin, R. C. (1974). The analysis of variance and the analysis of causes. *American Journal of Human Genetics* 26:400–411.

Lewontin, R. C. (1978). Adaptation. *Scientific American* 239:212–228.

Lewontin, R. C. (2001). Foreword. In *The Character Concept in Evolutionary Biology,* ed. G. P. Wagner, xvii–xiii. San Diego: Academic Press.

Lexell, J; C. C. Taylor; and M. Sjostrom (1988). What is the cause of the ageing atrophy? Total number, size and proportion of different fiber types studied in whole vastus lateralis muscle from 15- to 83-year-old men. *Journal of Neurological Sciences* 84:275–294.

Li, J.; K. C. Liu; F. Jin; M. M. Lu; and J. A. Epstein (1999). Transgenic rescue of congenital heart disease and spina bifida in *Splotch* mice. *Development* 126:2495–2503.

Li, P., and Johnston, M. O. (2000). Heterochrony in plant evolutionary studies through the twentieth century. *The Botanical Review* 66:57–88.

Li, X., and McGinnis, W. (1999). Activity regulation of Hox proteins, a mechanism for altering functional specificity in development and evolution. *Proceedings of the National Academy of Sciences, USA* 96:6802–6807.

Lieberman, D. E. (2000). Ontogeny, homology and phylogeny in the hominid craniofacial skeleton: The problem of the browridge. In *Development, Growth and Evolution: Implications for the Study of the Hominid Skeleton,* eds. P. O'Higgins and M. Cohn, 85–122. San Diego: Academic Press.

Liem, K. F. (1974). Evolutionary strategies and morphological innovations: Cichlid pharyngeal jaws. *Systematic Zoology* 22: 425–441.

Liem, K. F. (1990). Key evolutionary innovations, differential diversity, and symecomorphosis. In *Evolutionary Innovations,* ed. M. H. Nitecki, 147–170. Chicago: University of Chicago Press.

Liljegren, S. J.; C. Gustafson-Brown; A. Pinyopich; G. S. Ditta; and M. F. Yanofsky (1999). Interactions among *APETALA1, LEAFY,* and *TERMINAL FLOWER1* specify meristem fate. *Plant Cell* 11:1007–1018.

Lillie, F. R. (1898). Adaptation in cleavage. In *Biological Lectures from the Marine Biology Laboratories, Woods Hole, Massachusetts,* 43–67. Boston: Ginn.

Lloyd, E. (1988). *The Structure and Confirmation of Evolutionary Biology.* New York: Greenwood Press.

Lloyd, E. A., and Gould, S. J. (1993). Species selection on variability. *Proceedings of the National Academy of Sciences, USA* 90:595–599.

Loeb, J., and Northrop, J. H. (1916). Is there a temperature coefficient for the duration of life? *Proceedings of the National Academy of Sciences, USA* 2:456–457.

Lohmann, G. P. (1983). Eigenshape analysis of microfossils: A general morphometric procedure for describing changes in shape. *Mathematical Geology* 15:659–672.

Lohmann, G. P., and Schweitzer, P. N. (1990). On eigenshape analysis. In *Proceedings of the Michigan Morphometrics Workshop,* eds. F. J. Rohlf and F. L. Bookstein, 147–166. Ann Arbor, Mich.: The University of Michigan Museum of Zoology.

Lohmann, J. U.; R. L. Hong; M. Hobe; M. A. Busch; F. Parcy; R. Simon; and D. Weigel (2001). A molecular link between stem cell regulation and floral patterning in *Arabidopsis. Cell* 105:793–803.

Lohr, U.; M. Yussa; and L. Pick (2001). *Drosophila* fushi tarazu: A gene on the border of homeotic function. *Current Biology* 11:1403–1412.

Loredo, G. A.; A. Brukman; M. P. Harris; D. Kagle; E. E. LeClair; R. Gutman; E. Denney; E. Henkelman; B. P. Murray; J. F. Fallon; R. S. Tuan; and S. F. Gilbert (2001). Development of an evolutionarily novel structure: Fibroblast growth factor expression in the carapacial ridge of turtle embryos. *Journal of Experimental Zoology (Molecular and Developmental Evolution)* 291:274–281.

Losos, J. B.; D. A. Creer; D. Glossip; R. Goellner; A. Hampton; G. Roberts; N. Haskell; P. Taylor; and J. Ettling (2000). Evolutionary implications of phenotypic plasticity in the hindlimb of the lizard *Anolis sagrei. Evolution* 54:301–305.

Lovejoy, C. O.; M. J. Cohn; and T. D. White (1999). Morphological analysis of the mammalian postcranium: A developmental perspective. *Proceedings of the National Academy of Sciences, USA* 96:13247–13252.

Løvtrup, S. (1974). *Epigenetics.* London: John Wiley & Sons.

Løvtrup, S. (1987). *Darwinism: Refutation of a Myth.* Beckenham, England: Croom Helm.

Lowe, C. J., and Wray, G. A. (1997). Radical alterations in the roles of homeobox genes during echinoderm evolution. *Nature* 389:718–721.

Löwer, R.; J. Löwer; and R. Kurth (1999). The viruses in all of us: Characteristics and biological significance of human endogenous retrovirus sequences. *Proceedings of the National Academy of Sciences, USA* 93:5177–5184.

Ludwig, M. Z.; C. Bergman; N. H. Patel; and M. Kreitman (2000). Evidence for stabilizing selection in a eukaryotic enhancer element. *Nature* 403:564–567.

Lumsden, A. G. (1988). Spatial organization of the epithelium and the role of neural crest cells in the initiation of the mammalian tooth germ. *Development* 103 (Suppl): 155–169.

Luna, I. G., and Prokopy, R. J. (1995). Behavioral differences between hawthorn-origin and apple-origin *Rhagoletis pomonella* flies in patches of host trees. *Entomologia Experimentalis et Applicata* 74:277–282.

Lwoff, A. (1950). *Problems of Morphogenesis in Ciliates: The Kinetosomes in Development, Reproduction and Evolution.* New York: John Wiley & Sons.

Lwoff, A. (1990). L'Organization du cortex chez les cilés: Un exemple d'hérédité de caractère acquis. *Comptes Rendu de l'Académie des Sciences, Paris.* 310:109–111.

Lynch, M. (1999). The age and relationships of the major animal phyla. *Evolution* 53:319–325.

Lynch, M., and Force, A. (2000). The origin of interspecific genomic incompatibility via gene duplication. *American Naturalist* 156:590–605.

Lynch, M.; M. O'Hely; B. Walsh; and A. Force (2001). The probability of preservation of a newly arisen gene duplicate. *Genetics* 159:1789–1804.

Lyndon, R. F. (1990). *Plant Development.* London: Unwin Hyman.

Lyndon, R. F. (1998). *The Shoot Apical Meristem.* Cambridge: Cambridge University Press.

Lyon, J. D., and Vacquier, V. D. (1999). Interspecies chimeric sperm lysins identify regions mediating species-specific recognition of the abalone egg vitelline envelope. *Developmental Biology* 214:151–159.

Lyon, M. F. (1995). The history of X-chromosome inactivation and relation of recent findings to understanding of human X-linked conditions. *Turkish Journal of Pediatrics* 37:125–140.

Ma, H. (1994). The unfolding drama of flower development: Recent results from genetic and molecular analyses. *Genes & Development* 8:745–756.

Ma, H. (1998). To be, or not to be, a flower: Control of floral meristem identity. *Trends in Genetics* 14:26–32.

Ma, H., and dePamphilis, C. (2000). The ABCs of floral evolution. *Cell* 101:5–8.

Ma, H.; M. F. Yanofsky; and E. M. Meyerowitz (1991). *AGL1-AGL6,* an *Arabidopsis* gene family with similarity to floral homeotic and transcription factor genes. *Genes & Development* 5:484–495.

Mabee, P. M. (1993). Phylogenetic interpretation of ontogenetic change: Sorting out the actual from the artefactual in an empirical case study of centrarchid fishes. *Zoological Journal of the Linnean Society* 107:175–291.

Mabee, P. M. (2000). Developmental data and phylogenetic systematics: Evolution of the vertebrate limb. *American Zoologist* 40:789–800.
Mabee, P. M., and Humphries, J. (1993). Coding polymorphic data: Examples from allozymes and ontogeny. *Systematic Biology* 42:166–181.
MacDonald, M. E., and Hall, B. K. (2001). Altered timing of the extracellular-matrix-mediated epithelial-mesenchymal interaction that initiates mandibular skeletogenesis in three inbred strains of mice: Development, heterochrony, and evolutionary change in morphology. *Journal of Experimental Zoology (Molecular and Developmental Evolution)* 291:258–273.
MacLeod, N., and Rose, K. (1993). Inferring locomotor behavior in Paleogene mammals via Eigenshape analysis. *American Journal of Science* 293-A:300–355.
Maconochie, M.; S. Nonchev; A. Morrison; and R. Krumlauf (1996). Paralogous Hox genes: Function and regulation. *Annual Review of Genetics* 30:529–556.
Maddison, W. P., and Maddison, D. R. (1992). *MacClade: Analysis of Phylogeny and Character Evolution,* ver. 3. Sunderland, Mass.: Sinauer.
Maderson, P. F. A. (1971). On how an archosaurian scale might have given rise of an avian feather. *American Naturalist* 106:424–428.
Madsen, O.; M. Scally; C. J. Douady; D. J. Kao; R. W. Debry; R. Adkins; H. M. Amrine; M. J. Stanhope; W. W. de Jong; and M. S. Springer (2001). Parallel adaptive radiations in two major clades of placental mammals. *Nature* 409:610–614.
Maienschein, J. (1986). Preformation or new formation—or neither or both? In *A History of Embryology,* eds. T. H. Horder, J. A. Witkowsky, and C. C. Wylie, 73–108. Cambridge: Cambridge University Press.
Maillette, L.; R. J. N. Emery; C. C. Chinnappa; and N. K. Kimm (2000). Module demography does not mirror differentiation among populations of *Stellaria longipes* along an elevational gradient. *Plant Ecology* 149:143–156.
Malicki, J.; K. Schughart; and W. McGinnis (1990). Mouse Hox-2.2 specifies thoracic segmental identity in *Drosophila* embryos and larvae. *Cell* 63:961–967.
Mandel, M. A., and Yanofsky, M. F. (1998). The *Arabidopsis AGL9* MADS box gene is expressed in young flower primordia. *Sexual Plant Reproduction* 11:22–28.
Manley, N.; J. Barrow; T. Zhang; and M. R. Capecchi (2001). Hoxb2 and Hoxb4 act together to specify ventral body wall formation. *Developmental Biology* 237:130–144.
Mann, R., and Affolter, M. (1998). Hox proteins meet more partners. *Current Opinion in Genetics and Development* 8:423–429.
Mannervik, M. (1999). Target genes of homeodomain proteins. *BioEssays* 21:267–270.
Marden, J. H., and Kramer, M. G. (1994). Sufrace-skimming stone flies: A possible intermediate in insect flight evolution. *Science* 266:427–430.
Margulis, L. (1993). *Symbiosis in Cell Evolution,* 2nd ed. New York: W. H. Freeman.
Margulis, L., and Fester, R., eds. (1991). *Symbiosis as a Source of Evolutionary Innovation, Speciation and Morphogenesis.* Cambridge: MIT Press.

Mark Welch, D. B. (2000). Evidence from a protein-coding gene that acanthocephalans are rotifers. *Invertebrate Biology* 119:17–26.

Maroto, M., and Pourquié, O. (2001). A molecular clock involved in somite segmentation. *Current Topics in Developmental Biology* 51:221–248.

Marshall, C. R.; E. C. Raff; and R. A. Raff (1994). Dollo's law and the death and resurrection of genes. *Proceedings of the National Academy of Sciences, USA* 91:12283–12287.

Martin, G. R. (1987). Nomenclature for homoeobox-containing genes. *Nature* 325:21–22.

Martindale, M. Q., and Henry, J. Q. (1992). Evolutionary changes in the program of spiralian embryogensis: Fates of early blastomeres in a direct-developing nemertean worm. *American Zoologist* 32:79A.

Martindale, M. Q., and Henry, J. Q. (1999). Intracellular fate mapping in a basal metazoan, the ctenophore *Mnemiopsis leidyi,* reveals the origins of mesoderm and the existence of indeterminate cell lineages. *Developmental Biology* 214:243–257.

Martínez, D. E. (1998). Mortality patterns suggest lack of senescence in hydra. *Experimental Gerontology* 33:217.

Martins, E. P. (1996). Phylogenies, spatial autoregression, and the comparative method: A computer simulation test. *Evolution* 50:1750–1765.

Mathews, M. B.; N. Sonenberg; and J. W. B. Hershey (2000). Origins and principles of translational control. In *Translational Control of Gene Expression,* eds. N. Sonenberg, J. W. B. Hershey, and M. B. Mathews, 1–31. Cold Spring Harbor, N.Y.: Cold Spring Harbor Laboratory Press.

Matsuda, J. J.; R. F. Zernicke; A. C. Vailas; V. A. Pedrini; A. Pedrini-Mille; and J. A. Maynard (1986). Structural and mechanical adaptation of immature bone to strenuous exercise. *Journal of Applied Physiology* 60:2028–2034.

Matzke, M. A.; M. F. Mette; W. Aufsatz; J. Jakowitsch; and A. J. Matzke (1999). Host defenses to parasitic sequences and the evolution of epigenetic control mechanisms. *Genetica* 107:271–287.

Mayer, K. F.; H. Schoof; A. Haecker; M. Lenhard; G. Jurgens; and T. Laux (1998). Role of *WUSCHEL* in regulating stem cell fate in the *Arabidopsis* shoot meristem. *Cell* 95:805–815.

Maynard Smith, J. (1983). Evolution and development. In *Development and Evolution,* eds. B. C. Goodwin, N. Holder, and C. C. Wylie, 33–46. Cambridge: Cambridge University Press.

Maynard Smith, J. (1987). *Evolutionary Genetics.* Oxford: Oxford University Press.

Maynard Smith, J. (1990). Models of a dual inheritance system. *Journal of Theoretical Biology* 143:41–53.

Maynard Smith, J., and Szathmáry, E. (1995). *The Major Transitions in Evolution.* Oxford: W. H. Freeman.

Maynard Smith, J., and Szathmáry, E. (1999). *The Origins of Life: From the Birth of Life to the Origin of Language.* New York: Oxford University Press.

Maynard Smith, J.; R. M. Burian; S. A. Kauffman; P. Alberch; J. H. Campbell; B. Goodwin; R. Lande; D. M. Raup; and L. Wolpert (1985). Developmental con-

straints and evolution: A perspective from the Mountain Lake conference on development and evolution. *Quarterly Review of Biology* 60:265–287.
Mayr, E. (1942). *Systematics and the Origin of Species.* New York: Columbia University Press.
Mayr, E. (1958). Behavior and systematics. In *Behavior and Evolution,* eds. A. E. Roe and G. G. Simpson, 341–362. New Haven: Yale University Press.
Mayr, E. (1960). The emergence of evolutionary novelties. In *Evolution after Darwin,* ed. S. Tax, 349–380. Cambridge, Mass.: Harvard University Press.
Mayr, E. (1961). Cause and effect in biology. *Science* 134:1501–1506.
Mayr, E. (1963). *Animal Species and Evolution.* Cambridge, Mass.: Harvard University Press.
Mayr, E. (1966). *Animal Species and Evolution.* Cambridge, Mass.: Harvard University Press.
Mayr, E. (1980). Prologue: Some thoughts on the history of the evolutionary synthesis. In *The Evolutionary Synthesis: Perspectives on the Unification of Biology,* eds. E. Mayr and W. B. Provine, 1–48. Cambridge: Harvard University Press.
Mayr, E. (1982). *The Growth of Biological Thought: Diversity, Evolution, and Inheritance.* Cambridge, Mass.: Harvard University Press.
Mayr, E. (1983). How to carry out the adaptationist program? *American Naturalist* 121:324–334.
Mayr, E., and Provine, W. B., eds. (1980). *The Evolutionary Synthesis.* Cambridge, Mass.: Harvard University Press.
McCune, A. R. (1989). Evolutionary novelty and atavism in the *Semionotus* complex: Relaxed selection during colonization of an expanding lake. *Evolution* 44:71–85.
McDavid, C. N., and Poethig, R. S. (1988). Cell lineage patterns in the shoot apical meristem of the germinating maize embryo. *Planta* 175:13–22.
McEdward, L. R., ed. (1995). *Ecology of Marine Invertebrate Larvae.* Boca Raton, FL: CRC Press.
McEdward, L. R., and Janies, D. A. (1993). Life cycle evolution in asteroids: What is a larva? *Biological Bulletin* (Woods Hole) 184:255–268.
McEdward, L. R., and Miner, B. G. (2001). Larval and life-cycle patterns in echinoderms. *Canadian Journal of Zoology* 79:1125–1170.
McEdward, L. R., and Strathmann, R. R. (1987). The body plan of the cyphonautes larva of bryozoans prevents high clearance rates: Comparison with the pluteus and a growth model. *Biological Bulletin* 172:30–45.
McFall-Ngai, M. (2002). Unseen forces: The influences of bacteria on mammalian development. *Developmental Biology* 242:1–14.
McGinnis, W., and Krumlauf, R. (1992). Homeobox genes and axial patterning. *Cell* 68:283–302.
McGinnis, W.; M. S. Levine; E. Hafen; A. Kuroiwa; and W. J. Gehring (1984). A conserved DNA sequence in homeotic genes of the *Drosophila* Antennapedia and bithorax complexes. *Nature* 308:428–433.
McHugh, D. (1997). Molecular evidence that echiurans and pogonophorans are derived annelids. *Proceedings of the National Academy of Sciences, USA* 94:8006–8009.

McKay, R. (1997). Stem cells in the central nervous system. *Science* 276:66–71.

McKinney, M. L. (1986). Ecological causation of heterochrony: A test and implications for evolutionary theory. *Paleobiology* 12:282–289.

McKinney, M. L., ed. (1988). *Heterochrony in Evolution: A Multidisciplinary Approach.* New York: Plenum.

McKinney, M. L. (1999). Heterochrony: Beyond words. *Paleobiology* 25:149–153.

McKinney, M. L., and McNamara, K. J. (1991). *Heterochrony: The Evolution of Ontogeny.* New York: Plenum.

McKitrick, M. C. (1993). Phylogenetic constraint in evolutionary biology: Has it any explanatory power? *Annual Review of Ecology and Systematics* 24:307–330.

McNamara, J. M., and Houston, A. I. (1996). State-dependent life histories. *Nature* 380:215–221.

McNamara, K. (1995). *Evolutionary Change and Heterochrony.* New York: John Wiley & Sons.

McNamara, K. (1997). *Shapes of Time: The Evolution of Growth and Development.* Baltimore: Johns Hopkins University Press.

McShea, D. W. (1991). Complexity and evolution: What everybody knows. *Biology and Philosophy* 6:303–324.

Mead, K. S., and Epel, D. (1995). Beakers versus breakers: How fertilization in the laboratory differs from fertilization in nature. *Zygote* 3:95–99.

Medina-Martinez, O.; A. Bradley; and R. Ramirez-Solis (2000). A large targeted deletion of Hoxb1-Hoxb9 produces a series of single-segment anterior homeotic transformations. *Developmental Biology* 222:71–83.

Meinhardt, H. (1996). Models of biological pattern formation: Common mechanism in plant and animal development. *International Journal of Developmental Biology* 40:123–134.

Meinhardt, H. (1998). *The Algorithmic Beauty of Sea Shells.* Heidelberg: Springer.

Meinhardt, H., and Geirer, A. (2000). Pattern formation by local self-activation and lateral inhibition. *BioEssays* 22:753–760.

Mengal, P., ed. (1993). *Histoire du concept de Recapitulation—ntogenèse et Phylogenèse en Biologie et Sciences Humaines.* Paris: Masson et Cie.

Metcalfe, N. B., and Monaghan, P. (2001). Compensation for a bad start: Grow now, pay later? *Trends in Ecology and Evolution* 16:254–260.

Metchnikoff, E. (1879). Spongiologische Studien. *Zeitschrift für wissenschaftliche Zoologie* 33:349–387.

Metz, E. C., and Palumbi, S. R. (1996). Positive selection and sequence rearrangements generate extensive polymorphisms in the gamete recognition protein bindin. *Molecular Biology and Evolution* 13:397–406.

Meyer, A.; T. Kocher; P. Basasibwaki; and A. Wilson (1990). Monophyletic origin of Lake Victoria cichlid fishes suggested by mitochondrial DNA sequences. *Nature* 347:550–553.

Meyer, A. (1999). Homology and homoplasy: The retention of genetic programmes. In *Homology,* eds. G. R. Bock and G. Cardew, 141–157. Chichester: John Wiley & Sons.

Meyerowitz, E. M. (1996). Plant development: Local control, global patterning. *Current Opinion in Genetics and Development* 6:475–479.
Meyerowitz, E. M. (1999). Plants, animals and the logic of development. *Trends in Genetics* 15 (Millennium issue): M65–M68.
Meyerowitz, E. M.; D. R. Smyth; and J. L. Bowman (1989). Abnormal flowers and pattern formation in floral development. *Development* 106:209–217.
Miaud, C., and Guyetant, R. (1998). Plasticity and selection on life-history traits in a complex life cycle organism, the common frog *Rana temporaria* (Amphibia; Anura). *Bulletin Société Zoologique de France* 123:325–344.
Michod, R. E. (1999). *Darwinian Dynamics: Evolutionary Transitions in Fitness and Individuality.* Princeton: Princeton University Press.
Mikhailov, A. S. (1990). *Foundations of Synergetics,* vol. 1. Berlin: Springer-Verlag.
Miles, D. B., and Dunham, A. E. (1993). Historical perspectives in ecology and evolutionary biology: The use of phylogenetic comparative analyses. *Annual Review of Ecology and Systematics* 24:587–619.
Millar, J. S. (1977). Adaptive features of mammalian reproduction. *Evolution* 31:370–386.
Mindell, D. P., and Meyer, A. (2001). Homology evolving. *Trends in Ecology and Evolution* 16:434–440.
Minelli, A. (1998). Molecules, developmental modules, and phenotypes: A combinatorial approach to homology. *Molecular Phylogenetics and Evolution* 9:340–347.
Minelli, A., and Schram, F. R. (1994). Owen revisited: A reappraisal of morphology in evolutionary biology. *Bijdragen tot de Dierkunde* 64:65–74.
Minokawa, T.; K. Yagi; K. W. Makabe; and H. Nishida (2001). Binary specification of nerve cord and notochord cell fates in ascidian embryos. *Development* 128:2007–2017.
Mishler, B. D. (1994). The cladistic analysis of molecular and morphological data. *American Journal of Physical Anthropology* 94:143–156.
Mishler, B. D. (1995). Plant systematics and conservation: Science and society. *Madroño* 42:103–113.
Mishler, B. D. (2000). Deep phylogenetic relationships among "plants" and their implications for classification. *Taxon* 49:661–683.
Mivart, St. G. J. (1871). *On the Genesis of Species.* London: Macmillan.
Mizukami, Y., and Ma, H. (1992). Ectopic expression of the floral homeotic gene *AGAMOUS* in transgenic *Arabidopsis* plants alters floral organ identity. *Cell* 71:119–131.
Mizukami, Y., and Ma, H. (1997). Determination of *Arabidopsis* floral meristem identity by *AGAMOUS. Plant Cell* 9:393–408.
Monaghan, P., and Metcalfe, N. B. (2000). Genome size and longevity. *Trends in Genetics* 16:331–332.
Monk, M. (1988). Genetic imprinting. *Genes & Development* 2:921–925.
Moore, S. W. (1994). A fiber optic system for measuring dynamic mechanical properties of embryonic tissues. *IEEE Transactions on Biomedical Engineering* 41:45–50.

Moore, S. W.; R. E. Keller; and M. A. R. Koehl (1995). The dorsal involuting marginal zone stiffens anisotropically during its convergent extension in the gastrula of *Xenopus laevis*. *Development* 121:3131–3140.

Moran, N. A. (1992). The evolutionary maintenance of alternative phenotypes. *American Naturalist* 139:971–989.

Moran, N. A. (1994). Adaptation and constraint in the complex life cycles of animals. *Annual Review of Ecology and Systematics* 25:573–600.

Morgan, C. Lloyd. (1896). *Habit and Instinct.* Arnold: London.

Morgan, T. H. (1901). *Regeneration.* New York: Macmillan.

Morschhauser, J; G. Kohler; W. Ziebuhr; G. Blum-Oehler; U. Dobrindt; and J. Hacker (2000). Evolution of microbial pathogens. *Philosophical Transactions of the Royal Society of London (B)* 355:695–704.

Moscona, A. A. (1974). *The Cell Surface in Development.* New York: John Wiley & Sons.

Mosimann, J. (1979). Size allometry: Size and shape variables with characterizations of the lognormal and generalized gamma distributions. *Journal of the American Statistical Association* 63:930–978.

Mosimann, J., and James, F. (1979). New statistical methods for allometry with application to Florida re-winged blackbirds. *Evolution* 33:444–459.

Moss, L. (2001). Deconstructing the gene and reconstructing molecular developmental systems. In *Cycles of Contingency: Developmental Systems and Evolution,* eds. S. Oyama, P. E. Griffiths, and R. D. Gray, 85–97. Cambridge, Mass.: MIT Press.

Moss, M., and Salentijn, L. (1969). The primary role of functional matrices in facial growth. *American Journal of Orthodontics* 55:566–577.

Mousseau, T. A., and Fox, C. W. (1998). *Maternal Effects as Adaptations.* New York: Oxford University Press.

Müller, G. B. (1986). Effects of skeletal change on muscle pattern formation. *Bibliotheca Anatomica* 29:91–108.

Müller, G. B. (1989). Ancestral patterns in bird limb development: A new look at Hampé's experiment. *Journal of Evolutionary Biology* 2:31–47.

Müller, G. B. (1990). Developmental mechanisms at the origin of morphological novelty: A side-effect hypothesis. In *Evolutionary Innovations,* ed. M. H. Nitecki, 99–130. Chicago: University of Chicago Press.

Müller, G. B. (2002). Novelty and key innovations. In *Encyclopedia of Evolution,* ed. M. Pagel. Vol. 2, pp. 827–830. Oxford: Oxford University Press.

Müller, G. B., and Newman, S. A. (1999). Generation, integration, autonomy: Three steps in the evolution of homology. *Novartis Foundation Symposium* 222: 65–73.

Müller, G. B., and Newman, S. A., eds. (2002). *Origination of Organismal Form.* Cambridge, Mass.: MIT Press.

Müller, G. B., and Wagner, G. P. (1991). Novelty in evolution: Restructuring the concept. *Annual Reviews in Ecology and Systematics* 22:229–256.

Müller, H. J. (1942). Isolating mechanisms, speciation, and temperature. *Biological Symposia* 6:71–125.

Muller, M. M.; A. E. Carrasco; and E. M. De Robertis (1984). A homeo-box-containing gene expressed during oogenesis in *Xenopus*. *Cell* 39:157–162.

Mundel, P. (1979). The centipedes (Chilopoda) of the Mazon Creek. In *Mazon Creek Fossils,* ed. M. H. Nitecki, 361–378. New York: Academic Press.

Murphy, W. J.; E. Eizirik; S. J. O'Brien; O. Madsen; M. Scally; C. J. Douady; E. Teeling; O. A. Ryder; M. J. Stanhope; W. W. de Jong; and M. S. Springer (2001). Resolution of the early placental mammal radiation using Bayesian phylogenetics. *Science* 294:2348–2351.

Murray, J. (1998). *Mathematical Biology,* 2nd ed. Berlin: Springer-Verlag.

Nagy, L. M. (1994). Insect segmentation. A glance posterior. *Current Biology* 4:811–814.

Nakatani, Y., and Nishida, H. (1999). Duration of competence and inducing capacity of blastomeres in notochord induction during ascidian embryogenesis. *Development, Growth & Differentiation* 41:449–453.

Nanney, D. L. (1980). *Experimental Ciliatology.* New York: John Wiley & Sons.

Needham, A. E. (1952). *Regeneration and Wound-Healing.* New York: John Wiley & Sons.

Nei, M., and Kumar, S. (2000). *Molecular Evolution and Phylogenetics.* Oxford: Oxford University Press.

Nelson, C. E.; B. A. Morgan; A. C. Burke; E. Laufer; E. DiMambro; L. C. Murtaugh; E. Gonzales; L. Tessarollo; L. F. Parada; and C. J. Tabin (1996). Analysis of hox gene expression in the chick limb bud. *Development* 122:1449–1466.

Neumann, C., and Cohen, S. (1997). Morphogens and pattern formation. *BioEssays* 19:721–729.

Neumayr, M. (1873). Die Fauna der Schichten mit *Aspidoceras acanthicum. Herausgegeben von der koeniglinke geologischen Reichsanstalt* Band V, h. 6:141–257.

Nevo, E. (1977). Genetic variation in natural populations: Patterns and theory. *Theoretical Population Biology* 13:121–177.

Newman, S. A. (1993). Is segmentation generic? *BioEssays* 15:277–283.

Newman, S. A. (1994). Generic physical mechanisms of tissue morphogenesis: A common basis for development and evolution. *Journal of Evolutionary Biology* 7:467–488.

Newman, S. A., and Müller, G. B. (2000). Epigenetic mechanisms of character origination. *Journal of Experimental Zoology (Molecular and Developmental Evolution)* 288:304–317.

Newmark, P., and Sánchez, A. A. (2000). Bromodeoxyuridine specifically labels the regenerative stem cells of planarians. *Developmental Biology* 220:142–153.

Ng, H-H., and Bird, A. (1999). DNA methylation and chromatin modification. *Current Opinion in Genetics and Development* 9:158–163.

Ng, M., and Yanofsky, M. F. (2001). Function and evolution of the plant MADS-box gene family. *Nature Reviews, Genetics* 2:186–195.

Nichols, J.; B. Zevnik; K. Anastassiadis; H. Niura; D. Klowe-Nebenius; I. Chambers; H. Scholer; and A. Smith (1998). Formation of pluripotent stem cells in the

mammalian embryo depends on the POU transcription factor Oct 4. *Cell* 95:379–391.
Nielsen, C. (1998). Origin and evolution of animal life cycles. *Biological Reviews* 73:125–155.
Nielsen, C., and Nœrrevang, A. (1985). The trochaea theory: An example of life cycle phylogeny. In *The Origins and Relationships of Lower Invertebrates,* eds. S. Conway Morris, J. D. George, R. Gibson, and H. M. Platt, 28–41. Oxford: Oxford University Press.
Nieuwkoop, P. D., and Sutasurya, L. (1979). *Primordial Germ Cells in the Chordates.* Cambridge: Cambridge University Press.
Nieuwkoop, P. D., and Sutasurya, L. (1981). *Primordial Germ Cells in the Invertebrates.* Cambridge: Cambridge University Press.
Nijhout, H. F. (1991). *The Development and Evolution of Butterfly Wing Patterns.* Washington: Smithsonian Institution Press.
Nijhout, H. F. (1999). Control mechanisms of polyphenic development in insects. *BioScience* 49:181–192.
Nijhout, H. F. (2001). Origin of butterfly wing patterns. In *The Character Concept in Evolutionary Biology,* ed. G. P. Wagner, 511–529. San Diego: Academic Press.
Nijhout, H. F., and Emlen, D. J. (1998). Competition among body parts in the development and evolution of insect morphology. *Proceedings of the National Academy of Sciences, USA* 95:3685–3689.
Nijhout, H. F., and Paulsen, S. M. (1997). Developmental models and polygenic characters. *American Naturalist* 149:394–405.
Niklas, K. J. (1994). *Plant Allometry.* Chicago: University of Chicago Press.
Niklas, K. J. (2000). The evolution of plant body plans: A biomechanical perspective. *Annals of Botany* 85:411–438.
Niklas, K. J., and Enquist, B. J. (2001). Invariant scaling relationships for interspecific plant biomass production rates and body size. *Proceedings of the National Academy of Sciences, USA* 98:2928–2933.
Ninomiya H.; Q. Zhang; and R. P. Elinson (2001). Mesoderm formation in *Eleutherodactylus coqui:* Body patterning in a frog with a large egg. *Developmental Biology* 236:109–123.
Nishida, H., and Sawada, K. (2001). macho-1 encodes a localized mRNA in ascidian eggs that specifies muscle fate during embryogenesis. *Nature* 409:724–729.
Nitecki, M. H., ed. (1990). *Evolutionary Innovations.* Chicago: University of Chicago Press.
Northcutt, R. G., and Gans, C. (1983). The genesis of neural crest and epidermal placodes: A reinterpretation of vertebrate origins. *Quarterly Review of Biology* 58:1–28.
Novoplansky, A.; D. Cohen; and T. Sachs (1989). Ecological implications of correlative inhibition between plant shoots. *Physiologia Plantarum* 77:136–140.
Nübler-Jung, K., and Arendt, D. (1994). Is ventral in insects dorsal in vertebrates: A history of embryological arguments favoring axis inversion in chordate ancestors. *Roux Archives of Developmental Biology* 203:357–366.

Nyhart, L. K. (1995). *Biology Takes Form: Animal Morphology and the German Universities; 1800–1900.* Chicago: University of Chicago Press.

O'Higgins, P. (2000). Quantitative approaches to the study of craniofacial growth and evolution: Advances in morphometric techniques. In *Development, Growth and Evolution,* eds. P. O'Higgins and M. Cohn, 164–185. San Diego: Academic Press.

Oberpriller, J., and Oberpriller, J. (1974). Response of the adult newt ventricle to injury. *Journal of Experimental Zoology* 187:249–253.

Odelberg, S.; A. Kollhoff; and M. Keating (2000). Dedifferentiation of mammalian myotubes induced by msx1. *Cell* 103:1099–1109.

Odell, G. M.; G. F. Oster; P. Alberch; and B. Burnside (1981). The mechanical basis of morphogenesis. *Developmental Biology* 85:446–462.

Oelgeschlager, M.; J. Larrain; D. Geissert; and E. M. De Robertis (2000). The evolutionarily conserved BMP-binding protein Twisted gastrulation promotes BMP signalling. *Nature* 405:757–763.

Ohlsson, R.; K. Hall; and M. Ritzen (1995). *Genomic Imprinting: Causes and consequences.* Cambridge: Cambridge University Press.

Ohlsson, T., and Smith, H. G. (2001). Early nutrition causes persistent effects on pheasant morphology. *Physiological and Biochemical Zoology* 74:212–218.

Ohno, S. (1970). *Evolution by Gene Duplication.* New York: Springer-Verlag.

Okada, T. S. (2000). Lens studies continue to provide landmarks of embryology (developmental biology). *Journal of Bioscience* 25:133–144.

Olóriz, F., and Rodríguez-Tovar, F. J., eds. (1999) *Advancing Research on Living and Fossil Cephalopods.* New York: Kluwer Academic Plenum.

Olson, E., and Miller, R. (1958). *Morphological Integration.* Chicago: University of Chicago Press.

Opitz, J. M., and Gilbert, S. F. (1993). Developmental field theory and the molecular analysis of morphogenesis: A comment on Dr. Slavkin's observations. *American Journal of Medical Genetics* 47:687–688.

Orgel, L. E., and Crick, F. H. C. (1980). Selfish DNA: The ultimate parasite. *Nature* 284:604–607.

Orr, H. A. (1996). Dobzhansky, Bateson, and the genetics of speciation. *Genetics* 144:1331–1335.

Orr, H. A., and Turelli, M. (2001). The evolution of postzygotic isolation: Accumulating Dobzhansky-Müller incompatibilities. *Evolution* 55:1085–1094.

Osborn, H. F. (1896). A mode of evolution requiring neither natural selection nor the inheritance of acquired characters. *Transactions of the New York Academy of Science* 15:141–142, 148.

Owen, R. (1843). *Lectures on the Comparative Anatomy and Physiology of the Invertebrate Animals, delivered at the Royal College of Surgeons, in 1843.* From notes taken by William White Cooper, M. R. C. S. and revised by Professor Owen. London: Longman, Brown, Green, and Longmans.

Owen, R. (1846). On the archetype and homologies of the vertebrate skeleton. *Reports of the British Association for the Advancement of Science* 1846:169–340.

Owen, R. (1849). *On the Nature of Limbs.* London: J. van Voorst.

Oyama, S. (1981). What does the phenocopy copy? *Psychological Reports* 48:571–581.

Oyama, S. (1985). *The Ontogeny of Information: Developmental Systems and Evolution.* Cambridge: Cambridge University Press.

Oyama, S. (1999). Locating development: Locating developmental systems. In *Conceptual Development: Piaget's Legacy,* eds. E. K. Scholnick, K. Nelson, S. A. Gelman, and P. H. Miller, 185–208. Mahwah, NJ: Lawrence Erlbaum Associates.

Oyama, S. (2000a). *Evolution's Eye: A Systems View of the Biology-Culture Divide.* Durham: Duke University Press.

Oyama, S. (2000b). *The Ontogeny of Information: Developmental Systems and Evolution,* rev. ed. Durham: Duke University Press.

Oyama, S.; P. E. Griffiths; and R. D. Gray, eds. (2001). *Cycles of Contingency: Developmental Systems and Evolution.* Cambridge: MIT Press.

Padgett, R. W.; J. M. Wozney; and W. M. Gelbart (1993). Human BMP sequences can confer normal dorsal-ventral patterning in the *Drosophila* embryo. *Proceedings of the National Academy of Sciences, USA* 90:2905–2909.

Padian, K. (1995). A missing Hunterian Lecture on vertebrae by Richard Owen, 1837. *Journal of the History of Biology* 28:333–368.

Padian, K. (1997). The rehabilitation of Sir Richard Owen. *BioScience* 47:446–453.

Padian, K., and Chiappe, L. M. (1997). Bird origins. In *Encyclopedia of Dinosaurs,* eds. P. J. Currie and K. Padian, 71–79. San Diego: Academic Press.

Padian, K.; A. J. de Ricqles; and J. R. Horner (2001). Dinosaurian growth rates and bird origins. *Nature* 412:405–408.

Page, L. R. (2000). Inflated protoconchs and internally dissolved, coiled protoconchs of nudibranch larvae: Different developmental trajectories achieve the same morphological result. *Invertebrate Biology* 119:278–286.

Painter, K. J.; P. K. Maini; and H. G. Othmer (1999). Stripe formation in juvenile *Pomacanthus* explained by a generalized Turing mechanism with chemotaxis. *Proceedings of the National Academy of Sciences, USA* 96:5549–5554.

Paley, W. (1803). *Natural Theology.* Printed for R. Faulder, 1803.

Palladino, M. J.; L. P. Keegan; M. A. O'Connell; and R. A. Reenan (2000a). dADAR, a *Drosophila* double-stranded RNA-specific adenosine deaminase is highly developmentally regulated and is itself a target for RNA editing. *RNA* 6:1004–1018.

Palladino, M. J.; L. P. Keegan; M. A. O'Connell; and R. A. Reenan (2000b). A-to-I pre-mRNA editing in *Drosophila* is primarily involved in adult nervous system function and integrity. *Cell* 102:437–449.

Palmer, T.; J. Takahashi; and F. Gage (1997). The adult rat hippocampus contains primordial neural stem cells. *Molecular and Cellular Neurosciences* 8:389–404.

Panchen, A. L. (1994). Richard Owen and the concept of homology. In *Homology: The Hierarchical Basis of Comparative Biology,* ed. B. K. Hall, 21–62. San Diego: Academic Press.

Parcy, F.; O. Nilsson; M. A. Busch; I. Lee; and D. Weigel (1998). A genetic framework for floral patterning. *Nature* 395:561–566.

Parichy, D. M. (2001). Pigment patterns of ectothermic vertebrates: Heterochronic vs. nonheterochronic models for pigment pattern evolution. In *Beyond Heterochrony: The Evolution of Development,* ed. M. L. Zelditch, 229–269. New York: John Wiley & Sons.

Park, J.-K., and Foighil, D. O. (2000). Sphaeriid and corbiculid clams represent separate heterodont bivalve radiations into freshwater environments. *Molecular Phylogenetics & Evolution* 14:75–88.

Parra-Olea, G., and Wake, D. B. (2001). Extreme morphological and ecological homoplasy in tropical salamanders. *Proceedings of the National Academy of Sciences, USA* 98:7888–7891.

Parsons, P. A. (1981). Habitat selection and speciation in *Drosophila.* In *Evolution and Speciation: Essays in Honor of M.J.D. White,* eds. W. R. Atchley and D. S. Woodruff, 219–240. Cambridge: Cambridge University Press.

Pasquinelli, A. E.; B. J. Reinhart; F. Slack; M. Q. Martindale; M. I. Kuroda; B. Maller; D. C. Hayward; E. E. Ball; B. Degnan; P. Müller; J. Spring; A. Srinivasan; M. Fishman; J. Finnerty; J. Corbo; M. Levine; P. Leahy; E. Davidson; and G. Ruvkun (2000). Conservation of the sequence and temporal expression of *let-7* heterochronic regulatory RNA. *Nature* 408:86–89.

Patel, N. H.; T. B. Kornberg; and C. S. Goodman (1989). Expression of engrailed during segmentation in grasshopper and crayfish. *Development* 107:201–212.

Patterson, C. (1988). Homology in classical and molecular biology. *Molecular Biology and Evolution* 5:603–625.

Pearl, R. (1928). *The Rate of Living.* New York: Knopf.

Pelaz, S.; G. S. Ditta; E. Baumann; E. Wisman; and M. F. Yanofsky (2000). B and C floral organ identity functions require *SEPALLATA* MADS-box genes. *Nature* 405:200–203.

Pelaz, S.; C. Gustafson-Brown; S. E. Kohalmi; W. L. Crosby; and M. F. Yanofsky (2001a). *APETALA1* and *SEPALLATA3* interact to promote flower development. *Plant Journal* 26:385–394.

Pelaz, S.; R. Tapia-Lopez; E. R. Alvarez-Buylla; and M. F. Yanofsky (2001b). Conversion of leaves into petals in *Arabidopsis. Current Biology* 11:182–184.

Pen, I., and Weissing, F. J. (2000). Towards a unified theory of cooperative breeding: The role of ecology and life history re-examined. *Proceedings of the Royal Society, Biological Science Series B* 267:2411–2418.

Penny, D., and Hendy, M. D. (1985). The use of tree comparison metrics. *Systematic Zoology* 34:75–82.

Perez-Campo, R; M. Lopez-Torres; S. Cadenas; C. Rojas; and G. Barja (1998). The rate of free radical production as a determinant of the rate of aging: Evidence from the comparative approach. *Journal of Comparative Physiology B* 168:149–158.

Pesce, M; M. K. Gross; and H. R. Schöler (1998). In line with our ancestors: *Oct-4* and the mammalian *germ. BioEssays* 20:722–732.

Peters, H., and Balling, R. (1999). Teeth, where and how to make them. *Trends in Genetics* 15:59–65.

Peterson, K. J., and Eernisse, D. J. (2001). Animal phylogeny and the ancestry of bilaterians: Inferences from morphology and 18S rDNA gene sequences. *Evolution & Development* 3:170–205.

Peterson, K. J.; C. Arenas-Mena; and E. H. Davidson (2000). The A/P axis in echinoderm ontogeny and evolution: Evidence from fossils and molecules. *Evolution & Development* 2:93–101.

Petrov, D. A. (2001). Evolution of genome size: New approaches to an old problem. *Trends in Genetics* 17:23–28.

Petrov, D. A., and Hartl, D. L. (1997). Trash DNA is what gets thrown away: High rate of DNA loss in *Drosophila. Gene* 205:279–289.

Petrov, D. A., and Hartl, D. L. (2000). Pseudogene evolution and natural selection for a compact genome. *The Journal of Heredity* 91:221–227.

Petrov, D. A.; T. A. Sangster; J. S. Johnston; D. L. Hartl; and K. L. Shaw (2000). Evidence for DNA loss as a determinant of genome size. *Science* 287:1060–1062.

Pfennig, D. W., and Murphy, P. J. (2000). Character displacement in polyphenic tadpoles. *Evolution* 54:1738–1749.

Philipson, W. R.; J. M. Ward; and B. G. Butterfield (1971). *The Vascular Cambium.* London: Chapman and Hall.

Piaget, J. (1978). *Behavior and Evolution.* New York: Pantheon Books.

Pigliucci, M. (1996). Reply from M. Pigliucci. *Trends in Ecology and Evolution* 11:384.

Pigliucci, M. (2001). Characters and environments. In *The Character Concept in Evolutionary Biology,* ed. G. P. Wagner, 363–388. San Diego: Academic Press.

Pineda, D.; J. Gonzalez; P. Callaerts; K. Ikeo; W. J. Gehring; and E. Salo (2000). Searching for the prototypic eye genetic network: Sine oculis is essential for eye regeneration in planarians. *Proceedings of the National Academy of Sciences, USA* 97:4525–4529.

Pineda, D.; J. Gonzalez-Linares; M. Marsal; and E. Salo (2001). Functional analysis of the eye genetic network in planarian regeneration. *International Journal of Developmental Biology* 45:S123–124.

Pittenger, M. F.; A. M. Mackay; S. C. Beck; R. K. Jaiswal; R. Douglas; J. D. Mosca; M. A. Moorman; D. W. Simonetti; S. Craig; and D. R. Marshak (1999). Multilineage potential of adult human mesenchymal stem cells. *Science* 284:143–147.

Plotkin, H. C., ed. (1988). *The Role of Behavior in Evolution.* Cambridge: MIT Press.

Poethig, R. S. (1997). Leaf morphogenesis in flowering plants. *Plant Cell* 9:1077–1087.

Pol, D., and Siddall, M. E. (2001). Biases in maximum likelihood and parsimony: A simulation approach to a 10-taxon case. *Cladistics* 17:266–281.

Pollard, S. L., and Holland, P. (2000). Evidence for 14 homeobox gene clusters in human genome ancestry. *Current Biology* 10:1059–1062.

Ponder, W. F., and Lindberg, D. R. (1997). Towards a phylogeny of gastropod molluscs: A preliminary analysis using morphological characters. *Zoological Journal of the Linnean Society* 119:83–265.

Pourquie, O. (2001). Developmental biology: A macho way to make muscles. *Nature* 409:679–680.

Prince, V. E.; L. Joly; M. Ekker; and R. K. Ho (1998). Zebrafish Hox genes:

Genomic organization and modified colinear expression patterns in the trunk. *Development* 125:407–420.

Prockop, D. J. (1997). Marrow stromal cells as stem cells for nonhematopoietic tissues. *Science* 276:71–74.

Prokopy, R., and Bush, G. (1993). Evolution in an orchard. *Natural History* 102:4–10.

Prokopy, R.; S. R. Diehl; and S. S. Cooley (1988). Behavioral evidence for host races in *Rhagoletis pomonella* flies. *Oecologia* 76:138–147.

Provine, W. B. (1971). *The Origins of Theoretical Population Genetics.* Chicago: University of Chicago Press.

Prum, R. O. (1999). Development and evolutionary origin of feathers. *Journal of Experimental Zoology (Molecular and Developmental Evolution)* 285:291–306

Prum, R. O., and Brush, A. H. (2002). The evolutionary origin and diversification of feathers. *Quarterly Review of Biology* 77:261–295.

Qualls, C. P., and Shine, R. (1996). Reconstructing ancestral reaction norms: An example using the evolution of reptilian viviparity. *Functional Ecology* 10:688–697.

Quesada, H.; C. Gallagher; D. A. G. Skibinski; and D. O. F. Skibinski (1998). Patterns of polymorphism and gene flow of gender-associated mitochondrial DNA lineages in European mussel populations. *Molecular Ecology* 7:1041–1051.

Quiring, R.; U. Walldorf; U. Kloter; and W. J. Gehring (1994). Homology of the Eyeless gene of *Drosophila* to the Small Eye gene in mice and Aniridia in humans. *Science* 265:785–789.

Rachootin, S. P., and Thomson, K. S. (1981). Epigenetics, paleontology, and evolution. In *Evolution Today: Proceedings of the Second International Congress of Systematics and Evolutionary Biology,* eds. G. G. E. Scudder and J. L. Reveal, 181–193. Pittsburgh: Hunt Institute for Botanical Documentation, Carnegie-Mellon University.

Raff, E. C.; E. M. Popodi; B. J. Sly; F. R. Turner; J. T. Villinski; and R. A. Raff (1999). A novel ontogenetic pathway in hybrid embryos between species with different modes of development. *Development* 126:1937–1945.

Raff, M. C.; B. A. Barres; J. F. Burne; H. S. Coles; Y. Ishizaki; and M. D. Jacobson (1993). Programmed cell death and the control of cell survival: Lessons from the nervous system. *Science* 262:695–700.

Raff, R. A. (1992). Direct-developing sea urchins and the evolutionary reorganization of early development. *BioEssays* 14:211–218.

Raff, R. A. (1994). Developmental mechanisms in the evolution of animal form: Origins and evolvability of body plans. In *Early Life on Earth,* ed. S. Bengston, 489–500. New York: Columbia University Press.

Raff, R. A. (1996). *The Shape of Life: Genes, Development, and the Evolution of Animal Form.* Chicago: University of Chicago Press.

Raff, R. A. (2000). Evo-devo: The evolution of a new discipline. *Nature Reviews, Genetics* 1:74–9.

Raff, R. A., and Kaufman, T. C. (1983). *Embryos, Genes, and Evolution.* New York: Macmillan.

Raff, R. A., and Sly, B. J. (2000). Modularity and dissociation in the evolution of gene expression territories in development. *Evolution & Development* 2:102–113.

Raff, R. A., and Wray, G. A. (1989). Heterochrony: Developmental mechanisms and evolutionary results. *Journal of Evolutionary Biology* 2:409–434.

Raible, D. W., and Eisen, J. S. (1994). Restriction of neural crest cell fate in the trunk of the embryonic zebrafish. *Development* 120:495–503.

Raikow, R. J.; S. R. Boreky; and S. L. Berman (1979). The evolutionary re-establishment of a lost ancestral muscle in the bowerbird assemblage. *Condor* 81:203–206.

Rapaport, D. H.; S. L. Patheal; and W. A. Harris (2001). Cellular competence plays a role in photoreceptor differentiation in the developing *Xenopus* retina. *Journal of Neurobiology* 49:129–141.

Ratcliffe, O. J.; D. J. Bradley; and E. S. Coen (1999). Separation of shoot and floral identity in *Arabidopsis. Development* 126:1109–1120.

Reed, D. H., and Frankham, R. (2001). How closely correlated are molecular and quantitative measures of genetic varition? A meta-analysis. *Evolution* 55:1095–1103.

Reenan, R. A. (2001). The RNA world meets behavior: A→I pre-mRNA editing in animals. *Trends in Genetics* 17:53–56.

Reeve, H. K., and Sherman, P. W. (1992). Adaptation and the goals of evolutionary research. *The Quarterly Review of Biology* 68:1–32.

Reid, R. G. B. (1985). *Evolutionary Theory: The Unfinished Synthesis.* New York: Cornell University Press.

Reik, W. (1992). Genomic imprinting. In *Transgenic Animals,* eds. F. Grosveld and G. Kollias, 99–126. London: Academic Press.

Reik, W., and Walter, J. (2001). Evolution of imprinting mechanisms: The battle of the sexes begins in the zygote. *Nature Genetics* 27:255–256.

Reik, W.; W. Dean; and J. Walter (2001). Epigenetic reprogramming in mammalian development. *Science* 293:1089–1093.

Reilly, S. M., and Lauder, G. V. (1988). Atavisms and the homology of the hypobranchial elements in lower vertebrates. *Journal of Morphology* 195:237–245.

Reiss, J. O. (1989). The meaning of developmental time: A metric for comparative embryology. *American Naturalist* 134:170–189.

Remane, A. (1952). *Die Grundlagen des natürlichen Systems der vergleichenden Anatomie und der Phylogenetik.* Leipzig: Akademische Verlag.

Rendel, J. M. (1959). Canalization of the *scute* phenotype of *Drosophila. Evolution* 13:425–439.

Rendel, J. M. (1967). *Canalisation and Gene Control.* London: Logos Press.

Renfree, M., and Shaw, G. (2001). Germ cells, gonads and sex reversal in marsupials. *International Journal of Developmental Biology* 45:556–567.

Renner, M. J., and Rosenzweig, M. R. (1987). *Enriched and Impoverished Environments.* New York: Springer.

Rensch, B. (1959). *Evolution Above the Species Level.* London: Methuen.

Reyer, R. W. (1954). Regeneration in the lens in the amphibian eye. *Quarterly Review of Biology* 29:1–46.

Rheingold, H. L. (1985). Development as the acquisition of familiarity. *Annual Review of Psychology* 36:1–17.

Rice, S. H. (1997). The analysis of ontogenetic trajectories: When a change in size or shape is not heterochrony. *Proceedings of the National Academy of Sciences, USA* 94:907–912.

Rice, T. K., and Borecki, I. B. (1999). Familial resemblance and heritability. In *Genetic Dissection of Complex Traits,* eds. D. C. Rao and M. A. Province, 35–44. New York: Academic Press.

Richards, R. J. (1992). *The Meaning of Evolution: The Morphological Construction and Ideological Reconstruction of Darwin's Theory.* Chicago: University of Chicago Press.

Richardson, M. K. (1995). Heterochrony and the phylotypic period. *Developmental Biology* 172: 412–421.

Richardson, M. K. (1999). Vertebrate evolution: The developmental origins of adult variation. *BioEssays* 21:604–613.

Richardson, M. K.; J. Hanken; M. L. Gooneratne; C. Pieau; A. Raynaud; L. Selwood; and G. M. Wright (1997). There is no highly conserved embryonic stage in the vertebrates: Implications for current theories of evolution and development. *Anatomy and Embryology* 196:91–106.

Richner, H. (1998). Host-parasite interactions and life-history evolution. *Zoology* 101:333–344.

Richtsmeier, J. T., and Cheverud, J. (1986). Finite element scaling analysis of normal growth of the human craniofacial complex. *Journal of Craniofacial Genetics and Developmental Biology* 6:289–323.

Richtsmeier, J. T., and Lele, S. (1993). A coordinate-free approach to the analysis of growth patterns: Models and theoretical considerations. *Biological Reviews* 68:381–411.

Richtsmeier, J. T.; J. M. Cheverud; S. E. Danahey; B. D. Corner; and S. Lele (1993). Sexual dimorphism of ontogeny in the crab eating macaque *(Macaca fascicularis). Journal of Human Evolution* 25:1–30.

Richtsmeier, J. T.; T. M. Cole; G. Krovitz; C. J. Valeri; and S. Lele (1998). Preoperative morphology and development in sagittal synostosis. *Journal of Craniofacial Genetics and Developmental Biology* 18:64–78.

Richtsmeier, J. T.; V. DeLeon; and S. Lele (2002). The promise of geometric morphometrics. *Yearbook of Physical Anthropology* (in press).

Ricklefs, R. E.; J. M. Starck; and I. Choi (1994). Inverse relationship between functional maturity and exponential growth rate of avian skeletal muscle: A constraint on evolutionary response. *Evolution* 48:1080–1088.

Ricqlès, A. de. (1993). Some remarks on paleohistology from a comparative evolutionary point of view. In *Histology of Ancient Human Bone: Methods and Diagnosis,* eds. G. Grupe and A. N. Garland, 37–77. Berlin: Springer-Verlag.

Ricqlès, A. de; K. Padian; and J. R. Horner (2001). The bone histology of basal birds in phylogenetic and ontogenetic perspectives. In *New Perspectives on the Origin and Early Evolution of Birds,* eds. J. A. Gauthier and L. F. Gall, 411–428. New Haven: Yale University Press.

Ridley, M. (1983). *The Explanation of Organic Diversity: The Comparative Method and Adaptations for Mating.* New York: Oxford University Press.

Ridley, M. (1999). *Genome: The Autobiography of a Species in 23 Chapters.* New York: Harper Collins.

Riechmann, J. L., and Meyerowitz, E. M. (1997). MADS domain proteins in plant development. *Biological Chemistry* 378:1079–1101.

Riedl, R. (1975). *Die Ordnung des Lebendigen. Systembedingungen der Evolution.* Hamburg: Parey.

Riedl, R. (1977). A systems-analytical approach to macro-evolutionary phenomena. *The Quarterly Review of Biology* 52:351–370.

Rieger, R. M.; S. Tyler; J. P. S. Smith III; and G. E. Rieger (1991). Platyhelminthes: Turbellaria. In *Microscopic Anatomy of Invertebrates.* Vol. 3: *Platyhelminthes and Nemertina,* eds. F. W. Harrison and B. J. Bogitsh, 7–140. New York: Wiley-Liss.

Rieppel, O. (1989). Character incongruence: Noise or data? *Abhandlung der Naturwissenschaft Hamburg* (NF) 28:53–62.

Rieppel, O. (1994). Homology, topology, and typology: The history of modern debates. In *Homology: The Hierarchical Basis of Comparative Biology,* ed. B. K. Hall, 63–100. San Diego: Academic Press.

Riessen, H. P. (1992). Cost-benefit model for the induction of an antipredator defense. *American Naturalist.* 140:349–362.

Riggs, A. D. (1990). DNA methylation and late replication probably aid cell memory, and type I reeling could aid chromosome folding and enhancer function. *Philosophical Transactions of the Royal Society of London,* Series B 326:285–297.

Riggs, A. D., and Pfeifer, G. P. (1992). X-chromosome inactivation and cell memory. *Trends in Genetics* 8:169–174.

Robert, J. S.; B. K. Hall; and W. M. Olson (2001). Bridging the gap between developmental systems theory and evolutionary developmental biology. *BioEssays* 23:954–962.

Robertson, R. (1993). Snail handedness. *National Geographic Research and Exploration* 9:104–119.

Roff, D. A. (1997). *Evolutionary Quantitative Genetics.* New York: Chapman & Hall.

Roff, D. A., and Mousseau, T. A. (1987). Quantitative genetics and fitness: Lessons from *Drosophila. Heredity* 73:544–548.

Rohlf, F. (1993). Relative warp analysis and an example of its application to mosquito wings. In *Contributions to Morphometrics,* eds. L. Marcus, E. Bello, and A. Garcia-Valdecasas, 131–159. Madrid: Museo Nacional de Ciencias Naturales (CSIC).

Rongo C.; H. Broihier; L. Moore; M. Van Doren; A. Forbes; and R. Lehmann (1997). Germ plasm assembly and germ cell migration in *Drosophila. Cold Spring Harbor Symposium on Quantitative Biology* 62:1–11.

Ronshaugen, M.; N. McGinnis; and W. McGinnis (2002). Hox protein mutation and macroevolution of the insect body plan. *Nature* 415:914–917.

Roonwal, M. L. (1953). A new evolutionary phenomenon: The sharp increase of intraspecific variation in minimum populations, as evidences by the desert locust. *Records of the Indian Museum* 51:481–533.

Rosania, G.; Y. Chang; O. Perez; D. Sutherlin; H. Dong; D. Lockhart; and P. Schultz (2000). Myoseverin, a microtubule-binding molecule with novel cellular effects. *Nature Biotechnology* 18:304–308.

Rosenzweig, M. L. (1995). *Species Diversity in Space and Time.* Cambridge: Cambridge University Press.

Rosenzweig, M. L., and McCord, R. D. (1991). Incumbent replacement: Evidence for long-term evolutionary progress. *Paleobiology* 17:202–213.

Roskam, J. C., and Brakefield, P. M. (1996). A comparison of temperature-induced polyphenism in African *Bicyclus* butterflies from a seasonal savannah-rainforest ecotone. *Evolution* 50:2360–2372.

Roskam, J. C., and Brakefield, P. M. (1999). Seasonal polyphenism in *Bicyclus* (Lepidoptera: Satyridae) butterflies: Different climates need different cues. *Biological Journal of the Linnean Society* 66:345–356.

Rostand, J. (1951). *Les origines de la biologie expérimentale et l'abbé Spallanzani.* Paris: Fasquelle Editeurs.

Roth, G., and Wake, D. B. (1985). Trends in the functional morphology and sensorimotor control of feeding behavior in salamanders: An example of the role of internal dynamics in evolution. *Acta Biotheoretica* 34:175–192.

Roth, G., and Wake, D. B. (1989). Conservatism and innovation in the evolution of feeding in vertebrates. In *Complex Organismal Functions: Integration and Evolution in Vertebrates,* eds. D. B. Wake and G. Roth, 7–21. Chichester: John Wiley & Sons.

Roth, G.; J. Blanke; and D. B. Wake (1994). Cell size predicts morphological complexity in the brains of frogs and salamanders. *Proceedings of the National Academy of Sciences, USA* 91:4796–4800.

Roth, V. L. (1988). The biological basis of homology. In *Ontogeny and Systematics,* ed. C. J. Humphries, 1–26. New York: Columbia University Press.

Roth, V. L. (1991). Homology and hierarchies: Problems solved and unresolved. *Journal of Evolutionary Biology* 4:167–194.

Rothstein, S. I. (1973). The niche-variation model: Is it valid? *American Naturalist* 107:598–620.

Rountree, D. B., and Nijhout, H. F. (1995a). Hormonal control of a seasonal polyphenism in *Precis coenia* (Lepidoptera: Nymphalidae). *Journal of Insect Physiology* 41:987–992.

Rountree, D. B., and Nijhout, H. F. (1995b). Genetic control of a seasonal morph in *Precis coenia* (Lepidoptera: Nymphalidae). *Journal of Insect Physiology* 41:1141–1145.

Roux, W. (1881). *Der Kampf der Theile im Organismus.* Leipzig: Engelmann.

Ruddle, F. H.; C. T. Amemiya; J. L. Carr; C. B. Kim; C. Ledje; C. S. Shashikant; and G. P. Wagner (1999). Evolution of chordate hox gene clusters. *Annals of the New York Academy of Science* 870:238–248.

Rudwick, M. J. S. (1995). *Scenes from Deep Time: Early Pictorial Representations of the Prehistoric World.* Chicago: University of Chicago Press.

Rupke, N. A. (1994). *Richard Owen, Victorian Naturalist.* New Haven: Yale University Press.

Russell, E. S. (1916). *Form and Function.* London: John Murray [reprinted 1982, University of Chicago Press.]

Russo, V. E. A.; R. A. Martienssen; and A. D. Riggs, eds. (1996). *Epigenetic Mechanisms of Gene Regulation.* Cold Spring Harbor, N.Y.: Cold Spring Harbor Laboratory Press.

Rutherford, S. L. (2000). From genotype to phenotype: Buffering mechanisms and the storage of genetic information. *BioEssays* 22:1095–1105.

Rutherford, S. L., and Lindquist, S. (1998). Hsp90 as a capacitor for morphological evolution. *Nature* 396:336–342.

Ruvkun, G. (2001). Glimpses of a tiny RNA world. *Science* 294:797–799.

Sachs, T. (1982). A morphogenetic basis for plant morphology. *Acta Biotheoretica* 31A:118–131.

Sachs, T. (1988). Ontogeny and phylogeny: Phytohormones as indicators of labile changes. In *Plant Evolutionary Biology,* eds. L. D. Gottlieb and S. K. Jain, 157–176. London: Chapman and Hall.

Sachs, T. (1991). *Pattern Formation in Plant Tissues.* Cambridge: Cambridge University Press.

Sachs, T. (1999). "Node Counting," an internal control of balanced vegetative and reproductive development. *Plant Cell and Environment* 22:757–766.

Sachs, T. (2000). Integrating cellular and organismic aspects of vascular differentiation. *Plant and Cell Physiology* 41:649–656.

Sachs, T. (2001). The future of plant morphology. *Phytomorphology* 51:1–17.

Sachs, T. (2002). Developmental selection. In *Encyclopedia of Evolution,* ed. M. Pagel, in press. New York: Oxford University Press.

Sachs, T.; A. Novoplansky; and D. Cohen (1993). Plants as competing populations of redundant organs. *Plant Cell and Environment* 16:765–770.

Saffman, E., and Lasko, P. (1999). Germ line development in vertebrates and invertebrates. *Cellular and Molecular Life Sciences* 55:1141–1163.

Saffo, M. B. (1991). Symbiosis in evolution. In *The Unity of Evolutionary Biology, Proceedings of the Fourth International Congress of Systematics and Evolutionary Biology.* ed. E. C. Dudley, 674–680. Portland Ore.: Dioscorides Press.

Salazar-Ciudad, I.; S. A. Newman; and R. Solé (2001a). Phenotypic and dynamical transitions in model genetic networks. I. Emergence of patterns and genotype-phenotype relationships. *Evolution & Development* 3:84–94.

Salazar-Ciudad, I.; R. Solé; and S. A. Newman (2001b). Phenotypic and dynamical transitions in model genetic networks. II. Application to the evolution of segmentation mechanisms. *Evolution & Development* 3:95–103.

Salser, S. J., and Kenyon, C. (1996). A *C. elegans Hox* gene switches on, off, on and off again to regulate proliferation, differentiation and morphogenesis. *Development* 122:1651–1661.

Salvini-Plawen, L. V., and Mayr, E. (1977). Evolution of photoreceptors and eyes. *Evolutionary Biology* 10:207–263.

Samadi, S.; P. David; and P. Jarne (2000). Variation of shell shape in the clonal snail *Melanoides tuberculata* and its consequences for the interpretation of fossil series. *Evolution* 54:492–502.

Sánchez Alvarado, A. (2000). Regeneration in the metazoans: Why does it happen? *BioEssays* 22:578–590.

Sánchez Alvarado, A., and Newmark, P. A. (1999). Double-stranded RNA specifically disrupts gene expression during planarian regeneration. *Proceedings of the National Academy of Sciences, USA* 96:5049–5054.

Sander, K. (1983). The evolution of patterning mechanisms: Gleaning from insect embryogenesis and spermatogenesis. In *Development and Evolution,* eds. B. C. Goodwin, N. Holder, and C. C. Wyllie, 137–159. Cambridge: Cambridge University Press.

Sanderson, M. J. (1997). A nonparametric approach to estimating divergence times in the absence of rate constancy. *Molecular Biology and Evolution* 14:1218–1231.

Sanderson, M. J., and Hufford, L., eds. (1996). *Homoplasy: Recurrence of Similarity in Evolution.* San Diego: Academic Press.

Sanderson, M. J., and Wojciechowski, M. F. (1996). Diversification rates in a temperate legume clade: Are there "so many species" of *Astragalus* (Fabaceae)? *American Journal of Botany* 83:1488–1502.

Sanson, B. (2001). Generating patterns from fields of cells: Examples from *Drosophila* segmentation. *EMBO Reports* 2:1083–1088.

Santos, F. V. (1929). Studies on transplantation in planarian. *Biological Bulletin* 57:188–197.

Santos, F. V. (1931). Studies on transplantation in planaria. *Physiological Zoology* 4:111–164.

Sapp, J. (1987). *Beyond the Gene: Cytoplasmic Inheritance and the Struggle for Authority in Genetics.* New York: Oxford University Press.

Sapp, J. (1994). *Evolution by Association: A History of Symbiosis.* New York: Oxford University Press.

Sapp, J. (1998). Freewheeling centrioles. *History and Philosophy of the Life Sciences* 20:255–290.

Sapp, J. (in press). *Genesis: The Evolution of Biology.* New York: Oxford University Press.

Sarkar, S. (1994). The selection of alleles and the additivity of variance. In *PSA 1994,* vol. 1, eds. D. Hull, M. Forbes, and R. Burian, 3–12. East Lansing, Mich.: Philosophy of Science Association.

Sasai, Y., and De Robertis, E. M. (1997). Ectodermal patterning in vertebrate embryos. *Developmental Biology* 182:5–20.

Satoh, N. (1994). *Developmental Biology of Ascidians.* Cambridge and New York: Cambridge University Press.

Sattler, R. (1994). Homology, homeosis, and process morphology in plants. In *Homology: The Hierarchical Basis of Comparative Biology,* ed. B. K. Hall, 423–475, San Diego: Academic Press.

Sattler, R., and Rutishauser, R. (1997). The fundamental relevance of morphology and morphogenesis to plant research. *Annals of Botany* 80:571–582.

Schank, J. C., and Wimsatt, W. C. (2001). Evolvability, adaptation, and modularity. In *Thinking About Evolution: Historical, Philosophical, and Political Perspectives,* eds. R. S. Singh, C. B. Krimbas, D. B. Paul, and J. Beatty, 322–335. Cambridge: Cambridge University Press.

Scharloo, W. (1991). Canalization: Genetic and developmental aspects. *Annual Reviews of Ecology and Systematics* 22:65–93.

Schatt, P., and Feral, J.-P. (1996). Completely direct development of *Abatus cordatus,* a brooding schizasterid (Echinodermata: Echinoidea) from Kerguelen, with description of perigastrulation, a hypothetical new mode of gastrulation. *Biological Bulletin* (Woods Hole) 190:24–44.

Scheiner, S. M. (1993). Genetics and evolution of reaction norms. *Annual Review of Ecology and Systematics* 24:35–68.

Scheiner, S. M., and Berrigan, D. (1998). The genetics of phenotypic plasticity. VIII. The cost of plasticity in *Daphnia pulex. Evolution* 52:368–378.

Scheres, B. (2001). Plant cell identity: The role of position and lineage. *Plant Physiology* 125:112–114.

Schierwater, B., and DeSalle, R. (2001). Current problems with the zootype and the early evolution of Hox genes. *Journal of Experimental Zoology (Molecular and Developmental Evolution)* 291:169–174.

Schierwater, B., and Kuhn, K. (1998). Homology of Hox genes and the zootype concept in early metazoan evolution. *Molecular Phylogenetics and Evolution* 9:375–381.

Schlegel, M. (1994). Molecular phylogeny of eukaryotes. *Trends in Ecology and Evolution* 9:330–335.

Schlichting, C. D., and Pigliucci, M. (1993). Control of phenotypic plasticity via regulatory genes. *American Naturalist* 142:366–370.

Schlichting, C. D., and Pigliucci, M. (1995). Gene regulation, quantitative genetics, and the evolution of reaction norms. *Evolutionary Ecology* 9:154–168.

Schlichting, C. D., and Pigliucci, M. (1998). *Phenotypic Evolution: A Reaction Norm Perspective.* Sunderland, Mass.: Sinauer.

Schliewen, U.; K. Rassmann; M. Markmann; J. Markert; T. Kocher; and D. Tautz (2001). Genetic and ecological divergence of a monophyletic cichlid species pair under fully sympatric conditions in Lake Ejagham, Cameroon. *Molecular Ecology* 10:1471–1488.

Schlosser, G. (2001). Using heterochrony plots to detect the dissociated coevolution of characters. *Journal of Experimental Zoology (Molecular and Developmental Evolution)* 291:282–304.

Schlosser, G., and Roth, G. (1997). Evolution of nerve development in frogs. II: Modified development of the peripheral nervous system in the direct-developing frog *Eleutherodactylus coqui* (Leptodactylidae). *Brain, Behavior & Evolution* 50:94–128.

Schlosser, G.; C. Kintner; and R. G. Northcutt. (1999). Loss of ectodermal competence for lateral line placode formation in the direct-developing frog *Eleutherodactylus coqui. Developmental Biology* 213:354–369.

Schluter, D. (1996). Ecological causes of adaptive radiation. *American Naturalist* 148:S40–S64.

Schluter, D. (1998). Ecological causes of speciation. In *Endless Forms: Species and Speciation,* eds. D. J. Howard and S. H. Berlocher, 114–129. New York: Oxford University Press.

Schluter, D. (2000). *The Ecology of Adaptive Radiation.* New York: Oxford University Press.

Schluter, D., and McPhail, J. D. (1993). Character displacement and replicate adaptive radiation. *Trends in Ecology and Evolution* 8:197–200.

Schmalhausen, I. I. (1949). *Factors of Evolution; The Theory of Stabilizing Selection,* trans. I. Dordick; ed. T. Dobzhansky. Philadelphia: Blakiston. [reprinted 1986, University of Chicago Press.]

Schmid-Ott, U. (2001). Different ways to make a head. *BioEssays* 23:8–11.

Schmitt, J.; A. C. McCormac; and H. Smith (1995). A test of the adaptive plasticity hypothesis using transgenic and mutant plants disabled in phytochrome-mediated elongation responses to neighbors. *American Naturalist* 146:937–953.

Schnabel, R., and Priess, J. R. (1997). Specification of cell fates in the early embryos. In *C. elegans II,* eds. D. L. Riddle, T. Blumenthal, B. J. Meyer, and J. R. Priess, 361–382. Cold Spring Harbor, N.Y.: Cold Spring Harbor Laboratory Press.

Schoof, H.; M. Lenhard; A. Haecker; K. F. Mayer; G. Jurgens; and T. Laux (2000). The stem cell population of *Arabidopsis* shoot meristems is maintained by a regulatory loop between the *CLAVATA* and *WUSCHEL* genes. *Cell* 100:635–644.

Schultz, J. W. (1986). The origin of the spinning apparatus in spiders. *Biological Reviews* 62:89–113.

Schütt, C., and Nöthiger, R. (2000). Structure, function and evolution of sex-determining systems in dipteran insects. *Development* 127:667–677.

Schwann, T. (1839). *Mikroscopische Untersuchungen über die Übereinstimmung in der Structur und dem Wachsthum der Thiere und Pflanzen.* Berlin, trans. H. Smith, 1969, *Microscopical Researches into the Accordance in the Structure of Growth of Animals and Plants,* New York, Kraus Reprint.

Schwarz-Sommer, Z.; P. Huijser; W. Nacken; H. Saedler; and H. Sommer (1990). Genetic control of flower development: Homeotic genes in *Antirrhinum majus. Science* 250:931–936.

Schwemmler, W. (1989). *Symbiogenesis: A Macro-Mechanism of Evolution. Progress Toward a Unified Theory of Evolution.* Berlin: Walter de Gruyter.

Schwenk, K. (1995). A utilitarian approach to evolutionary constraint. *Zoology* 98:251–262.

Schwenk, K. (2000). Functional units and their evolution. In *The Character Concept in Evolutionary Biology,* ed. G. P. Wagner, 165–198. San Diego: Academic Press.

Schwenk, K. (2002). Constraint. In *Encyclopedia of Evolution,* ed. M. Pagel. Oxford: Oxford University Press. Vol. 1, pp. 196–199.

Schwenk, K., and Wagner, G. P. (2001). Function and the evolution of phenotypic stability: Connecting pattern to process. *American Zoologist* 41:552–563.

Scott, M. P. (1992). Vertebrate Homeobox gene nomenclature. *Cell* 71:551–553.

Scott, M. P. (1999). Hox proteins reach out round DNA. *Nature* 397:649, 651.

Scott, M. P., and Weiner, A. J. (1984). Structural relationships among genes that control development: Sequence homology between the Antennapedia, Ultrabithorax, and fushi tarazu loci of *Drosophila. Proceedings of the National Academy of Sciences, USA* 81:4115–4119.

Seehausen, O.; J. J. M. van Alphen; and F. Witte (1997). Cichlid fish diversity threatened by eutrophication that curbs sexual selection. *Science* 277:1808–1811.

Seehausen, O.; P. J. Mayhew; and J. J. M. van Alphen (1999). Evolution of colour patterns in East African cichlid fish. *Evolutionary Biology* 12:514–534.

Seidel, F. (1960). Körpergrundgestalt und Keimstruktur. Eine Erörterung über die Grundlagen der vergleichenden und experimentellen Embryologie und deren Gültigkeit bei phylogenetischen Überlegungen. *Zoologische Anzeiger* 164:245–305.

Sengel, P. (1976). Morphogenesis of the skin. In *Developmental and Cell Biology Series,* eds. M. Abercrombie, D. R. Newth, and J. G. Torrey, 1–277. Cambridge: Cambridge University Press.

Serres, A. E. (1860). Anatomie comparée trasncendante, principes d'Embryogénie, de Zoogénie et de Tératogénie. *Mémoires de l'Académie des Sciences, Paris* 25: XV+943 p.

Sewertzoff, A. N. (1929). Directions of evolution. *Acta Zoologica (Stockholm)* 10:59–141.

Seydoux, G., and Strome, S. (1999). Launching the germ line in *Caenorhabditis elegans:* Regulation of gene expression in early germ cells. *Development* 126:3275–3283.

Shain, D. H.; F. A. Ramirez-Weber; J. Hsu; and D. A. Weisblat (1998). Gangliogenesis in leech: Morphogenetic processes leading to segmentation in the central nervous system. *Development, Genes & Evolution* 208:28–36.

Shankland, M., and Seaver, E. C. (2000). Evolution of the bilaterian body plan: What have we learned from annelids? *Proceedings of the National Academy of Sciences, USA* 97:4434–4437.

Shapiro, A. M. (1976). Seasonal polyphenism. *Evolutionary Biology* 9:259–333.

Shapiro, A. M. (1984). Experimental studies on the evolution of seasonal polyphenism. In *The Biology of Butterflies,* eds. R. I. Vane-Wright and P. R. Ackery, 297–307. London: Academic Press.

Sharman, A. C., and Holland, P. W. (1998). Estimation of Hox gene cluster number in lampreys. *International Journal of Developmental Biology* 42:617–620.

Sharpe, P. T. (2001). Fish scale development: Hair today, teeth and scales yesterday? *Current Biology* 11:R751–752.

Shashikant, C.; C. Bieberich; H-G. Belting; J. Wang; M. Borbely; and F. Ruddle (1995). Regulation of *Hoxc-8* during mouse embryonic development: Identification and characterization of critical elements in early neural tube expression. *Development* 121:4339–4347.

Shastry, B. S. (1995). Genetic knockouts in mice: An update. *Experientia* 51:1028–1039.

Shea, B. (1983). Allometry and heterochrony in the African apes. *American Journal of Physical Anthropology* 62:275–289.

Shima, D. T., and Warren, G. (1998). Inheritance of the cytoplasm during cell division. In *Dynamics of Cell Division,* eds. S. A. Endow and D. M. Glover, 248–269. New York: Oxford University Press.

Shipley, B., and Lechowicz, M. J. (2000). The functional co-ordination of leaf morphology, nitrogen concentration, and gas exchange in 40 wetland species. *EcoScience* 7:183–194.

Shubin, N., and Wake, D. B. (1996). Phylogeny, variation and morphological integration. *American Zoologist* 36:51–60.

Shubin, N.; D. B. Wake; and A. J. Crawford (1995). Morphological variation in the limbs of *Taricha granulosa* (Caudata: Salamandridae): Evolutionary and phylogenetic implications. *Evolution* 49:874–884.

Shubin, N.; C. Tabin; and S. Carroll (1997). Fossils, genes and the evolution of animal limbs. *Nature* 388:639–648.

Siegel, A., and Benson, R. (1982). A robust comparison of biological shapes. *Biometrics* 38:341–350.

Simons, A. M., and Johnston, M. O. (1997). Developmental instability as a bet-hedging strategy. *OIKOS* 80:401–406.

Simpson, G. G. (1944). *Tempo and Mode in Evolution.* New York: Columbia University Press.

Simpson, G. G. (1953). *The Major Features of Evolution.* New York: Columbia University Press.

Simpson, T. L. (1984). *The Cell Biology of Sponges.* New York: Springer-Verlag.

Sinervo, B. (1999). Mechanistic analysis of natural selection and a refinement of Lack's and Williams's principles. *American Naturalist* 154 (Suppl): S26–42.

Sinervo, B., and Doughty, P. (1996). Interactive effects of offspring size and timing of reproduction on offspring evolution: Experimental, maternal, and quantitative genetic aspects. *Evolution* 50:1314–1327.

Sinervo, B., and Lively, C. M. (1996). The rock-paper-scissors game and the evolution of alternative male strategies. *Nature* 380:240–243.

Sinervo, B., and Svensson, E. (1998). Mechanistic and selective causes of life history trade-offs and plasticity. *Oikos* 83:432–442.

Singer, S. R. (2000). Plant Development. In *Developmental Biology,* 6th ed., ed., S.F. Gilbert, 621–647. Sunderland, Mass.: Sinauer.

Slack, J. M. W. (1991). *From Egg to Embryo.* Cambridge: Cambridge University Press.

Slack, J. M. W. (2000). Stem cells in epithelial tissues. *Science* 287:1431–1433.

Slack, J. M. W. (2001). *Essential Developmental Biology.* Oxford: Blackwell Science.

Slack, J. M. W.; P. W. H. Holland; and C. F. Graham (1993). The zootype and the phylotypic stage. *Nature* 361:490–492.

Slatkin, M., and Lande, R. (1976). Niche width in a fluctuating environment: Density independent model. *The American Naturalist* 110:31–55.

Sloan, P. R. (1992). On the Edge of Evolution. Introduction to R. Owen, *The Hunterian Lectures in Comparative Anatomy, May–June 1837,* 3–72. Chicago: University of Chicago Press.

Slowinski, J. B., and Guyer, C. (1989). Testing the stochasticity of patterns of organismal diversity: An improved null model. *American Naturalist* 134:907–921.

Sluckin, W. (1965). *Imprinting and Early Learning.* Chicago: Aldine.

Smekens, M. J. E. (1998). On phenotypic plasticity in relation to salt tolerance in *Plantago coronopus.* Ph.D. thesis, University of Utrecht, the Netherlands.

Smith, J. C. (1995). Mesoderm-inducing factors and mesodermal patterning. *Current Opinion in Cell Biology* 7:856–861.

Smith, J. P. S. III, and Tyler, S. (1985). The acoel turbellarians: Kingpins of metazoan evolution or a specialized offshoot? In *The Origins and Relationships of Lower Invertebrates,* eds. S. Conway Morris; J. D. George; R. Gibson; and H. M. Platt, 123–142. Oxford: Clarendon Press.

Smith, K. K. (1997). Comparative patterns of craniofacial development in eutherian and metatherian mammals. *Evolution* 51:1663–1678.

Smith, K. K. (2001). Heterochrony revisited: The evolution of developmental sequences. *Biological Journal of the Linnean Society* 73:169–186.

Smith, L. G.; S. Hake; and A. W. Sylvester (1996). The *tangled-1* mutation alters cell division orientations throughout maize leaf development without altering leaf shape. *Development* 122:481–489.

Smits, J. D.; F. Witte; and F. G. Van Veen (1996). Functional changes in the anatomy of the pharyngeal jaw apparatus of *Astatoreochromis alluaudi* (Pisces, Cichlidae), and their effects on adjacent structures. *Biological Journal of the Linnean Society* 59:389–409.

Sneath, P. H. A. (1964). Comparative biochemical genetics in bacterial taxonomy. In *Taxonomic Biochemistry and Serology,* ed. C. A. Leone, 565–583. New York: Ronald.

Sneath, P. H. A. (1967). Trend-surface analysis of transformation grids. *Journal of Zoology, London* 151:65–122.

Sober, E. (1984). *The Nature of Selection.* Cambridge, Mass.: MIT Press.

Sober, E. (1988). *Reconstructing the Past.* Cambridge, Mass.: MIT. Press.

Soltis, D., and Soltis, P. (2000) Dicots. In *Plant Sciences,* ed. R. Robinson, 78–79. New York: MacMillan Reference.

Sonneborn, T. M. (1963). Does preformed cell structure play an essential role in cell heredity? In *The Nature of Biological Diversity,* ed. J. M. Allen, 165–221. New York: McGraw-Hill.

Sordino, P.; F. van der Hoeven; and D. Duboule (1995). Hox gene expression in teleost fins and the origin of vertebrate digits. *Nature* 375:678–681.

Sordino, P.; D. Duboule; and T. Kondo (1996). Zebrafish Hoxa and Evx-2 genes: Cloning, developmental expression and implications for the functional evolution of posterior Hox genes. *Mechanisms of Development* 59:165–175.

Soulé, M. E. (1982). Allomeric variation 1: The theory and some consequences. *The American Naturalist* 120:751–764.

Spemann, H. (1915). Zur Geschichte und Kritik des Begriffs der Homologie. In *Allgemeine Biologie,* eds. C. Chun and W. Johannsen, 63–86. Leipzig: Teubner.

Spemann, H., and Mangold, H. (1924). Induction of embryonic primordia by implantation of organizers from a different species. In *Foundations of Experimental Embryology*, eds. B. H. Willier and J. M. Oppenheimer, 144–184. Rept. 1964, New York: Hafner.

Stanley, S. M. (1979). *Macroevolution: Pattern and Process.* San Francisco: Freeman.

Stauber, M.; A. Prell; and U. Schmidt-Ott (2002). A single Hox3 gene with composite bicoid and zerknüllt expression characteristics in non-cyclorrhaphan flies. *Proceedings of the National Academy of Sciences, USA* 99:274–279.

Stearns, S. C. (1986). Natural selection and fitness, adaptation and constraint. In *Patterns and Processes in the History of Life,* eds. D. M. Raup and D. Jablonski, 23–44. Berlin: Springer-Verlag.

Stearns, S. C. (1989). The evolutionary significance of phenotypic plasticity. *BioScience* 39: 436–445.

Stearns, S. C. (1992). *The Evolution of Life Histories.* Oxford: Oxford University Press.

Stearns, S. C., and Kawecki, T. J. (1994). Fitness sensitivity and the canalization of life-history traits. *Evolution* 48:1438–1450.

Stearns, S. C.; C. M. Kaiser; and T. J. Kawecki (1995). The differential genetic and environmental canalization of fitness components in *Drosophila melanogaster. Journal of Evolutionary Biology* 8:539–557.

Steeves, T. A., and Sussex, I. M. (1989). *Patterns in Plant Development,* 2nd ed. Cambridge: Cambridge University Press.

Steiner, W. W. M. (1977). Niche width and genetic variation in Hawaiian *Drosophila. American Naturalist* 111:1037–1045.

Steinmann, T.; N. Geldner; M. Grebe; S. Mangold; C. L. Jackson; S. Paris; L. Galweiler; K. Palme; and G. Jürgens (1999). Coordinated polar localization of auxin efflux carrier PIN1 by GNOM ARF GEF. *Science* 286:316–318.

Sterelny, K. (2000). Development, evolution, and adaptation. *Philosophy of Science* 67:369–387.

Sterelny, K., and Kitcher, P. (1988). The return of the gene. *Journal of Philosophy* 85:339–361.

Stern, C. D., and Fraser, S. E. (2001). Tracing the lineage of tracing cell lineages. *Nature Cell Biology* 3:E216–218.

Stern, D. L. (1998). A role of *Ultrabithorax* in morphological differences between *Drosophila* species. *Nature* 396:463–466.

Stern, D. L. (2000). Perspective: Evolutionary developmental biology and the problem of variation. *Evolution* 54:1079–1091.

Stiassny, M. L. J. (1981). Phylogenetic versus convergent relationship between piscivorous cichlid fishes from Lakes Malawi and Tanganyika. *Bulletin of the British Museum of Natural History, Zoology* 40:67–101.

Stiassny, M. L. J. (1992). Atavisms, phylogenetic character reversals, and the origin of evolutionary novelties. *Netherlands Journal of Zoology* 42:260–276.

Strachan, T., and Read, A. P. (1999). *Human Molecular Genetics 2.* Oxford: Bios.

Strathmann, R. R. (1985). Feeding and nonfeeding larval development and life history evolution in marine invertebrates. *Annual Reviews of Ecology and Systematics* 16:339–361.

Straus, W. L., and Temkin, O. (1943). Vesalius and the problem of variability. *Bulletin of the History of Medicine* 14:609–633.

Strauss, R., and Bookstein, F. L. (1982). The truss: Body form reconstructions in morphometrics. *Systematic Zoology* 31:113–135.

Streicher, J., and Müller, G. B. (1992). Natural and experimental reduction of the avian fibula: Developmental thresholds and evolutionary constraint. *Journal of Morphology* 214:269–285.

Striedter, G. F., and Northcutt, R. G. (1991). Biological hierarchies and the concept of homology. *Brain Behavior and Evolution* 38:177–189.

Strigini, M., and Cohen, S. M. (1997). A Hedgehog activity gradient contributes to AP axial patterning of the *Drosophila* wing. *Development* 124:4697–4705.

Subramaniam, K., and Seydoux, G. (1999). nos-1 and nos-2, two genes related to *Drosophila* nanos, regulate primordial germ cell development and survival in *Caenorhabditis elegans. Development* 126:4861–4871.

Sucena, E., and Stern, D. L. (2000). Divergence of larval morphology between *Drosophila sechellia* and its sibling species caused by *cis*-regulatory evolution of ovoy shaven-baby. *Proceedings of the National Academy of Sciences, USA* 97:4530–4534.

Suemori, H., and Noguchi, S. (2000). Hox C cluster genes are dispensable for overall body plan of mouse embryonic development. *Developmental Biology* 220:333–342.

Sulston, J. E.; E. Schierenberg; J. G. White; and J. N. Thomson (1983). The embryonic cell lineage of the nematode *Caenorhabditis elegans. Developmental Biology* 100:64–119.

Sulston, J. E., and White, J. G. (1980). Regulation and cell autonomy during postembryonic development of *Caenorhabditis elegans. Developmental Biology* 78:577–597.

Sultan, S. E. (1996). Phenotypic plasticity for offspring traits in *Polygonum persicaria. Ecology* 77:1791–1807.

Sultan, S. E. (2000). Phenotypic plasticity for plant development, function and life history. *Trends in Plant Science* 5:537–542.

Surani, M. A. H.; S. C. Barton; and M. L. Norris (1984). Development of reconstituted mouse eggs suggests imprinting of the genome during gametogenesis. *Nature* 308:548–550.

Sussex, I. M., and Kerk, N. M. (2001). The evolution of plant architecture. *Current Opinion in Plant Biology* 4:33–37.

Sutasurja, L., and Nieuwkoop, P. D. (1974). The induction of the primordial germ cells in the urodeles. *Wilhelm Roux' Archiv* 175:199–220.

Suzuki, D. T.; A. J. F. Griffiths; J. H. Miller; and R. C. Lewontin (1986). *Introduction to Genetic Analysis.* New York: W. H. Freeman.

Svensson, E., and Sheldon, B. C. (1998). The social context of life history evolution. *Oikos* 83:466–477.

Svensson, E., and Sinervo, B. (1998). Mechanistic and selective causes of life history trade-offs and plasticity. *Oikos* 83:432–442.

Svensson, E., and Sinervo, B. (2000). Experimental excursions on adaptive landscapes: Density dependent selection on egg size. *Evolution* 54:1396–1403.

Swammerdam, J. (1685). *Historia insectorum generalis*. Holland: Luchtmans.

Szentesi, A., and Jermy, T. (1989). The role of experience in host plant choice by phytophagous insects. In *Insect-plant Interactions,* vol. 2, ed. E. A. Bernays, 39–74. Boca Raton, FL: CRC Press.

Szymkowiak, E. J., and Sussex, I. M. (1996). What chimeras tell us about plant development. *Annual Review Plant Physiology Plant Molecular Biology* 47:351–376.

Tabin, C. J.; S. B. Carroll; and G. Panganiban (1999). Out on a limb: Parallels in vertebrate and invertebrate limb patterning and the origin of appendages. *American Zoologist* 39:650–663.

Takahashi, H.; K. Hotta; A. Erives; A. Di Gregorio; R. W. Zeller; M. Levine; and N. Satoh (1999a). Brachyury downstream notochord differentiation in the ascidian embryo. *Genes & Development* 13:1519–1523.

Takahashi, H.; Y. Mitani; G. Satoh; and N. Satoh (1999b). Evolutionary alterations of the minimal promoter for notochord-specific *Brachyury* expression in ascidian embryos. *Development* 126:3725–3734.

Takimoto, G.; M. Higashi; and N. A. Yamamura (2000). A deterministic genetic model for sympatric speciation by sexual selection. *Evolution* 54:1870–1881.

Tam, P., and Zhou, S. (1996). The allocation of epiblast cells to ectodermal and germ-line lineages is influenced by the postion of the cells in gastrulating mouse embryos. *Developmental Biology* 178:124–132.

Tanaka, E., and Brockes, J. (1998). A target of thrombin activation promotes cell cycle re-entry by urodele muscle cells. *Wound Repair and Regeneration* 6:371–381.

Tanaka, E.; D. N. Drechsel; and J. P. Brockes (1999). Thrombin regulates S-phase re-entry by cultured newt myotubes. *Current Biology* 9:792–799.

Tartar, V. (1961). *The Biology of Stentor.* Oxford: Pergammon Press.

Taylor, J. S.; Y. Van de Peer; and A. Meyer (2001). Genome duplication, divergent resolution and speciation. *Trends in Genetics* 17:299–301.

Teeling, E. C.; O. Madsen; R. A. Van Den Bussche; W. W. de Jong; M. J. Stanhope; and M. S. Springer (2002). Microbat paraphyly and the convergent evolution of a key innovation in Old World rhinolophoid microbats. *Proceedings of the National Academy of Sciences, USA* 99:1431–1436.

Teleman, A. A.; M. Strigini; and S. M. Cohen (2001). Shaping morphogen gradients. *Cell* 105:559–562.

Telford, M. J. (2000). Evidence for the derivation of the *Drosophila* fushi tarazu gene from a Hox gene orthologous to lophotrochozoan Lox5. *Current Biology* 10:349–352.

Tepperman, J. M.; T. Zhu; H.-S. Chang; X. Wang; and P. H. Quail (2001). Multiple transcription-factor genes are early targets of phytochrome A signaling. *Proceedings of the National Academy of Sciences, USA* 98:9437–9442.

Theissen, G. (2001). Development of floral organ identity: Stories from the MADS house. *Current Opinion in Plant Biology* 4:75–85.

Thompson, D.'A. (1917). *On Growth and Form,* rev. ed. Rept., 1992, New York: Dover.

Thomson, K. (1988). *Morphogenesis and Evolution.* New York: Oxford University Press.

Tilghman, S. M. (1999). The sins of the fathers and mothers: Genomic imprinting in mammalian development. *Cell* 96:185–193.

Tort, P., ed. (1996). *Dictionnaire du Darwinisme et de l'Evolution.* 3 vols. Paris: Presses Universitaires de France.

Toulmin, S. E., and Goodfield, J. (1990). *The Discovery of Time.* Chicago: University of Chicago Press.

Townsend, D. S., and Stewart, M. M. (1985). Direct development in *Eleutherodactylus coqui* (Anura: Leptodactylidae): A staging table. *Copeia* 1985:423–436.

Trainor, P. A., and Krumlauf, R. (2000). Patterning the cranial neural crest: Hindbrain segmentation and Hox gene plasticity. *Nature Reviews, Neuroscience* 1:116–124.

Trouve, S.; P. Sasal; J. Jourdane; F. Renaud; and S. Morand (1998). The evolution of life-history traits in parasitic and free-living platyhelminthes: A new perspective. *Oecologia* (Berlin) 115:370–378.

True, J., and Haag, E. (2001). Developmental system drift and flexibility in evolutionary trajectories. *Evolution & Development* 3:109–119.

Trussell, G. C. (2000). Phenotypic clines, plasticity, and morphological trade-offs in an intertidal snail. *Evolution* 54:151–166.

Turner, G. F., and Burrows, M. T. (1995). A model of sympatric speciation by sexual selection. *Proceedings of the Royal Society of London, Series B* 260:287–292.

Urnov, F. D., and Wolffe, A. P. (2001). Above and within the genome: Epigenetics past and present. *Journal of Mammary Gland Biology and Neoplasia* 6:153–167.

Valentine, J. W. (1991). Major factors in the rapidity and extent of the metazoan radiation during the Proterozoic-Phanerozoic transition. In *The Early Evolution of Metazoa and the Significance of Problematic Taxa,* eds. A. M. Simonetta and S. Conway Morris, 11–13. Cambridge: Cambridge University Press.

Valentine, J. W. (1994). Late Precambrian bilaterians: Grades and clades. *Proceedings of the National Academy of Sciences, USA* 91:6751–6757.

Valentine, J. W. (1995). Why no new phyla after the Cambrian? Genome and ecospace hypotheses revisited. *Palaios* 10:190–194.

Valentine, J. W. (2000). Two genomic paths to the evolution of complexity in bodyplans. *Paleobiology* 26:522–528.

Valentine, J. W., and Collins, A. G. (2000). The significance of moulting in Ecdysozoan evolution. *Evolution & Development* 2:152–156.

Valentine, J. W.; A. G. Collins; and C. P. Meyer (1994). Morphological complexity increase in metazoans. *Paleobiology* 20:131–142.

Van Auken, K.; D. C. Weaver; L. G. Edgar; and W. B. Wood (2000). *Caenorhabditis elegans* embryonic axial patterning requires two recently discovered posterior-group Hox genes. *Proceedings of the National Academy of Sciences, USA* 97:4499–4503.

Van Buskirk, J., and Saxer, G. (2001). Delayed costs of an induced defense in tadpoles? Morphology, hopping, and development rate at metamorphosis. *Evolution* 55:821–829.

Van Buskirk, J.; S. A. McCollum; and E. E. Werner (1997). Natural selection for environmentally induced phenotypes in tadpoles. *Evolution* 51:1983–1192.

van den Akker, E.; C. Fromental-Ramain; W. Graff; H. Le Mouellic; P. Brulet; P. Chambon; and J. Deschamps (2001). Axial skeletal patterning in mice lacking all paralogous group 8 Hox genes. *Development* 128:1911–1921.

Van den Biggelaar, J. A. M., and Haszprunar, G. (1996). Cleavage patterns and mesentoblast formation in the Gastropoda: An evolutionary perspective. *Evolution* 50:1520–1540.

Van der Meulen, M. C. H., and Carter, D. R. (1995). Developmental mechanics determine long bone allometry. *Journal of Theoretical Biology* 172: 323–327.

van der Weele, C. (1999). *Images of Development: Environmental Causes in Ontogeny.* Albany: State University of New York Press.

Van Hinsberg, A. (1996). On phenotypic plasticity in *Plantago lanceolata:* Light quality and plant morphology. Ph.D. thesis, University of Utrecht, the Netherlands.

Van Noordwijk, A. J. (1989). Reaction norms in genetical ecology. *BioScience* 39:453–458.

Van Tienderen, P. H. (1991). Evolution of generalists and specialists in spatially heterogeneous environments. *Evolution* 45:1317–1331.

Van Tienderen, P. H. (1997). Generalists, specialists and the evolution of phenotypic plasticity in sympatric populations of distinct species. *Evolution* 51:1372–1380.

Van Valen, L. M. (1965). Morphological variation and width of ecological niche. *American Naturalist* 99:377–390.

Van Valen, L. M. (1971). Adaptive zones and the orders of mammals. *Evolution* 25:420–428.

Van Valen, L. M. (1974). Foreword. In *Variability of Mammals,* by A.V. Yablokov, iii–iv. New Dehli: Amerind.

Van Valen, L. M. (1978). The statistics of variation. *Evolutionary Theory* 4:33–3.

Van Valen, L. M. (1982). Homology and causes. *Journal of Morphology* 173:305–312.

Vanbrunt, J. (1991). Mitochondrial defects and disease. *Biotechnology* 9: 329.

Vasiliauskas, D., and Stern, C. D. (2001). Patterning the embryonic axis: FGF signaling and how vertebrate embryos measure time. *Cell* 106:133–136.

Vazquez-Martinez, R.; J. R. Peinado; J. L. Gonzalez De Aguilar; L. Desrues; M. C. Tonon; H. Vaudry; F. Gracia-Navarro; and M. M. Malagon (2001). Melanotrope cell plasticity: A key mechanism for the physiological adaptation to background color changes. *Endocrinology* 142:3060–3067.

Velhagen, W. A., Jr. (1997). Analyzing developmental sequences using sequence units. *Systematic Biology* 46:204–210.

Velloso, C.; A. Kumar; E. Tanaka; and J. Brockes (2000). Generation of mononucleate cells from post-mitotic myotubes proceeds in the absence of cell cycle progression. *Differentiation* 66:239–246.

Velloso, C.; A. Simon; and J. P. Brockes (2001). Mammalian postmitotic nuclei re-enter the cell cycle after serum stimulation in newt/mouse hybrid myotubes. *Current Biology* 11:855–858.

Ventner, J. C., + 283 co-authors. (2001). The sequence of the human genome. *Science* 291:1304–1351.

Verdonk, N. H., and van den Biggelaar, J. A. M. (1983). Early development and the formation of the germ layers. In *The Mollusca,* vol. 3, *Development,* eds. N. H. Verdonk, J. A. M. van den Biggelaar, and A. S. Tompa, 91–122. New York: Academic Press.

Vermeij, G. J. (1973). Adaptation, versatility and evolution. *Systematic Zoology* 22:466–477.

Via, S. (1993a). Adaptive phenotypic plasticity: Target or byproduct of selection in a variable environment? *American Naturalist* 142:352–365.

Via, S. (1993b). Regulatory genes and reaction norms. *American Naturalist* 142:374–378.

Via, S.; R. Gomulkiewicz; G. De Jong; S. M. Scheiner; C. D. Schlichting; and P. H. Van Tienderen (1995). Adaptive phenotypic plasticity: Consensus and controversy. *Trends in Ecology and Evolution* 10:212–217.

Viebahn, C. (1999). The anterior margin of the mammalian gastrula: Comparative and phylogenetic aspects of its role in axis formation and head induction. *Current Topics in Developmental Biology* 46:64–103.

Vincent, S., and Perrimon, N. (2001). Developmental biology: Fishing for morphogens. *Nature* 411:533, 535–536.

Vinogradov, A. E. (1995). Nucleotypic effect in homeotherms: Body-mass corrected basal metabolic rate of mammals is related to genome size. *Evolution* 49:1249–1259.

Vinogradov, A. E. (1997). Nucleotypic effect in homeotherms: Body-mass independent resting metabolic rate of passerine birds is related to genome size. *Evolution* 51:220–225.

Vinogradov, A. E. (1999). Intron-genome size relationship on a large evolutionary scale. *Journal of Molecular Evolution* 49:376–384.

Voesenek, L. A. C. J., and Blom, C. W. P. M. (1996). Plants and hormones: An ecophysiological view on timing and plasticity. *Journal of Ecology* 84:111–119.

Vogel, G. (2000). Can old cells learn new tricks? *Science* 287:1418–1419.

Vogl, C. (1993). Theoretical enhancements of finite element scaling (FESA) methodology. *Systematic Biology* 42:341–355.

von Baer, K. E. (1828). *Entwicklungsgeschichte der Thiere: Beobachtung und Reflexion.* Koenigsberg: Borntraeger.

von Dassow, G., and Munro, E. (1999). Modularity in animal development and evolution: Elements of a conceptual framework for evodevo. *Journal of Experimental Zoology (Molecular and Developmental Evolution)* 285:307–325.

von Dassow, G.; E. Meir; E. M. Munro; and G. M. Odell (2000). The segment polarity network is a robust developmental module. *Nature* 406:188–192.

Waagen, W. (1869). Die formenreihe des *Ammonites subradiatus*. Versuch einer paläontologischen Monographie. In *Geognostisch-paläontologischen Beiträgen, Bd. 2,* ed. E. W. Benecke, 179–256. Munchen: R. Oldenbourg.

Waddington, C. H. (1942). Canalization of development and the inheritance of acquired characters. *Nature* 150:563–565.

Waddington, C. H. (1953a). Genetic assimilation of an acquired character. *Evolution* 7:118–126.

Waddington, C. H. (1953b). Epigenetics and evolution. In *Evolution* (SEB Symposium VII), eds. R. Brown and J. F. Danielli, 186–199. Cambridge: Cambridge University Press.

Waddington, C. H. (1956a). *Principles of Embryology.* London: Allen and Unwin.

Waddington, C. H. (1956b). Genetic assimilation of the bithorax phenotype. *Evolution* 10:1–13.

Waddington, C. H. (1957). *The Strategy of the Genes.* New York: MacMillan.

Waddington, C. H. (1959). Canalisation of development and genetic assimilation of acquired characters. *Nature* 183:1654–1655.

Waddington, C. H. (1961). Genetic assimilation. *Advances in Genetics* 10:257–293.

Waddington, C. H. (1962). *New Patterns in Genetics and Development.* New York: Columbia University Press.

Waddington, C. H. (1975). *The Evolution of an Evolutionist.* Ithaca: Cornell University Press.

Wagner, A. (1999). Redundant gene functions and natural selection. *Journal of Evolutionary Biology* 12:1–16.

Wagner, A. (2000). Robustness against mutations in genetic networks of yeast. *Nature Genetics* 24:351–361.

Wagner, D.; R. W. Sablowski; and E. M. Meyerowitz (1999). Transcriptional activation of *APETALA1* by LEAFY. *Science* 285:582–584.

Wagner, G. P. (1988). The significance of developmental constraints for phenotypic evolution by natural selection. In *Population Genetics and Evolution,* ed. G. de Jong, 222–229. Berlin: Springer-Verlag.

Wagner, G. P. (1989). The biological homology concept. *Annual Review of Ecology and Systematics* 20:51–69.

Wagner, G. P. (1994). Homology and the mechanisms of development. In *Homology: The Hierarchical Basis of Comparative Biology,* ed. B. K. Hall, 273–299. San Diego: Academic Press.

Wagner, G. P. (1996). Homologues, natural kinds, and the evolution of modularity. *American Zoologist* 36:36–43.

Wagner, G. P. (1999). A research programme for testing the biological homology concept. In *Homology,* eds. G. R. Bock and G. Cardew, 125–140. Chichester: John Wiley & Sons.

Wagner, G. P., ed. (2001). *The Character Concept in Evolutionary Biology.* San Diego: Academic Press.

Wagner, G. P., and Altenberg, L. (1996). Complex adaptations and the evolution of evolvability. *Evolution* 50:967–976.
Wagner, G. P., and Laubichler, M. D. (2000). Character identification in evolutionary biology: The role of the organism. *Theory in Biosciences* 119:20–40.
Wagner, G. P., and Misof, B. (1993). How can a character be developmentally constrained despite variation in developmental pathways? *Journal of Evolutionary Biology* 6:449–455.
Wagner, G. P., and Müller, G. B. (2002). Evolutionary innovations overcome ancestral constraints: A re-examination of character evolution in male sepsid flies (Diptera: *Sepsidae*). *Evolution & Development* 4:1–6.
Wagner, G. P., and Schwenk, K. (2000). Evolutionarily stable configurations: Functional integration and the evolution of phenotypic stability. *Evolutionary Biology* 31:155–217.
Wagner, G. P.; G. Booth; and H. Bagheri-Chaichian (1997). A population genetic theory of canalization. *Evolution* 51:329–347.
Wake, D. B. (1982). Functional and evolutionary biology. *Perspectives in Biology and Medicine* 25:603–620.
Wake, D. B. (1985). Phylogenetic implications of ontogenetic data. *Geobios, memoire speciale* 12:369–378.
Wake, D. B. (1991). Homoplasy: The result of natural selection, or evidence of design limitations? *The American Naturalist* 138:543–567.
Wake, D. B. (1994). Comparative terminology. *Science* 265:268–269.
Wake, D. B. (1996). Evolutionary developmental biology: Prospects for an evolutionary synthesis at the developmental level. In *New Perspectives on the History of Life,* eds. M. T. Ghiselin and G. Pinna, 97–107. San Francisco: California Academy of Sciences.
Wake, D. B. (1999). Homoplasy, homology and the problem of "sameness" in biology. In *Homology,* eds. G. R. Bock and G. Cardew, 24–46. Chichester: John Wiley & Sons.
Wake, D. B., and Hanken, J. (1996). Direct development in the lungless salamanders: What are the consequences for developmental biology, evolution and phylogenesis? *International Journal of Developmental Biology* 40:859–869.
Wake, D. B. and Larson, A. (1987). Multidimensional analysis of an evolving lineage. *Science* 238:42–48.
Wake, D. B., and Roth, G. (1989). The linkage between ontogeny and phylogeny in the evolution of complex systems. In *Complex Organismal Functions: Integration and Evolution in Vertebrates,* eds. D. B. Wake and G. Roth, 361–377. Chichester: John Wiley & Sons.
Wake, M. H. (1989). Phylogenesis of direct development and viviparity. In *Complex Organismal Functions: Integration and Evolution in Vertebrates,* eds. D. B. Wake and G. Roth, 235–250. Chichester: John Wiley & Sons.
Wake, M. H. (1992). Morphology, the study of form and function, in modern evolutionary biology. In *Oxford Surveys in Evolutionary Biology,* eds. D. Futuyma and J. Antonovics, 289–346. New York: Oxford University Press.
Walter, J., and Biggin, M. D. (1996). DNA binding specificity of two homeodomain

proteins in vitro and in *Drosophila* embryos. *Proceedings of the National Academy of Sciences, USA* 93:2680–2685.

Wanntorp, H.-E.; D. R. Brooks; T. Nilsson; S. Nylin; F. Ronquist; S. C. Stearns; and N. Wedell (1990). Phylogenetic approaches in ecology. *Oikos* 57:119–132.

Waters, K. (1991). Tempered realism about the forces of selection. *Philosophy of Science* 58:417–435.

Watt, F. M., and Hogan, B. L. (2000). Out of Eden: Stem cells and their niches. *Science* 287:1427–1430.

Webb, J. F. (1999). Larvae in fish development and evolution. In *The Origin and Evolution of Larval Forms,* eds. B. K. Hall and M. H. Wake, 109–158. San Diego: Academic Press.

Webster, G., and Goodwin, B. (1996). *Form and Transformation: Generative and Relational Principles in Biology.* Cambridge: Cambridge University Press.

Weigel, D.; J. Alvarez; D. R. Smyth; M. F. Yanofsky; and E. M. Meyerowitz (1992). *LEAFY* controls floral meristem identity in *Arabidopsis. Cell* 69:843–859.

Weinig, C. (2000). Plasticity versus canalization: Population differences in the timing of shade-avoidance responses. *Evolution* 54:441–451.

Weinig, C., and Delph, L. F. (2001). Phenotypic plasticity early in life constrains developmental responses later. *Evolution* 55:930–936.

Weismann, A. (1893). *The Germ Plasm.* New York: Scribners. [Translation of Weismann, A. (1892). *Das Keimplasma: Eine Theorie der Vererbung.* Jena: Gustav Fischer.]

Weiss, K., and Fullerton, S. M. (2000). Phenogenetic drift and the evolution of genotype-phenotype relationships. *Theoretical Population Biology* 57:187–195.

Weiss, K.; D. Stock; and A. Zhao (1998). Dynamic interactions and the evolutionary genetics of dental patterning. *Critical Reviews of Oral Biology Medicine* 9:369–398.

Werren, J. H. (1997). Biology of *Wolbachia. Annual Review of Entomology* 42:587–609.

Werth, C. R., and Windham, M. D. (1991). A model for divergent, allopatric speciation of polyploid pteridophytes resulting from silencing of duplicate-gene expression. *American Naturalist* 137:515–526.

Wessells, N. K. (1977). *Tissue Interactions and Development.* Menlo Park: Benjamin/Cummings.

West, G.; J. Brown; and B. Enquist (2001). A general model for ontogenetic growth. *Nature* 413:628–631.

West, S. A.; C. M. Lively; and A. F. Read (1999). A pluralist approach to sex and recombination. *Journal of Evolutionary Biology* 12:1003–1012.

Wheeler, W. (2001). Homology and the optimization of DNA sequence data. *Cladistics* 17:S3–11.

White, M. M., and McLaren, I. A. (2000). Copepod development rates in relation to genome size and 18S rDNA copy number. *Genome* 43:750–755.

White, R. J. (2001). *Gene Transcription: Mechanisms and Control.* Oxford: Blackwell Science.

White, S., and Doebley, J. (1998). Of genes and genomes and the origin of maize. *Trends in Genetics* 14:327–332.
Whittington, H. B. (1985). *The Burgess Shale.* New Haven: Yale University Press.
Whittle, J. R. S. (1998). How is developmental stability sustained in the face of genetic variation? *International Journal of Developmental Biology* 42:495–499.
Whyte, L. L. (1965). *Internal Factors in Evolution.* New York: George Braziller.
Wickens, M.; E. B. Goodwin; J. Kimble; S. Strickland; and M. W. Hentze (2000). Translational control of developmental decisions. In *Translational Control of Gene Expression,* eds. N. Sonenberg, J. W. B. Hershey, and M. B. Mathews, 295–370. Cold Spring Harbor, N.Y.: Cold Spring Harbor Laboratory Press.
Wiedmann, J., and Kullmann, J. (1988). *Cephalopods Present and Past.* Stuttgart: Schweitzerbart'sche Verlagsbuchhandlung.
Wijngaarden, P. J., and Brakefield, P. M. (2000). The genetic basis of eyespot size in the butterfly *Bicyclus anynana:* An analysis of line crosses. *Heredity* 85:471–479.
Wijngaarden, P. J., and Brakefield, P. M. (2001). Lack of response to artificial selection on the slope of reaction norms for seasonal polyphenism in the butterfly *Bicyclus anynana. Heredity* 87:410–420.
Wijngaarden, P. J.; P. B. Koch; and P. M. Brakefield (2002). Artificial selection on the shape of reaction norms for eyespot size in the butterfly *Bicyclus anynana:* Direct and correlated responses. *Journal of Evolutionary Biology* 15:290–300.
Wiley, E. O. (1981). Phylogenetics: The theory and practice of phylogenetic systematics. New York: John Wiley & Sons.
Wilkins, A. S. (1997). Canalisation: A molecular genetic perspective. *BioEssays* 19:257–262.
Wilkins, A. S. (2002). *The Evolution of Developmental Pathways.* Sunderland, Mass.: Sinauer.
Williams, G. C. (1966). *Adaptation and Natural Selection.* Princeton: Princeton University Press.
Williams, G. C. (1992). *Natural Selection: Domains, Levels, and Challenges.* New York: Oxford University Press.
Williams, T. A. (1994). The nauplius larva of crustaceans- functional diversity and the phylotypic stage. *American Zoologist* 34:562–569.
Williamson, A., and Lehmann, R. (1996). Germ cell development in *Drosophila. Annual Review of Cellular and Developmental Biology* 12:365–391.
Willmer, P. (1990). *Invertebrate Relationships: Patterns in Animal Evolution.* Cambridge: Cambridge University Press.
Wills, M. A., and Fortey, R. A. (2000). The shape of life: How much is written in stone? *BioEssays* 22:1142–1152.
Wilson, A. B; K. Noack-Kunnmann; and A. Meyer (2000). Incipient speciation in sympatric Nicaraguan crater lake cichlid fishes: Sexual selection versus ecological diversification. *Proceedings of the Royal Society of London, Series B* 267:2133–2141.
Wilson, E. B. (1898). Cell lineage and ancestral reminiscence. In *Biological Lectures from the Marine Biology Laboratories, Woods Hole, Massachusetts,* 21–42. Boston: Ginn.

Wimsatt, W. C. (1980). Reductionistic research strategies and their biases in the unit of selection controversy. In *Scientific Discovery,* ed. T. Nickles, 213–249. Dordrecht, The Netherlands: Reidel.

Wimsatt, W. C. (1981). Units of selection and the structure of the multi-level genome. In *PSA 1980,* eds. P. Asquith and R. Giere, 122–183. East Lansing, Mich.: Philosophy of Science Association.

Wimsatt, W. C. (2001). Generative entrenchment and the developmental systems approach to evolutionary processes. In *Cycles of Contingency: Developmental Systems and Evolution,* eds. S. Oyama, P. E. Griffiths, and R. D. Gray, 219–237. Cambridge: MIT Press.

Windig, J. J.; P. M. Brakefield; N. Reitsma; and J. G. M. Wilson (1994). Seasonal polyphenism in the wild: Survey of wing patterns in five species of *Bicyclus* butterflies in Malawi. *Ecological Entomology* 19:285–298.

Winther, R. G. (2001). Varieties of modules: Kinds, levels, origins, and behaviors. *Journal of Experimental Zoology (Molecular and Developmental Evolution)* 291:116–129.

Wistow, G. (1993). Lens crystallins: Gene recruitment and evolutionary dynamism. *Trends in Biochemical Science* 18:301–306.

Witte, H., and Döring, D. (1999). Canalized pathways of change and constraints in the evolution of reproductive modes of microarthropods. *Experimental and Applied Acarology* 23:181–216.

Wochok, Z. S., and Sussex, I. M. (1974). Morphogenesis in *Selaginella.* II. Auxin transport in the root (rhizophore). *Plant Physiology* 53:738–741.

Wolf, U. (1995). Identical mutations and phenotypic variation. *Human Genetics* 100:305–321.

Wolff, C. F. (1764). *Theorie von der Generation.* Berlin: Birnstiel.

Wolpert, L. (1969). Positional information and the spatial pattern of cellular differentiation. *Journal of Theoretical Biology* 25:1–47.

Wolpert, L. (1992). *The Triumph of the Embryo.* Oxford: Oxford University Press.

Wolpert, L. (1994). The evolutionary origin of development: Cycles, patterning, privilege and continuity. *Development* (Cambridge) Supplement: 79–84.

Woltereck, R. (1909). Weitere experimentelle Untersuchungen über Artveränderung, speziell über das Wesen quantitativer Artunterschieden bei Daphniden. *Verhandlungen der Deutschen Zoologischen Gesellschaft* 19:110–172.

Woo, K.; J. Shih; and S. E. Fraser (1995). Fate maps of the zebrafish embryo. *Current Opinion in Genetics and Development* 5:439–443.

Woodger, J. H. (1945). On biological transformations. In *Essays on Growth and Form Presented to D'Arcy Wentworth Thompson,* eds. W. E. Le Gros Clark and P. B. Medawer, 95–120. Cambridge: Cambridge University Press.

Wray, G. A. (1994). The evolution of cell lineage in echinoderms. *American Zoologist* 34:353–363.

Wray, G. A. (1995). Punctuated evolution of embryos. *Science* 267:1115–1116.

Wray, G. A. (1999). Evolutionary dissociations between homologous genes and homologous structures. In *Homology, Novartis Foundation Symposium 222,* eds. G. R. Bock and G. Cardew, 189–206. Chichester: John Wiley & Sons.

Wray, G. A., and Raff, R. A. (1989). Evolutionary modification of cell lineage in the direct-developing sea urchin *Heliocidaris erythrogramma. Developmental Biology* 132:458–470.

Wray, G. A., and Raff, R. A. (1990). Novel origins of lineage founder cells in the direct-developing sea urchin *Heliocidaris erythrogramma. Developmental Biology* 141:41–54.

Wray, G. A.; J. S. Levinton; and L. M. Shapiro (1996). Molecular evidence for deep pre-Cambrian divergences among metazoan phyla. *Science* 214:568–573.

Wright, S. (1969). *Evolution and the Genetics of Populations,* vol. 2: *The Theory of Gene Frequences.* Chicago: University of Chicago Press.

Wu, L. H., and Lengyel, J. A. (1998). Role of caudal in hindgut specification and gastrulation suggests homology between *Drosophila* amnioproctodeal invagination and vertebrate blastopore. *Development* 125:2433–2442.

Wutz, A.; O. W. Smrza; N. Schweifer; K. Schellander; E. F. Wagner; and D. P. Barlow (1997). Imprinted expression of the Igf2r gene depends on an intronic CpG island. *Nature* 389:745–749.

Wyles, J. S.; J. G. Kunkel; and A. C. Wilson (1983). Birds, behavior, and anatomical evolution. *Proceedings of the National Academy of Sciences, USA* 80:4394–4397.

Wylie, C. (1999). Germ cells. *Cell* 96:165–174.

Wylie, C. (2000). Germ cells. *Current Opinion in Genetics and Development* 10:410–413.

Wyss, A. R. (1988). On "retrogression" in the evolution of the Phocinae and phylogenetic affinities of the Monk seals. *American Museum Novitates* 2924:1–38.

Xu, P.-X.; I. Woo; H. Her; D. R. Beier; and R. L. Maas (1997). Mouse *Eya* homologues of the *Drosophila eyes absent* gene require *Pax6* for expression in lens and nasal placode. *Development* 124:219–231.

Xu, X.; Z. Zhou; and R. O. Prum (2001). Branched integumental structures in *Sinornithosaurus* and the origin of feathers. *Nature* 410:200–204.

Yablokov, A. V. (1966). *Variability of Mammals.* New Delhi: Amerind.

Yanze, N.; J. Spring; C. Schmidli; and V. Schmid (2001). Conservation of Hox/ParaHox-related genes in the early development of a cnidarian. *Developmental Biology* 236:89–98.

Yuh, C.-H.; H. Bolouri; and E. H. Davidson (1998). Genomic *cis*-regulatory logic: Experimental and computational analysis of a sea urchin gene. *Science* 279:1896–1902.

Zajonc, R. B. (1971). Attraction, affiliation, and attachment. In *Man and Beast: Comparative Social Behavior,* eds. J. F. Eisenberg and W. S. Dillon, 143–179. Washington: Smithsonian Institution Press.

Zecca, M.; K. Basler; and G. Struhl (1996). Direct and long-range action of a wingless morphogen gradient. *Cell* 87:833–844.

Zeger, S., and Harlow, S. (1987). Mathematical models from laws of growth to tools for biological analysis: Fifty years of Growth. *Growth* 51:1–21.

Zelditch, M. L., ed. (2001). *Beyond Heterochrony: The Evolution of Development.* New York: John Wiley & Sons.

Zelditch, M. L., and Fink, W. L. (1996). Heterochrony and heterotopy: Stability and innovation in the evolution of form. *Paleobiology* 22:241–254.

Zelditch, M. L.; F. Bookstein; and B. Lundrigan (1992). Ontogeny of integrated skull growth in the cotton rat *Sigmodon fulviventer. Evolution* 46:1164–1180.

Zelditch, M. L.; H. D. Sheets; and W. L. Fink (2000). Spatiotemporal reorganization of growth rates in the evolution of ontogeny. *Evolution* 54:1363–1371.

Zelditch, M. L.; H. D. Sheets; and W. L. Fink (2001). The spatial complexity and evolutionary dynamics of growth. In *Beyond Heterochrony: The Evolution of Development,* ed. M. L. Zelditch, 145–194. New York: John Wiley & Sons.

Zhao, D.; Q. Yu; C. Chen; and H. Ma (2001a). Genetic control of reproductive meristems. In *Annual Plant Review: Meristematic Tissues in Plant Growth and Development,* eds. M. T. McManus and B. Veit, 89–142. Sheffield: Sheffield Academic Press.

Zhao, D.; Q. Yu; M. Chen; and H. Ma (2001b). The *ASK1* gene regulates B function gene expression in cooperation with *UFO* and *LEAFY* in *Arabidopsis. Development* 128:2735–2746.

Zimmer, E. A. (1994). Perspectives on future applications of experimental biology to evolution. In *Molecular Ecology and Evolution: Approaches and Applications,* eds. B. Schierwater, B. Streit, G. P. Wagner, and R. DeSalle, 607–616. Basel, Switzerland: Birkhäuser Verlag.

Zimmermann, W. (1953). *Evolution: Geschichte ihrer Probleme und Erkenntnisse.* Freiburg: Alber.

Zouros, E. (2000). The exceptional mitochondrial DNA system of the mussel family Mytilidae. *Genes & Genetic Systems* 75:313–318.

Zrzavy, J., and Stys, P. (1997). The basic body plan of arthropods: Insights from evolutionary morphology and development biology. *Journal of Evolutionary Biology* 10:353–367.

Contributors

Alejandro Sánchez Alvarado
Department of Neurobiology and Anatomy
University of Utah School of Medicine

Wallace Arthur
Ecology Centre
University of Sunderland, United Kingdom

Jessica A. Bolker
Department of Zoology
University of New Hampshire

John Tyler Bonner
Department of Ecology and Evolutionary Biology
Princeton University

Paul M. Brakefield
Institute for Evolution and Ecological Sciences
Leiden University, Netherlands

Susan J. Brown
Division of Biology
Kansas State University

Graham E. Budd
Department of Earth Sciences
Historical Geology and Palaeontology
University of Uppsala, Sweden

Richard M. Burian
Department of Philosophy
Virginia Technical University

Anne C. Burke
Department of Biology
Wesleyan University

Bruce M. Carlson
Institute of Gerontology
University of Michigan

Andres Collazo
House Ear Institute
Los Angeles, California

Frietson Galis
Institute of Evolutionary and Ecological Sciences
Leiden University, Netherlands

Gillian L. Gass
Department of Biology
Dalhousie University, Canada

John Gerhart
Department of Molecular and Cell Biology
University of California, Berkeley

Scott F. Gilbert
Department of Biology
Swarthmore College

Gilbert Gottlieb
Center for Developmental Science
University of North Carolina, Chapel Hill

Robert P. Guralnick
Department of Environmental, Population and Organismal Biology and University of Colorado Museum
University of Colorado, Boulder

Brian K. Hall
Department of Biology
Dalhousie University, Canada

Benedikt Hallgrímsson
Department of Cell Biology and Anatomy
University of Calgary, Canada

James Hanken
Museum of Comparative Zoology
Harvard University

Michael Hart
Department of Biology
Dalhousie University, Canada

Susan W. Herring
Department of Orthodontics
University of Washington

Jukka Jernvall
Institute of Biotechnology
University of Helsinki, Finland

Elizabeth L. Jockusch
Ecology and Evolutionary Biology
University of Connecticut

Mary Lou King
Department of Cell Biology and Anatomy
University of Miami School of Medicine

Marc Kirschner
Department of Cell Biology
Harvard Medical School

Loren Knapp
Department of Biological Sciences
University of South Carolina

Manfred D. Laublicher
Department of Biology
Arizona State University

David R. Lindberg
Department of Integrative Biology
University of California, Berkeley

Hong Ma
Department of Biology and
The Life Sciences Consortium
Penn State University

Norman Maclean
School of Biological Sciences
University of Southampton, United Kingdom

Brent Mishler
University Herbarium, Jepson Herbarium, and Department of Integrative Biology
University of California, Berkeley

Gerd B. Müller
Department of Anatomy
University of Vienna, Austria

Stuart A. Newman
Department of Cell Biology and Anatomy
New York Medical College

Karl J. Niklas
Department of Plant Biology
Cornell University

Wendy M. Olson
Department of Biology
Dalhousie University, Canada

Lennart Olsson
Institut für Spezielle Zoologie und Evolutionsbiologie
Friedrich-Schiller-Universität Jena, Germany

Kevin Padian
Department of Integrative Biology
University of California, Berkeley

Grace Panganiban
Department of Anatomy
University of Wisconsin Medical School

John O. Reiss
Department of Biological Sciences
Humboldt State University, California

Joan T. Richtsmeier
Department of Anthropology
Pennsylvania State University

Armand de Ricqlès
Historical Biology and Evolutionism, Collège de France and UMR 8570 CNRS, France

Jason Scott Robert
Department of Philosophy
Dalhousie University, Canada

Tsvi Sachs
Department of Plant Sciences
Hebrew University, Israel

Jan Sapp
Centre interuniversitaire de recherche sur la science et la technologie (CIRST)
Université du Québec à Montréal, Canada

Roger Sawyer
College of Science and Mathematics
University of South Carolina

Carl D. Schlichting
Department of Ecology and Evolutionary Biology
University of Connecticut

Kurt Schwenk
Department of Ecology and Evolutionary Biology
University of Connecticut

Barry Sinervo
Department of Ecology and Evolutionary Biology
University of California, Santa Cruz

Jonathan M. W. Slack
Department of Biology and Biochemistry
University of Bath, United Kingdom

David Stern
Department of Ecology and Evolutionary Biology
Princeton University

Melanie L. J. Stiassny
Department of Ichthyology
American Museum of Natural History, New York

Adam P. Summers
Department of Ecology and Evolutionary Biology
University of California, Irvine

James W. Valentine
Museum of Paleontology
University of California, Berkeley

Günter P. Wagner
Department of Ecology and Evolutionary Biology
Yale University

David B. Wake
Museum of Vertebrate Zoology
University of California, Berkeley

Marvalee H. Wake
Department of Integrative Biology
University of California, Berkeley

Kenneth M. Weiss
Departments of Anthropology and Biology
Pennsylvania State University

Pieter J. Wijngaarden
Laboratory of Genetics
Wageningen University, Netherlands

Adam Wilkins
Editorial Office, *BioEssays*
Cambridge, United Kingdom

Miriam Zelditch
Museum of Paleontology
University of Michigan

Acknowledgments

Paul Brakefield and Pieter Wijngaarden are grateful to Patrícia Beldade and the editors for valuable comments on their entry.

Andres Collazo's years teaching the Woods Hole Marine Biological Laboratory Embryology class have been invaluable for writing his entry. Numerous discussions with Bob Zeller, Joel Rothman, Richard Harland, and too many others to list have helped immensely in choosing the examples discussed. He wishes to thank Scott Fraser and Marianne Bronner-Fraser for their support, and Olivier Bricaud, Andy Groves, Sung-Hee Kil, and Bob Zeller for comments on the manuscript. He is supported by an R01 grant from the NIH and NIDCD.

Scott Gilbert and Richard Burian wish to thank Ron Amundson, Marjorie Grene, Sahotra Sarkar, Günter Wagner, and the editors for their critical comments on earlier drafts of their essay.

Brian Hall and Wendy Olson are delighted to thank the authors for their willingness to contribute and for the quality (and timeliness) of their entries. They also thank Harvard University Press, in particular Michael Fisher and Sara Davis, for their professionalism in the preparation of the book. Brian Hall is supported by NSERC (Canada) grant #A5056.

Hong Ma apologizes for not citing many early papers and reviews due to space constraints and thanks Z. Liu, W. Ni, and D. Soltis for valuable comments on the manuscript. The work in his laboratory is supported by grants from the U.S. National Science Foundation and the U.S. Department of Agriculture.

Norman Maclean is very grateful to Tom Fleming, Tom Papenbrock, and David Shepherd for critical comments on the original version of his entry.

John Reiss thanks the editors for their patience and editorial skill, and Pere Alberch for inspiring his thinking about time, what now seems a very long time ago; he is supported by NSF grant IBN-0092070.

Joan Richtsmeier acknowledges that many of the ideas expressed, especially with reference to the quantitative study of growth, were improved and clarified through discussions with Valerie Burke DeLeon, Kristina Aldridge, Tim Cole, and Subhash Lele; she thanks them for their collegial contributions to her entry and her education.

Index